PREFACE

This book is intended as a first course in complex analysis for students who take their mathematics seriously. The prerequisites for reading this book consist of some basic real analysis (often as motivation) and a minimal encounter with the language of modern mathematics. The text is suitable for students with a background of one year of analysis proper, that is one year beyond the usual preparatory informal courses on calculus and analytic geometry. There is sufficient material for a course varying from 40 to 60 lectures according to the background and needs of the students. Although the book is a development of several courses I have given to honours students of mathematics at Aberdeen University, I have tried to keep the exposition as simple as possible with the object of making some of the theory of complex analysis available to a fairly wide audience.

Since there are already many books on complex analysis it is necessary to say a word about the special features of the present book. This can best be done by a simple illustration. The central theorem of complex analysis is the famous Cauchy theorem and it has long been considered a deep and difficult theorem. This is certainly true of the most general case; but the theorem admits a straightforward proof for the case in which the given domain is starlike. This latter case is adequate for almost all the results of complex function theory. Since I believe that the student has every right to demand a proper treatment of the version of the Cauchy theorem which is to be used for subsequent results, I have chosen to prove the Cauchy theorem only for the starlike case. This means that I am able to give complete (and straightforward) proofs of all the subsequent theorems even though the theorems are not always the most general possible. I indicate how the theorems may be extended for the rare occasions on which this is necessary. This then describes the aim of the book—to give an introductory course in complex analysis that is rigorous and yet amenable to the serious undergraduate student. I hope that the course is also enjoyable.

I have freely borrowed from other sources the ideas that I consider to be the best and the simplest. Certain sources will be transparent to the discerning reader. For example Section 9.2 is wholly inspired by the corresponding account in the more advanced book *Theory of Analytic Functions of One or Several Variables*, by Professor H. Cartan. On the other hand I think that occasionally some of the details are new as in parts of Chapters 5 and 10.

It is a pleasure to record several personal acknowledgements. I owe much indirectly to Professor F. F. Bonsall; any lucidity of style that may occur in this book arises largely from his personal example as a supervisor and friend. Some of the work for this book was carried out while I was visiting Yale University where I was fortunate to have many fruitful conversations with Professor E. Lee Stout. I am grateful to my Aberdeen colleagues Eric E. Morrison and Alan J. White who read parts of the manuscript and gave encouragement and constructive criticisms. I am also indebted to the secretaries and typists of the mathematics departments both at Aberdeen and Yale who prepared various editions of the manuscript. Finally I wish to express my thanks to the editorial staff of John Wiley & Sons for their advice and encouragement.

Aberdeen University, 1968 JOHN DUNCAN

CONTENTS

1

METRIC SPACE PRELIMINARIES

1.1 Set theoretic notation and terminology

The basic language of contemporary mathematics is the language of set theory. We shall assume that the reader is acquainted with the intuitive ideas of set theory. Since the notation of set theory varies from book to book we shall begin by listing briefly the notation which we shall employ in this book.

We denote set membership by $\in$; thus if E is a set and x is an element (or member) of E, we write $x \in E$. If x is not an element of E, we write $x \notin E$. The set of all elements of E with property P is denoted by

$$\{x: x \in E, P(x)\}.$$

When there is no danger of confusion we shall write this set more simply as $\{x: P(x)\}$. When we wish to indicate that property P holds for each x in E we write

$$P(x) \qquad (x \in E).$$

Singletons (i.e. sets consisting of exactly one element) are denoted by $\{a\}$. The empty set is denoted by $\varnothing$.

Set inclusion is denoted by $\subset$; thus if A and B are subsets of E we write $A \subset B$ if every element of A is also an element of B. If $A \subset B$ and $B \subset A$ then $A = B$. If $A \subset B$ and $A \neq B$ we say that A is a proper subset of B. We use no special symbol for proper inclusion since it is rarely necessary to exclude the possibility of equality.

Union, intersection and complementation are denoted by $\cup$, $\cap$, $\setminus$ respectively; thus if A and B are subsets of E we write

$$A \cup B = \{x: x \in A \text{ or } x \in B\}$$
$$A \cap B = \{x: x \in A \text{ and } x \in B\}$$
$$A \setminus B = \{x: x \in A \text{ and } x \notin B\}.$$

More generally if Λ is some index set and A_λ is a subset of E for each $\lambda \in \Lambda$ we write

$$\bigcup \{A_\lambda : \lambda \in \Lambda\} = \{x : x \in A_\lambda \text{ for some } \lambda\}$$
$$\bigcap \{A_\lambda : \lambda \in \Lambda\} = \{x : x \in A_\lambda \ (\lambda \in \Lambda)\}.$$

If A and B are two non-empty sets we denote their Cartesian product by $A \times B$; thus

$$A \times B = \{(x, y) : x \in A, y \in B\}.$$

When $B = A$ we sometimes write $A \times A$ more briefly as A^2. More generally we write A^{n+1} for the Cartesian product of A with itself n times.

The sets consisting of the integers, the positive integers (i.e. 1, 2, 3, ...), the rationals and the reals are denoted respectively by **Z**, **P**, **Q**, **R**. The symbols **Z**, **P**, **Q**, **R** will be used exclusively for the above sets. We assume that the reader is acquainted with some of the basic properties of these sets. In particular we assume that the reader knows that the real numbers form a complete ordered field.

Given non-empty sets X and Y, a function (or mapping) from X to Y is a subset f of $X \times Y$, such that

(i) given $x \in X$ there is $y \in Y$, such that $(x, y) \in f$
(ii) $(x, y) \in f, (x, z) \in f \Rightarrow y = z$.

We denote the mapping symbolically by

$$f: X \to Y.$$

As usual we write $f(x)$ for the image of x under f. Given $S \subset X, T \subset Y$ we write

$$f(S) = \{f(x) : x \in S\}$$
$$f^{-1}(T) = \{x : x \in X, f(x) \in T\}.$$

The set $f(X)$ is called the range of f. We say that the mapping f is onto* if $f(X) = Y$. We say that the mapping f is one-to-one if

$$f(x) = f(y) \Rightarrow x = y.$$

If f is one-to-one the inverse function $f^{-1}: f(X) \to X$ is defined by

$$f^{-1}(y) = x \quad \text{if} \quad x = f(y).$$

The reader should be careful to distinguish the possible meanings of f^{-1} at each occurrence.

* We write 'onto' as one word since we regard it as a special mathematical term.

If E is a non-empty subset of X we define the restriction of f to E by

$$f|_E(x) = f(x) \qquad (x \in E)$$

so that $f|_E$ is a function from E to Y. When there is no possibility of confusion we shall often write $f|_E$ more simply as f.

We shall use $\circ$ to denote the composition of functions; more precisely, given functions $f: X \to Y$, $g: Y \to Z$ we define $g \circ f: X \to Z$ by

$$(g \circ f)(x) = g(f(x)) \qquad (x \in X).$$

A sequence in a set X is a function from **P** into X. We shall follow the standard practice of writing a sequence as $\{x_n\}$. Moreover we shall also use $\{x_n\}$ to denote the range of the sequence, i.e. the set $\{x_n : n \in \mathbf{P}\}$. This is the only case in which we identify a function with its range. A subsequence of $\{x_n\}$ is a sequence of the form $\{x_{\varphi(n)}\}$ where $\varphi: \mathbf{P} \to \mathbf{P}$ is a function satisfying

$$\varphi(m) > \varphi(n) \qquad (m > n).$$

Thus a subsequence is a function of the form $x \circ \varphi$ where φ is a function as above.

We shall sometimes use the common abbreviation 'iff' for 'if and only if'; we shall also use the symbol $\Leftrightarrow$ for equivalent statements. The symbol $\Rightarrow$ is used for 'implies', when appropriate.

1.2 Elementary properties of metric spaces

For a satisfactory attack on the theory of complex analysis the student must be equipped with several topological tools. Most of these tools find their setting in the general theory of topological spaces, but in this book we need only the special case of metric spaces. Our main interest is with very special examples of metric spaces, namely subsets of the complex plane; but we shall also be considering metric spaces whose elements are functions. We therefore set out in this chapter those results from the general theory of metric spaces that we shall need in subsequent chapters. We believe that the student will find it easier to understand metric space theory in the general case than in the case of, say, subsets of the complex plane; for in an abstract metric space we have stripped away all the additional structure which, in special cases, often obscures the essence of the theory.

Much of the material in this chapter is probably familiar to many of our readers. Accordingly we shall give but little motivation and few illustrative examples.

A *metric space* is a pair (X, d) where X is a non-empty set and d is a *metric* on X, i.e. d is a real valued function on $X \times X$ satisfying

(i) $d(x, y) \geqslant 0$ $(x, y \in X)$; $d(x, y) = 0$ iff $x = y$,
(ii) $d(x, y) = d(y, x)$ $(x, y \in X)$,
(iii) $d(x, z) \leqslant d(x, y) + d(y, z)$ $(x, y, z \in X)$.

When no confusion is likely we denote a metric space by X for short. It is of course helpful to think of $d(x, y)$ as the 'distance' from the point x to the point y. For this reason, condition (iii) is referred to as the 'triangle inequality'. The student may also find it helpful on occasions to draw rough geometrical pictures to facilitate his understanding.

The fundamental example of a metric space is obtained by taking $X = \mathbf{R}$ and

$$d(x, y) = |x - y| \qquad (x, y \in \mathbf{R}).$$

It is a simple exercise to verify that d is indeed a metric on $\mathbf{R}$. We say that d is the *usual metric* on $\mathbf{R}$, and whenever we speak of the metric space $\mathbf{R}$ we mean that $\mathbf{R}$ has the usual metric.

Every metric space gives rise to other metric spaces in the following manner. Let (X, d) be any metric space and let E be a non-empty subset of X. Let ρ be the restriction of the function d to $E \times E$. It is trivially verified that ρ is a metric on E. We say that ρ is the *relative metric* induced on E by the metric d. Given any metric space X and any non-empty subset E of X, the set E with the relative metric will often simply be called the metric space E. The metric space E is often referred to as a *subspace* of E. We shall subsequently see that the properties of subspaces may be quite different from the properties of the original metric space. If E is a non-empty subset of $\mathbf{R}$, by the metric space E we mean the set E together with the relative metric induced on E by the usual metric on $\mathbf{R}$.

Another standard method for producing new metric spaces from old ones is by taking Cartesian products. We illustrate the method by an example. Given $n \in \mathbf{P}$ let $X = \mathbf{R}^n$ and define d on $\mathbf{R}^n \times \mathbf{R}^n$ by

$$d((x_1, x_2, \ldots, x_n), (y_1, y_2, \ldots, y_n)) = \left\{ \sum_{r=1}^{n} (x_r - y_r)^2 \right\}^{\frac{1}{2}}.$$

The student may verify that d is a metric on $\mathbf{R}^n$ (only the triangle inequality needs any thought); we say that d is the *usual metric* on $\mathbf{R}^n$. Other examples of metrics on $\mathbf{R}^n$ may be found in Problem 1.2.

The following metric space concepts are motivated by elementary geometry and the case of the metric space $\mathbf{R}$. Let (X, d) be any metric space. Given $p \in X$ and $\epsilon > 0$, the set

$$\{x \colon x \in X, d(p, x) < \epsilon\}$$

is called the *open ball* with *centre* p and *radius* ϵ and is denoted by $B(p, \epsilon)$. When we wish to specify which metric is involved we shall write $B_d(p, \epsilon)$. The set

$$\{x: x \in X, d(p, x) \leqslant \epsilon\}$$

is called the *closed ball* with *centre* p and *radius* ϵ. A *neighbourhood* of p is any subset of X that contains some open ball with centre p. A subset E of X is said to be *bounded* if it is contained in some ball (open or closed). Observe that $B(p, \epsilon)$ is thus a bounded neighbourhood of p for each $\epsilon > 0$. Given a sequence $\{x_n\}$ in X we say that $\{x_n\}$ is *convergent* if there is $x \in X$ such that for every $\epsilon > 0$ there is $N(\epsilon) \in \mathbf{P}$ such that

$$x_n \in B(x, \epsilon) \qquad (n > N(\epsilon)).$$

We call x a *limit* of the sequence $\{x_n\}$ and say that $\{x_n\}$ *converges* to x. We write this as

$$x_n \to x \quad \text{as} \quad n \to \infty$$

or

$$\lim_{n \to \infty} x_n = x.$$

It is convenient at this point to make some simple observations about convergent sequences. For the metric space $\mathbf{R}$ the above definition coincides exactly with the classical definition of the convergence of a sequence of real numbers. For any metric space it is simple to verify that

$$\lim_{n \to \infty} x_n = x \Leftrightarrow \lim_{n \to \infty} d(x, x_n) = 0.$$

This indicates how convergence in the metric space $\mathbf{R}$ is fundamental to convergence in any metric space. If $\{x_n\}$ converges to x it is clear that any subsequence of $\{x_n\}$ also converges to x. In the metric space $\mathbf{R}^2$ it is easy to see that $\{(x_n, y_n)\}$ converges to (x, y) if and only if $\{x_n\}$ converges to x and $\{y_n\}$ converges to y; there is an obvious generalization to sequences in $\mathbf{R}^m$ for any $m > 2$. At this point we have still to determine how many limits a convergent sequence may have. We ought to suspect that a convergent sequence in any metric space can have only one limit. This is indeed true—but it requires to be proved.

Proposition 1.1. *Let (X, d) be a metric space and let $\{x_n\}$ be a sequence in X such that $\{x_n\}$ converges to x and to y. Then $x = y$.*

Proof. We have

$$\lim_{n \to \infty} d(x, x_n) = 0 = \lim_{n \to \infty} d(y, x_n).$$

Given $\epsilon > 0$ we thus have some $N_1(\epsilon) \in \mathbf{P}$, such that

$$d(x, x_n) < \epsilon \qquad (n > N_1(\epsilon)).$$

Similarly, there is $N_2(\epsilon) \in \mathbf{P}$, such that

$$d(y, x_n) < \epsilon \qquad (n > N_2(\epsilon)).$$

Choose any integer $n > \max(N_1(\epsilon), N_2(\epsilon))$ and we have

$$\begin{aligned} d(x, y) &\leqslant d(x, x_n) + d(x_n, y) \\ &= d(x, x_n) + d(y, x_n) \\ &< 2\epsilon. \end{aligned}$$

We have now shown that $0 \leqslant d(x, y) < 2\epsilon$ for every $\epsilon > 0$. It follows that $d(x, y) = 0$, and so $x = y$.

We are now entitled to speak about *the* limit of a convergent sequence in a metric space. We consider next some special classes of subsets of a metric space. Given a metric space X, a subset O of X is said to be *open* if each point x of O has a neighbourhood that is contained in O. Thus O is open in X if and only if for each $x \in O$ there is $\epsilon > 0$ such that $B(x, \epsilon) \subset O$. Given $A \subset X$ and $p \in X$, we say that p is a *cluster point* of A if every neighbourhood of p contains a point of A other than p. Thus p is a cluster point of A if and only if for each $\epsilon > 0$ there is $q \in A \cap B(p, \epsilon)$ with $q \neq p$. The set of all cluster points of A is denoted by A'. Given $A \subset X$ we say that A is *closed* if it contains all its cluster points, i.e. if $A' \subset A$. Since both open and closed subsets of a metric space are of particular significance we shall now prove several results about them.

Proposition 1.2. *The union of any family of open subsets of a metric space is open; the intersection of a finite family of open subsets is open.*

Proof. Let $\{O_\lambda : \lambda \in \Lambda\}$ be a family of open subsets of a metric space and let $O = \bigcup \{O_\lambda : \lambda \in \Lambda\}$. Given $p \in O$ there is some $\lambda \in \Lambda$ such that $p \in O_\lambda$. Since O_λ is open there is $\epsilon > 0$ such that $B(p, \epsilon) \subset O_\lambda$. Since $O_\lambda \subset O$ we have $B(p, \epsilon) \subset O$ and therefore O is open.

Let $O_1, O_2, \ldots, O_n$ be open sets and let $O = \bigcap \{O_r : r = 1, 2, \ldots, n\}$. Given $p \in O$ we have $p \in O_r$ $(r = 1, 2, \ldots, n)$. Since each O_r is open there is $\epsilon_r > 0$ such that

$$B(p, \epsilon_r) \subset O_r \qquad (r = 1, 2, \ldots, n).$$

Let $\epsilon = \min\{\epsilon_r : r = 1, 2, \ldots, n\}$. Then $\epsilon > 0$ and

$$B(p, \epsilon) \subset B(p, \epsilon_r) \subset O_r \qquad (r = 1, 2, \ldots, n).$$

It follows that $B(p, \epsilon) \subset O$ and so O is open.

We remark at this point that Proposition 1.2 serves as the definition for the more general objects known as topological spaces. More precisely a *topology* on a non-empty set X is defined to be a family $\mathscr{T}$ of subsets of X that contains $\varnothing$ and X and is closed under the operations of arbitrary unions and finite intersections. A *topological space* is then a pair $(X, \mathscr{T})$ where X is a non-empty set and $\mathscr{T}$ is a topology on X.

Proposition 1.3. *A subset F of a metric space is closed if and only if $X \setminus F$ is open.*

Proof. Let F be closed, so that $F' \subset F$. Let $p \in X \setminus F$. Since $p \notin F'$ some neighbourhood of p must be contained in $X \setminus F$. Therefore $X \setminus F$ is open.

Let $X \setminus F$ be open and let $p \in X \setminus F$. Then some neighbourhood of p is contained in $X \setminus F$, so that p cannot be a cluster point of F. This shows that $F' \cap (X \setminus F) = \varnothing$ and so $F' \subset F$, i.e. F is closed.

Proposition 1.4. *The union of a finite family of closed subsets of a metric space is closed; the intersection of any family of closed sets is closed.*

Proof. This follows easily from Propositions 1.2 and 1.3, and the de Morgan laws for complementation of sets.

The next result gives a very useful characterization of closed sets in terms of convergent sequences.

Proposition 1.5. *Given a subset F of a metric space X, the following statements are equivalent.*

(i) *F is closed.*

(ii) $\{x_n\} \subset F,\ x = \lim_{n\to\infty} x_n \quad \Rightarrow \quad x \in F.$

Proof. (i) $\Rightarrow$ (ii). Let F be closed and let $\{x_n\}$ be any convergent sequence such that $\{x_n\} \subset F$ and $x = \lim_{n\to\infty} x_n$. Suppose $x \in X \setminus F$. Given $\epsilon > 0$ we may choose $N \in \mathbf{P}$ such that $x_n \in B(x, \epsilon)$ $(n > N)$. This means that every neighbourhood of x contains points of $\{x_n\}$, and so of F, other than x. Therefore $x \in F'$. Since F is closed we have $F' \subset F$ and so $x \in F$. This is a contradiction and therefore $x \notin X \setminus F$, i.e. $x \in F$ and so condition (ii) is satisfied.

(ii) $\Rightarrow$ (i). Let condition (ii) hold and let $p \in F'$. Then every neighbourhood of p contains points of F. In particular, for each $n \in \mathbf{P}$ we may choose $p_n \in F \cap B\left(p, \dfrac{1}{n}\right)$. We thus have $\{p_n\} \subset F$ and $\lim_{n\to\infty} d(p, p_n) = 0$ so that $\lim_{n\to\infty} p_n = p$. It now follows from condition (ii) that $p \in F$. Thus $F' \subset F$ and so F is closed.

The basic examples of open sets and closed sets are given by open balls and closed balls. Let $B(x, \epsilon)$ be an open ball in a metric space and let $y \in B(x, \epsilon)$. To show that $B(x, \epsilon)$ is open we must produce some $\delta > 0$ such that $B(y, \delta) \subset B(x, \epsilon)$. Let $\delta = \epsilon - d(x, y)$ so that $\delta > 0$. Given $p \in B(y, \delta)$ we have

$$\begin{aligned} d(x, p) &\leqslant d(x, y) + d(y, p) \\ &< d(x, y) + \delta = \epsilon. \end{aligned}$$

Therefore $B(y, \delta) \subset B(x, \epsilon)$ and so $B(x, \epsilon)$ is an open set. We leave the student to prove that a closed ball is a closed set. As a further example of closed sets we see from Proposition 1.5 that any singleton is a closed set. It follows from Proposition 1.4 that any finite subset of a metric space is closed.

The unwary student may be tempted to think that any given subset of a metric space is either open or closed. In fact a subset of a metric space may be neither open nor closed—or it may be both open and closed! For example, in the metric space $\mathbf{R}$ it is easy to see that

$$E = \{x : x \in \mathbf{R}, 0 \leqslant x < 1\}$$

is neither open nor closed. Indeed there is no neighbourhood of 0 contained in E, so that E is not open. Moreover 1 is a cluster point of E which does not belong to E. On the other hand it is trivial to verify that in any metric space X the set X itself is both open and closed. Suppose now that we give the above set E the relative metric ρ. Then the set E is both open and closed in the metric space (E, ρ) even though it is neither open nor closed in the metric space $\mathbf{R}$. This situation is often a source of confusion to the student. Indeed much of the common misunderstanding of metric spaces revolves around a misunderstanding of the relative metric. In subsequent chapters we shall frequently be working with relative metrics. It is therefore essential for the student to understand what happens to open sets, closed sets and convergent sequences when the relative metric comes into play.

Proposition 1.6. *Let (X, d) be a metric space and let E be a non-empty subset of X with relative metric ρ. A subset V of E is open in (E, ρ) iff it is of the form $V = E \cap U$ where U is open in (X, d).*

Proof. Given $x \in E$ we have

$$\begin{aligned} B_\rho(x, \epsilon) &= \{y : y \in E, \rho(x, y) < \epsilon\} \\ &= \{y : y \in E, d(x, y) < \epsilon\} \\ &= E \cap B_d(x, \epsilon). \end{aligned}$$

Let V be open in (E, ρ). For each $x \in V$ there is $\epsilon(x) > 0$ such that $B_\rho(x, \epsilon(x)) \subset V$. It follows that

$$\begin{aligned} V &= \bigcup \{B_\rho(x, \epsilon(x)) : x \in V\} \\ &= \bigcup \{E \cap B_d(x, \epsilon(x)) : x \in V\} \\ &= E \cap U \end{aligned}$$

where $U = \bigcup \{B_d(x, \epsilon(x)) : x \in V\}$. Since each $B_d(x, \epsilon(x))$ is open in (X, d) it follows from Proposition 1.2 that U is open in (X, d) and so V has the required form.

Conversely let $V = E \cap U$ where U is open in (X, d). Given $x \in V$ we have $x \in U$. Since U is open there is $\epsilon > 0$ such that $B_d(x, \epsilon) \subset U$. Therefore

$$\begin{aligned} B_\rho(x, \epsilon) &= E \cap B_d(x, \epsilon) \\ &\subset E \cap U = V. \end{aligned}$$

This shows that V is open in (E, ρ) as required.

Corollary. *Let E be an open subset of a metric space (X, d). Then $V \subset E$ is open in (E, ρ) iff it is open in (X, d).*

Proof. Let $V \subset E$ be open in (E, ρ). Then $V = E \cap U$, where U is open in (X, d). Since E is open in (X, d), we have V open in (X, d) by Proposition 1.2. Conversely, if $V \subset E$ is open in (X, d), then $V = E \cap V$ is open in (E, ρ) as required.

Proposition 1.7. *Let (X, d) be a metric space and let E be a non-empty subset of X with relative metric ρ. A subset F of E is closed in (E, ρ) iff it is of the form $F = E \cap H$, where H is closed in (X, d).*

Proof. This follows readily from Propositions 1.3 and 1.6.

Corollary. *Let E be a closed subset of a metric space (X, d). Then $F \subset E$ is closed in (E, ρ) iff it is closed in (X, d).*

Proof. Similar to the last corollary.

To illustrate the above two results let $X = \mathbf{R}$ and

$$E = \{x : 0 \leqslant x \leqslant 1\} \cup \{x : 2 < x < 3\}.$$

Then $V = \{x : 0 \leqslant x \leqslant 1\}$ is open in (E, ρ) but not in $\mathbf{R}$; moreover $F = \{x : 2 < x < 3\}$ is closed in (E, ρ) but not in $\mathbf{R}$.

Proposition 1.8. *Let (X, d) be a metric space and let E be a non-empty subset of X with relative metric ρ. If $\{x_n\}$ is a convergent sequence in (E, ρ) then it is a convergent sequence in (X, d) (with the same limit). If $\{x_n\} \subset E$*

is a convergent sequence in (X, d) then it is a convergent sequence in (E, ρ) iff its limit is in E.

Proof. Let $\{x_n\}$ be a convergent sequence in (E, ρ). Then there is $x \in E$ such that $\lim_{n\to\infty} \rho(x, x_n) = 0$. Therefore $\lim_{n\to\infty} d(x, x_n) = 0$ and so $\{x_n\}$ converges to x in (X, d).

Let $\{x_n\} \subset E$ be a convergent sequence in (X, d) with limit x. Suppose that $x \in E$. Then

$$\lim_{n\to\infty} \rho(x, x_n) = \lim_{n\to\infty} d(x, x_n) = 0$$

and so $\{x_n\}$ converges to x in (E, ρ). Suppose now that $x \notin E$ and that $\{x_n\}$ converges to y in (E, ρ). By the first part of this proposition we have that $\{x_n\}$ converges to y in (X, d). Proposition 1.1 now gives $x = y$. This contradiction completes the proof.

To illustrate the above result, observe that the sequence $\left\{\frac{1}{n}\right\}$ converges in $\mathbf{R}$ but fails to converge in the metric space $\{x: 0 < x \leqslant 1\}$.

We have already remarked that a subset of a metric space may be neither open nor closed. It is, however, possible, in a natural way, to associate with each subset of a metric space a corresponding open set and a corresponding closed set. This association will be useful in subsequent chapters. Let (X, d) be any metric space and let $E \subset X$. The *interior* of E is the set of all points of E that have some neighbourhood contained in E; it is denoted* by E^0. We show below that the interior of a set is open. The *closure* of E is the union of E and E'; it is denoted by E^-. We say that a subset E of X is *dense* in X if $E^- = X$. The *boundary* of E is the intersection of E^- and $(X \setminus E)^-$; it is denoted by $b(E)$. We show below that the closure of a set is closed and so the boundary of a set is also closed.

Proposition 1.9. *Let (X, d) be a metric space and let $E \subset X$.*

(i) *E^0 is the largest open set contained in E.*
(ii) *E^- is the smallest closed set containing E.*
(iii) $b(E) = b(X \setminus E) = E^- \setminus E^0$.

Proof. (i) Given $p \in E^0$ there is $\epsilon > 0$ such that $B(p, \epsilon) \subset E$. Let $x \in B(p, \epsilon)$. Since $B(p, \epsilon)$ is open there is $\delta > 0$ such that $B(x, \delta) \subset B(p, \epsilon)$.

* In some texts the interior of E is denoted by int E. We shall use the symbol *int* for a different purpose in Chapter 5 and subsequently.

Thus $B(x, \delta) \subset E$ and so $x \in E^0$. This shows that $B(p, \epsilon) \subset E^0$ and so E^0 is an open set. Let U be any open set such that $U \subset E$. Given $x \in U$ there is $\epsilon > 0$ such that $B(x, \epsilon) \subset U \subset E$. Thus $x \in E^0$ and so $U \subset E^0$. In other words E^0 is the largest open set contained in E.

(ii) We observe that $x \in X \setminus E^-$ iff some neighbourhood of x is contained in $X \setminus E$, i.e. iff $x \in (X \setminus E)^0$. It follows from (i) above that $X \setminus E^-$ is open and so E^- is closed by Proposition 1.3. Let F be any closed set such that $F \supset E$. Then $X \setminus F$ is open and $X \setminus F \subset X \setminus E$. It follows from (i) above that $X \setminus F \subset (X \setminus E)^0 = X \setminus E^-$. Therefore $F \supset E^-$ and so E^- is the smallest closed set containing E.

(iii) It is immediate from the definition of boundary that $b(X \setminus E) = (X \setminus E)^- \cap (X \setminus (X \setminus E))^- = (X \setminus E)^- \cap E^- = b(E)$. Recall from the proof of (ii) that

$$X \setminus (X \setminus E)^- = (X \setminus (X \setminus E))^0 = E^0$$

and so $(X \setminus E)^- = X \setminus E^0$. Therefore

$$b(E) = E^- \cap (X \setminus E^0) = E^- \setminus E^0.$$

Observe from the above result that E is open iff $E = E^0$ and E is closed {iff $E = E^-$. To illustrate the above result further let $X = \mathbf{R}$ and $E = x: 0 \leqslant x < 1\}$. We may easily check that

$$E^0 = \{x: 0 < x < 1\}, \qquad E^- = \{x: 0 \leqslant x \leqslant 1\}, \qquad b(E) = \{0, 1\}.$$

On the other hand if X is the metric space $\{x: 0 \leqslant x \leqslant 1\}$ and $E = \{x: 0 \leqslant x < 1\}$, then

$$E^0 = E, \qquad E^- = X, \qquad b(E) = \{1\}.$$

In other words, care is again required when working with the relative metric.

PROBLEMS 1

1. Let X be any non-empty set. Define d on $X \times X$ by

$$d(x, y) = \begin{cases} 1 & \text{if } x \neq y \\ 0 & \text{if } x = y. \end{cases}$$

Show that d is a metric on X (it is called the *discrete metric*). Which subsets of X are (i) open, (ii) closed, (iii) bounded? Which sequences in X converge?

2. Let $n \in \mathbf{P}$ and let $X = \mathbf{R}^n$. Define d_1 and d_∞ on $X \times X$ by

$$d_1((x_1, x_2, \ldots, x_n), (y_1, y_2, \ldots, y_n)) = \sum_{r=1}^{n} |x_r - y_r|$$

$$d_\infty((x_1, x_2, \ldots, x_n), (y_1, y_2, \ldots, y_n)) = \max \{|x_r - y_r| : r = 1, 2, \ldots, n\}.$$

Show that d_1 and d_∞ are metrics on $\mathbf{R}^n$. Show also that $\mathbf{R}^n$ with the usual metric, $(\mathbf{R}^n, d_1)$ and $(\mathbf{R}^n, d_\infty)$ have the same open sets. Now let (X_r, ρ_r) $(r = 1, 2, \ldots, n)$ be metric spaces and let $X = X_1 \times X_2 \times \cdots \times X_n$. Generalize the above results by defining analogous metrics on X.

3. Let S be any non-empty set and let $\mathscr{B}_{\mathbf{R}}(S)$ be the set of all bounded real functions on S. Define d on $\mathscr{B}_{\mathbf{R}}(S) \times \mathscr{B}_{\mathbf{R}}(S)$ by

$$d(f, g) = \sup \{|f(s) - g(s)| : s \in S\}.$$

Show that d is a metric on $\mathscr{B}_{\mathbf{R}}(S)$.

4. Show that any convergent sequence in a metric space is bounded.

5. Let E be a bounded subset of a metric space X.

(i) Given $p \in X$ show that there is $\epsilon > 0$ such that $E \subset B(p, \epsilon)$.
(ii) Show that E^- is bounded.

6. Let E be a subset of a metric space.

(i) Given $p \in E'$ show that every neighbourhood of p contains an infinite subset of E. Show also that there is a sequence $\{p_n\}$ of distinct points of E such that $p = \lim_{n\to\infty} p_n$.
(ii) Show that E' is closed. Is $(E')' = E'$?
(iii) Given $p \in E^-$ show that there is a sequence $\{p_n\}$ in E such that $p = \lim_{n\to\infty} p_n$. If $\{p_n\}$ is any convergent sequence such that $\{p_n\} \subset E$ show that its limit is in E^-.

7. Let E be a subset of a metric space. Show that E^- is the intersection of all the closed sets that contain E, and E^0 is the union of all the open sets that are contained in E. Show also that

$$E^- = E \cup b(E), \qquad E^0 = E \setminus b(E).$$

8. Given subsets A, B of a metric space show that

(i) $(A^0)^0 = A^0$; $A \subset B \Rightarrow A^0 \subset B^0$; $(A \cap B)^0 = A^0 \cap B^0$,
(ii) $(A^-)^- = A^-$; $A \subset B \Rightarrow A^- \subset B^-$; $(A \cup B)^- = A^- \cup B^-$.

9. Let Y be a subspace of a metric space X and let $E \subset Y$. Relate the interior, closure, and boundary of E in Y to those in X.

10. Given any open ball $B(x, \epsilon)$ in a metric space determine $B(x, \epsilon)^-$ and $b(B(x, \epsilon))$. Can $b(B(x, \epsilon))$ be empty?

11. In the metric space **R** determine $\mathbf{Q}^0$, $\mathbf{Q}^-$, $b(\mathbf{Q})$. Show also that $\mathbf{Q}^2$ is dense in $\mathbf{R}^2$ (with the usual metric). Generalize this last result.

12. Let E be a bounded subset of **R**. Show that $\sup E$ and $\inf E$ belong to E^-.

1.3 Continuous functions on metric spaces

In this section we shall consider the important concept of the continuity of mappings between metric spaces and we shall show how it is related to the concepts of open sets, closed sets, and convergence.

Let (X, d), (Y, e) be metric spaces, let f be a function from X to Y, and let $x \in X$. We say that f is *continuous at* x if for each $\epsilon > 0$ there is $\delta(\epsilon) > 0$ such that

$$f(B_d(x, \delta(\epsilon))) \subset B_e(f(x), \epsilon)$$

i.e.

$$d(x, y) < \delta(\epsilon) \quad \Rightarrow \quad e(f(x), f(y)) < \epsilon.$$

We then write

$$\lim_{y \to x} f(y) = f(x).$$

In the case when $X = Y = \mathbf{R}$ this definition coincides exactly with the classical definition of continuity at a point. If E is a subset of X we say that f is *continuous on* E if f is continuous at each point of E. If f is continuous on X we say more simply that f is *continuous*. If f maps X one-to-one onto Y and if both f and f^{-1} are continuous we say that f is a *homeomorphism* of X with Y; we then say that X is *homeomorphic* with Y. We shall denote by *Hom* (X, Y) the set of all homeomorphisms of X with Y. If f maps X onto Y and if

$$e(f(x), f(y)) = d(x, y) \qquad (x, y \in X)$$

we say that f is an *isometry* of X with Y; we then say that X is *isometric* with Y. It is immediate that an isometry is one-to-one; for if $f(x) = f(y)$ then $d(x, y) = e(f(x), f(y)) = 0$ and so $x = y$. We leave to the student the elementary exercise of proving that an isometry is a homeomorphism. The student should also find an example of a homeomorphism that is not an isometry.

By way of illustration we show how to produce continuous real functions on any metric space (X, d). Let $x_0 \in X$ and define $f\colon X \to \mathbf{R}$ by

$$f(x) = d(x, x_0) \qquad (x \in X).$$

Given $x, y \in X$ we have

$$d(x, x_0) \leqslant d(y, x_0) + d(x, y)$$
$$d(y, x_0) \leqslant d(x, x_0) + d(x, y).$$

It follows that

$$|f(x) - f(y)| = |d(x, x_0) - d(y, x_0)| \leqslant d(x, y).$$

Given $\epsilon > 0$ let $\delta = \epsilon$. Then

$$d(x, y) < \delta \quad \Rightarrow \quad |f(x) - f(y)| < \epsilon$$

and so f is continuous at x. Since x was arbitrary f is thus continuous.

It is extremely useful to have several characterizations of when a given function is continuous. The next two results are fundamental in this connection.

Proposition 1.10. *Let (X, d), (Y, e) be metric spaces, let $f: X \to Y$ and let $x \in X$. The following statements are equivalent.*

(i) *f is continuous at x.*
(ii) *$\{f(x_n)\}$ converges to $f(x)$ whenever $\{x_n\}$ converges to x.*

Proof. (i) $\Rightarrow$ (ii). Let f be continuous at x and let $\{x_n\}$ converge to x. Given $\epsilon > 0$ there is $\delta > 0$ such that $f(B_d(x, \delta)) \subset B_e(f(x), \epsilon)$. Since $\{x_n\}$ converges to x there is $N \in \mathbf{P}$ such that

$$x_n \in B_d(x, \delta) \qquad (n > N)$$

and hence

$$f(x_n) \in B_e(f(x), \epsilon) \qquad (n > N).$$

Therefore $\{f(x_n)\}$ converges to $f(x)$ as required.

(ii) $\Rightarrow$ (i). Let condition (ii) hold and suppose that f is not continuous at x. Then there is some $\epsilon_0 > 0$ such that for every $\delta > 0$ we have

$$f(B_d(x, \delta)) \not\subset B_e(f(x), \epsilon_0).$$

In particular, if we choose $\delta = \dfrac{1}{n}$ $(n \in \mathbf{P})$ there is a corresponding x_n in $B_d\left(x, \dfrac{1}{n}\right)$ such that $f(x_n) \notin B_e(f(x), \epsilon_0)$. It follows that $\{x_n\}$ converges to x and

$$e(f(x), f(x_n)) \geqslant \epsilon_0 \qquad (n \in \mathbf{P})$$

so that $\{f(x_n)\}$ does not converge to $f(x)$. This contradiction completes the proof.

Proposition 1.11. *Let (X, d), (Y, e) be metric spaces and let $f: X \to Y$. The following statements are equivalent.*

(i) *f is continuous.*
(ii) *$f^{-1}(F)$ is closed in X whenever F is closed in Y.*
(iii) *$f^{-1}(O)$ is open in X whenever O is open in Y.*

Proof. (i) $\Rightarrow$ (ii). Let f be continuous and let F be closed in Y. Let $\{x_n\} \subset f^{-1}(F)$ with $\lim_{n\to\infty} x_n = x$. Since f is continuous it follows from Proposition 1.10 that $\lim_{n\to\infty} f(x_n) = f(x)$. Since $\{f(x_n)\} \subset F$ and F is closed, it follows from Proposition 1.5 that $f(x) \in F$ and so $x \in f^{-1}(F)$. We conclude from Proposition 1.5 that $f^{-1}(F)$ is closed, i.e. condition (ii) holds.

(ii) $\Rightarrow$ (iii). Let condition (ii) hold and let O be open in Y. By Proposition 1.3 $Y \setminus O$ is closed in Y and so $f^{-1}(Y \setminus O)$ is closed in X by condition (ii). Since

$$f^{-1}(Y \setminus O) = X \setminus f^{-1}(O)$$

it follows that $X \setminus f^{-1}(O)$ is closed in X and therefore $f^{-1}(O)$ is open in X by Proposition 1.3. Thus condition (iii) holds.

(iii) $\Rightarrow$ (i). Let condition (iii) hold and let $x \in X$. Given $\epsilon > 0$ we have that $B_e(f(x), \epsilon)$ is open in Y and so by condition (iii) $f^{-1}(B_e(f(x), \epsilon)$ is an open set in X which contains x. Hence there is $\delta > 0$ such that

$$B_d(x, \delta) \subset f^{-1}(B_e(f(x), \epsilon)).$$

Therefore

$$f(B_d(x, \delta)) \subset B_e(f(x), \epsilon).$$

and so f is continuous at x. Since x was arbitrary we have that f is continuous. The proof is complete.

Continuity has the pleasant property of being preserved under composition. This statement is made precise in the next result.

Proposition 1.12. *Let X, Y, Z be metric spaces and let $f: X \to Y$, $g: Y \to Z$. If f is continuous at x and g is continuous at $f(x)$, then $g \circ f$ is continuous at x. If f and g are continuous, then $g \circ f$ is continuous.*

Proof. Let $\{x_n\} \subset X$ with $\lim_{n\to\infty} x_n = x$. Since f is continuous at x it follows from Proposition 1.10 that

$$\lim_{n\to\infty} f(x_n) = f(x).$$

Since g is continuous at $f(x)$ it then follows that

$$\lim_{n\to\infty} g(f(x_n)) = g(f(x)).$$

We conclude from Proposition 1.10 that $g \circ f$ is continuous at x. The final statement of the proposition is now obvious.

Corollary. *If $f \in Hom(X, Y)$, $g \in Hom(Y, Z)$, then $g \circ f \in Hom(X, Z)$.*

Proof. It is trivial to verify that $g \circ f$ maps X one-to-one onto Z. By the above result we have that $g \circ f$ is continuous. Since $(g \circ f)^{-1} = f^{-1} \circ g^{-1}$ we also have that $(g \circ f)^{-1}$ is continuous.

It is also important to study continuity in relation to the relative metric.

Proposition 1.13. *Let (X, d), (Y, e) be metric spaces, let $f: X \to Y$, and let E be a non-empty subset of X with relative metric ρ. If f is continuous on E then $f|_E$ is continuous on the metric space (E, ρ). If $f|_E$ is continuous on the metric space (E, ρ) and if E is open in (X, d), then f is continuous on E.*

Proof. Let f be continuous on E and let $x \in E$. Given $\epsilon > 0$ there is $\delta > 0$ such that $f(B_d(x, \delta)) \subset B_e(f(x), \epsilon)$. Since $B_\rho(x, \delta) \subset B_d(x, \delta)$ we have

$$f(B_\rho(x, \delta)) \subset f(B_d(x, \delta)) \subset B_e(f(x), \epsilon).$$

Since x was arbitrary this shows that $f|_E$ is continuous on the metric space (E, ρ).

Let $f|_E$ be continuous on (E, ρ) and let E be open in (X, d). Given $x \in E$ there is $\delta_1 > 0$ such that $B_d(x, \delta_1) \subset E$, and so for $0 < \delta \leqslant \delta_1$ we have

$$B_\rho(x, \delta) = E \cap B_d(x, \delta) = B_d(x, \delta).$$

Since $f|_E$ is continuous at x, given $\epsilon > 0$ there is $\delta_2 > 0$ such that $f(B_\rho(x, \delta_2)) \subset B_e(f(x), \epsilon)$. Let $\delta = \min(\delta_1, \delta_2)$ so that $\delta > 0$. We now have

$$f(B_d(x, \delta)) = f(B_\rho(x, \delta)) \subset f(B_\rho(x, \delta_2)) \subset B_e(f(x), \epsilon)$$

and so f is continuous at x. Since x was arbitrary in E the proof is now complete.

This last result will often appear tacitly in subsequent chapters. To illustrate the result, let $a, b \in \mathbf{R}$ with $a < b$, let $E = \{x: x \in \mathbf{R}, a \leqslant x \leqslant b\}$, and let f be a continuous real function on the metric space E. In classical language the continuity at a and b is 'one-sided' continuity; we follow the classical notation by writing

$$\lim_{x \to a+} f(x) = f(a), \qquad \lim_{x \to b-} f(x) = f(b).$$

Now let $g: \mathbf{R} \to \mathbf{R}$ be defined by

$$g(x) = \begin{cases} 1 & \text{if} \quad x \in E \\ 0 & \text{if} \quad x \notin E. \end{cases}$$

It is obvious that $g|_E$ is continuous on the metric space E. On the other hand g itself is not continuous on E since it clearly fails to be continuous at a and b. It is of course true that g is continuous on $\{x: a < x < b\}$.

In subsequent chapters we shall frequently discuss sequences of functions with values in a metric space. For such sequences there are two common concepts of convergence, namely pointwise and uniform. More precisely, let X be any non-empty set, let $E \subset X$, and let (Y, e) be a metric space. Let f and f_n $(n \in \mathbf{P})$ be functions from X to Y. We say that $\{f_n\}$ converges to f *pointwise on E* if given $x \in E$, $\epsilon > 0$ there is $N(x, \epsilon) \in \mathbf{P}$ such that

$$f_n(x) \in B_e(f(x), \epsilon) \qquad (n > N(x, \epsilon))$$

i.e. if for each $x \in E$, $\{f_n(x)\}$ converges to $f(x)$. We write

$$\lim_{n\to\infty} f_n = f \qquad \text{(pointwise on } E).$$

We say that $\{f_n\}$ converges to f *uniformly on E* if given $\epsilon > 0$ there is $N(\epsilon) \in \mathbf{P}$ such that

$$f_n(x) \in B_e(f(x), \epsilon) \qquad (n > N(\epsilon), x \in E).$$

We write

$$\lim_{n\to\infty} f_n = f \qquad \text{(uniformly on } E).$$

Roughly speaking, in uniform convergence the same N satisfies the required condition for each of the points of E.

Special interest occurs when X is itself a metric space and the functions f_n are continuous. We may then ask which types of convergence preserve continuity. Pointwise convergence fails to have this property. For example let $X = \{x: x \in \mathbf{R}, 0 \leqslant x \leqslant 1\}$, $Y = \mathbf{R}$,

$$f_n(x) = x^n \qquad (x \in X, n \in \mathbf{P})$$

$$f(x) = \begin{cases} 0 & \text{if} \quad 0 \leqslant x < 1 \\ 1 & \text{if} \quad x = 1. \end{cases}$$

It is easy to show that each f_n is continuous on X and that

$$\lim_{n\to\infty} f_n = f \qquad \text{(pointwise on } X).$$

But it is clear that f is not continuous on X since it is not continuous at 1. The situation is much more satisfactory for uniform convergence.

Proposition 1.14. *Let (X, d), (Y, e) be metric spaces. Let $f_n\colon X \to Y$ be continuous for each $n \in \mathbf{P}$ and let $\{f_n\}$ converge to f uniformly on X. Then f is continuous.*

Proof. Let $p \in X$ and let $\epsilon > 0$. Since $\{f_n\}$ converges to f uniformly on X, we may certainly choose some $N \in \mathbf{P}$ such that

$$f_N(x) \in B_e(f(x), \tfrac{1}{3}\epsilon) \qquad (x \in X).$$

Since f_N is continuous at p, we may now choose $\delta > 0$ such that

$$f_N(B_d(p, \delta)) \subset B_e(f_N(p), \tfrac{1}{3}\epsilon).$$

If $d(p, x) < \delta$ we now have

$$\begin{aligned} e(f(p), f(x)) &\leqslant e(f(p), f_N(p)) + e(f_N(p), f_N(x)) + e(f_N(x), f(x)) \\ &< \tfrac{1}{3}\epsilon + \tfrac{1}{3}\epsilon + \tfrac{1}{3}\epsilon = \epsilon. \end{aligned}$$

This shows that f is continuous at p. Since p was arbitrary f is thus continuous as required.

We shall further discuss continuity in relation to the topics to be introduced in §§1.4, 1.6.

PROBLEMS 1

13. Let (X, d), (Y, e) be metric spaces and let $f\colon X \to Y$. Show that f is continuous as a mapping into the metric space Y if and only if it is continuous as a mapping onto the metric space $f(X)$.

14. Let (X, d) be a discrete metric space (i.e. d is the discrete metric on X as defined in Problem 1.1) and let (Y, e) be any metric space. Which functions $f\colon X \to Y$ are continuous? Which functions $g\colon Y \to X$ are continuous? Give an example of a one-to-one function from one metric space onto another which is continuous but is not a homeomorphism.

15. Let (X, d) be a metric space, and let f and g be real functions on X. Define the functions $f + g, fg, f \vee g, f \wedge g, |f|$ on X by

$$\begin{aligned} (f + g)(x) &= f(x) + g(x) \\ (fg)(x) &= f(x)\, g(x) \\ (f \vee g)(x) &= \max (f(x), g(x)) \\ (f \wedge g)(x) &= \min (f(x), g(x)) \\ |f|(x) &= |f(x)|. \end{aligned}$$

If f and g are continuous show that the above functions are also continuous.

16. Given $f: \mathbf{R} \to \mathbf{R}$, $g: \mathbf{R} \to \mathbf{R}$ define $h: \mathbf{R}^2 \to \mathbf{R}$ by

$$h(x, y) = (f(x), g(y)).$$

Show that h is continuous iff f and g are continuous.

17. Let X be a non-empty set and let $\{f_n\}$, $\{g_n\}$ be sequences of real functions on X such that

$$\lim_{n\to\infty} f_n = f \qquad \text{(uniformly on } X\text{)}$$
$$\lim_{n\to\infty} g_n = g \qquad \text{(uniformly on } X\text{)}.$$

Show that

$$\lim_{n\to\infty} (f_n + g_n) = f + g \qquad \text{(uniformly on } X\text{)}$$
$$\lim_{n\to\infty} (f_n g_n) = fg \qquad \text{(pointwise on } X\text{)}$$
$$\lim_{n\to\infty} (f_n \vee g_n) = f \vee g \qquad \text{(uniformly on } X\text{)}$$
$$\lim_{n\to\infty} (f_n \wedge g_n) = f \wedge g \qquad \text{(uniformly on } X\text{)}$$
$$\lim_{n\to\infty} |f_n| = |f| \qquad \text{(uniformly on } X\text{)}.$$

Give an example in which $\{f_n g_n\}$ fails to converge to fg uniformly on X.

18. Let (X, d) be a metric space. Given $p \in X$ let

$$f_p(x) = d(x, p) \qquad (x \in X).$$

If $\{p_n\} \subset X$ and $\lim_{n\to\infty} p_n = p$ show that

$$\lim_{n\to\infty} f_{p_n} = f_p \qquad \text{(uniformly on } X\text{)}.$$

19. Given a metric space (X, d) define e on $X \times X$ by

$$e(x, y) = \frac{d(x, y)}{1 + d(x, y)}.$$

Show that (X, e) is a metric space and that the identity mapping is a homeomorphism of (X, d) with (X, e). Note that $X = B_e(x, 1)$ for each $x \in X$.

20. Show that the relation of 'being homeomorphic' is an equivalence relation on the class of all metric spaces.

1.4 Compactness

In this section we shall discuss another special class of subsets of a metric space, namely compact sets. The concept of compactness is one of the most important ideas in topology. Many of the results that we obtain in later chapters depend upon compactness arguments. We shall soon see that the technical significance of compactness revolves around the fact that it allows many problems to be treated by finiteness arguments.

There are several equivalent definitions of compactness. The following definition will be most convenient for our purposes. Let (X, d) be a metric space and let $K \subset X$. We say that $\{U_\lambda: \lambda \in \Lambda\}$ is an *open cover* of K if each U_λ is open and $K \subset \bigcup\{U_\lambda: \lambda \in \Lambda\}$. We say that K is *compact* if every open cover of K contains a finite subcover, i.e. if every open cover contains a finite number of sets whose union contains K. We may observe immediately that every finite subset of a metric space is compact, so that compactness may be regarded as a generalization of finiteness. We say that X is a *compact metric space* if the set X itself is compact. We now have two possible concepts of compactness for a subset K of X. K may be a compact subset of X or K may be a compact metric space with the relative metric. Happily the two concepts are equivalent and so there will be no need to distinguish between them.

Proposition 1.15. *Let (X, d) be a metric space and let K be a non-empty subset of X with relative metric ρ. Then K is compact in (X, d) if and only if (K, ρ) is a compact metric space.*

Proof. Let K be a compact subset of (X, d) and let $\{V_\lambda: \lambda \in \Lambda\}$ be an open cover for K in the metric space (K, ρ). It follows from Proposition 1.6 that for each $\lambda \in \Lambda$ we may choose an open set U_λ in (X, d) such that $V_\lambda = K \cap U_\lambda$. It is clear that $\{U_\lambda: \lambda \in \Lambda\}$ is now an open cover for K in (X, d). Since K is compact in (X, d) there exist $\lambda_1, \lambda_2, \ldots, \lambda_n \in \Lambda$ such that

$$K \subset \bigcup\{U_{\lambda_j}: j = 1, 2, \ldots, n\}.$$

It follows that

$$\begin{aligned} K &\subset K \cap [\bigcup\{U_{\lambda_j}: j = 1, 2, \ldots, n\}] \\ &= \bigcup\{K \cap U_{\lambda_j}: j = 1, 2, \ldots, n\} \\ &= \bigcup\{V_{\lambda_j}: j = 1, 2, \ldots, n\} \end{aligned}$$

so that $\{V_{\lambda_j}: j = 1, 2, \ldots, n\}$ is an open cover for K in (K, ρ). Therefore (K, ρ) is a compact metric space.

Suppose now that (K, ρ) is a compact metric space and let $\{U_\lambda: \lambda \in \Lambda\}$ be an open cover for K in (X, d). By Proposition 1.6, each set $K \cap U_\lambda$ is open in (K, ρ) and hence $\{K \cap U_\lambda: \lambda \in \Lambda\}$ is an open cover for K in (K, ρ). Since (K, ρ) is compact there exist $\lambda_1, \lambda_2, \ldots, \lambda_n \in \Lambda$ such that

$$\begin{aligned} K &\subset \bigcup\{K \cap U_{\lambda_j}: j = 1, 2, \ldots, n\} \\ &\subset \bigcup\{U_{\lambda_j}: j = 1, 2, \ldots, n\}. \end{aligned}$$

Thus $\{U_{\lambda_j}: j = 1, 2, \ldots, n\}$ is an open cover for K in (X, d) and so K is compact in (X, d) as required.

Compact subsets of metric spaces have many pleasant properties. The next result begins our study of these properties.

Proposition 1.16. *A compact subset of a metric space is closed and bounded; a closed subset of a compact set is compact.*

Proof. Let K be a compact subset of a metric space (X, d) and let $p \in X \setminus K$. For each $x \in K$ let $\epsilon(x) = \frac{1}{2}d(x, p)$. We then have

$$B(x, \epsilon(x)) \cap B(p, \epsilon(x)) = \varnothing \quad (x \in K).$$

Clearly $\{B(x, \epsilon(x)): x \in K\}$ is an open cover of the compact set K and so has a finite subcover, say

$$\{B(x_j, \epsilon(x_j)): j = 1, 2, \ldots, n\}.$$

Let $\delta = \min\{\epsilon(x_j): j = 1, 2, \ldots, n\}$ so that $\delta > 0$. We now have

$$\begin{aligned} B(p, \delta) \cap K &\subset B(p, \delta) \cap [\bigcup\{B(x_j, \epsilon(x_j)): j = 1, 2, \ldots, n\}] \\ &\subset \bigcup\{B(p, \epsilon(x_j)) \cap B(x, \epsilon(x_j)): j = 1, 2, \ldots, n\} \\ &= \varnothing. \end{aligned}$$

It follows that $B(p, \delta) \subset X \setminus K$. Since p was arbitrary in $X \setminus K$ this shows that $X \setminus K$ is open and so K is closed. To see that K is bounded we observe that $\{B(x, 1): x \in K\}$ is an open cover of K and so has a finite subcover, say $\{B(x_j, 1): j = 1, 2, \ldots, n\}$. Let $\delta = \max\{d(x_1, x_j): j = 1, 2, \ldots, n\}$. Given $x \in K$ there is some j such that $x \in B(x_j, 1)$. Then

$$\begin{aligned} d(x_1, x) &\leqslant d(x_1, x_j) + d(x_j, x) \\ &< \delta + 1. \end{aligned}$$

Therefore $K \subset B(x_1, \delta + 1)$ and so K is bounded.

Suppose now that F is closed and $F \subset K$. We then have that $X \setminus F$ is open. Let $\{U_\lambda: \lambda \in \Lambda\}$ be any open cover of F. Then $\{U_\lambda: \lambda \in \Lambda\} \cup \{X \setminus F\}$ is an open cover for K and so has a finite subcover since K is compact. If we exclude $X \setminus F$ from the finite subcover (if necessary) the remaining sets must still cover F (since $X \setminus F$ can cover no part of F). This proves that F is compact.

In the final part of the above result we took F to be closed in X. There would be no difference if we took F to be closed in K. To see this we note that K is closed in X and so by Proposition 1.7 a subset of K is closed in (X, d) if and only if it is closed in (K, ρ).

It is not true in general that every closed and bounded subset of a metric space is compact (see Problem 1.25). On the other hand the above statement is true for the metric spaces $\mathbf{R}^n$ ($n \in \mathbf{P}$). This fact, which we prove below, is extremely useful since it enables us to tell almost at a glance when a subset of $\mathbf{R}^n$ is compact.

Theorem 1.17. *A subset of* $\mathbf{R}^n$ *is compact iff it is closed and bounded.*

Proof. In view of the last proposition we need only prove that a closed and bounded subset of $\mathbf{R}^n$ is compact. We shall give the proof for the case $n = 2$ (which is the most significant for us); it will be clear that our proof can be adapted to deal with the general case. Let K be closed and bounded in $\mathbf{R}^2$. Then K is contained in some ball in $\mathbf{R}^2$ and hence K is contained in some closed square S. (A *closed square* in $\mathbf{R}^2$ is of course a subset of the form $\{(x, y): a \leqslant x \leqslant c, b \leqslant y \leqslant d\}$ where $a < c, b < d$ and $c - a = d - b$.) We shall show below that S is compact and it will then follow from Proposition 1.16 that K is compact.

Suppose that S is not compact. Then there is an open cover $\{U_\lambda: \lambda \in \Lambda\}$ of S which contains no finite subcover. Suppose that S has sides length l. Join the mid points of opposite sides of S and so divide S into four closed squares with sides length $\frac{1}{2}l$. At least one of these four squares must fail to be covered by a finite subcollection of $\{U_\lambda: \lambda \in \Lambda\}$, else S itself would be covered by a finite subcollection of $\{U_\lambda: \lambda \in \Lambda\}$. Let S_1 be one such square and let its bottom left vertex be (a_1, b_1). We now repeat the process on S_1 and so obtain a closed square S_2 with sides length $\dfrac{1}{2^2}\, l$ and bottom left vertex (a_2, b_2) such that no finite subcollection of $\{U_\lambda: \lambda \in \Lambda\}$ covers S_2. We now proceed by induction and so obtain for each $n \in \mathbf{P}$ a closed square S_n with sides length $\dfrac{1}{2^n}\, l$ and bottom left vertex (a_n, b_n) such that no finite subcollection of $\{U_\lambda: \lambda \in \Lambda\}$ covers S_n. It follows from the method of construction of the squares that the real sequences $\{a_n\}$, $\{b_n\}$ are monotonic and bounded, and so are convergent. Thus there exist $a, b \in \mathbf{R}$ such that

$$\lim_{n \to \infty} a_n = a, \qquad \lim_{n \to \infty} b_n = b.$$

It follows that

$$\lim_{n \to \infty} (a_n, b_n) = (a, b).$$

Since $(a_n, b_n) \in S$ ($n \in \mathbf{P}$) and S is closed it follows from Proposition 1.5 that $(a, b) \in S$. Since $(a, b) \in S$ there is $\lambda_0 \in \Lambda$ such that $(a, b) \in U_{\lambda_0}$. Since

U_{λ_0} is open there is $\epsilon > 0$ such that $B((a, b), \epsilon) \subset U_{\lambda_0}$. We may now choose $m \in \mathbf{P}$ such that

$$\frac{\sqrt{2}l}{2^m} < \tfrac{1}{2}\epsilon \quad \text{and} \quad d((a, b), (a_m, b_m)) < \tfrac{1}{2}\epsilon.$$

Given $(x, y) \in S_m$ we now have

$$\begin{aligned} d((a, b), (x, y)) &\leqslant d((a, b), (a_m, b_m)) + d((a_m, b_m), (x, y)) \\ &< \tfrac{1}{2}\epsilon + \frac{\sqrt{2}l}{2^m} \\ &< \epsilon. \end{aligned}$$

This shows that $S_m \subset B((a, b), \epsilon) \subset U_{\lambda_0}$, so that the singleton $\{U_{\lambda_0}\}$ is a cover of S_m. This contradicts the definition of S_m. Therefore S must be compact as required.

The above technique of repeated dissection is a useful argument in analysis; we shall use a similar form of argument in Chapter 6. The student should have little difficulty in using the above technique to show that every infinite bounded subset of $\mathbf{R}^n$ has a cluster point.

We now investigate compactness in relation to continuity. The first part of the next result shows that the continuous image of a compact set is compact. The second part will be very useful in later chapters, especially in Chapter 5.

Proposition 1.18. *Let X be a compact metric space and let f be a continuous mapping from X to a metric space Y. Then $f(X)$ is compact in Y. If, further, f is one-to-one and onto, then f^{-1} is also continuous, i.e. $f \in Hom(X, Y)$.*

Proof. Let $\{U_\lambda : \lambda \in \Lambda\}$ be an open cover of $f(X)$. Since f is continuous it follows from Proposition 1.11 that $f^{-1}(U_\lambda)$ is open in X for each $\lambda \in \Lambda$. We now have that $\{f^{-1}(U_\lambda) : \lambda \in \Lambda\}$ is an open cover of X. Since X is compact, there is a finite subcover $\{f^{-1}(U_{\lambda_j}) : j = 1, 2, \ldots, n\}$ of X. It is readily verified that $\{U_{\lambda_j}; j = 1, 2, \ldots, n\}$ now forms a cover of $f(X)$. Therefore $f(X)$ is compact.

Suppose now that f is also one-to-one and onto. Let F be any closed subset of X. Since X is compact we have by Proposition 1.16 that F is compact. By the first part above we now have that $(f^{-1})^{-1}(F) = f(F)$ is also compact. Using Proposition 1.16 again we have that $(f^{-1})^{-1}(F)$ is closed. It follows from Proposition 1.11 that f^{-1} is continuous. The proof is complete.

We now consider a stronger concept of continuity. Let (X, d), (Y, e) be metric spaces, let $f: X \to Y$, and let E be a non-empty subset of X. We say that f is *uniformly continuous on E* if for every $\epsilon > 0$ there is $\delta(\epsilon) > 0$ such that

$$f(B_d(x, \delta(\epsilon))) \subset B_e(f(x), \epsilon) \qquad (x \in E).$$

If f is uniformly continuous on X we say more simply that f is *uniformly continuous*. The student should note the parallels between continuity and uniform continuity, and pointwise and uniform convergence. It is clear that uniform continuity implies continuity, but the converse is false as is easily seen by taking

$$f(x) = x^2 \qquad (x \in \mathbf{R}).$$

On the other hand, when X is compact the two concepts coincide.

Proposition 1.19. *Let (X, d), (Y, e) be metric spaces with X compact, and let $f: X \to Y$ be continuous. Then f is uniformly continuous.*

Proof. Let $\epsilon > 0$. Since f is continuous, given $x \in X$ there is $\delta(\epsilon, x) > 0$ such that

$$f(B_d(x, 2\delta(\epsilon, x))) \subset B_e(f(x), \tfrac{1}{2}\epsilon).$$

The collection $\{B_d(x, \delta(\epsilon, x)): x \in X\}$ is clearly an open cover for the compact space X and so has a finite subcover, say $\{B_d(x_j, \delta(\epsilon, x_j)): j = 1, 2, \ldots, n\}$. Let

$$\eta(\epsilon) = \min \{\delta(\epsilon, x_j): j = 1, 2, \ldots, n\}$$

so that $\eta(\epsilon) > 0$. Given $x, y \in X$ with $d(x, y) < \eta(\epsilon)$, there is some j such that $x \in B_d(x_j, \delta(\epsilon, x_j))$. Since $\eta(\epsilon) \leqslant \delta(\epsilon, x_j)$ we have

$$d(y, x_j) \leqslant d(y, x) + d(x, x_j) < 2\delta(\epsilon, x_j).$$

Thus $x, y \in B_d(x_j, 2\delta(\epsilon, x_j))$ and so

$$\begin{aligned} e(f(x), f(y)) &\leqslant e(f(x_j), f(x)) + e(f(x_j), f(y)) \\ &< \tfrac{1}{2}\epsilon + \tfrac{1}{2}\epsilon = \epsilon. \end{aligned}$$

This shows that

$$f(B_d(x, \eta(\epsilon))) \subset B_e(f(x), \epsilon) \qquad (x \in X)$$

and so f is uniformly continuous as required.

Let X, Y be metric spaces and let $f: X \to Y$. We say that f is a *bounded* function if $f(X)$ is bounded in Y. Suppose now that $Y = \mathbf{R}$. We say that $x_1 \in X$ is an *absolute minimum point* of f if

$$f(x) \geqslant f(x_1) \qquad (x \in X).$$

We say that $x_2 \in X$ is an *absolute maximum point* of f if

$$f(x) \leqslant f(x_2) \qquad (x \in X).$$

Proposition 1.20. *Let X, Y be metric spaces with X compact and let $f: X \to Y$ be continuous. Then f is bounded, and if $Y = \mathbf{R}$ then f has absolute maximum and minimum points in X.*

Proof. Since X is compact and f is continuous, $f(X)$ is compact by Proposition 1.18 and so bounded and closed by Proposition 1.16. In particular f is bounded. Suppose now that $Y = \mathbf{R}$. Since $f(X)$ is bounded and closed it follows that $\sup f(X)$ and $\inf f(X)$ are elements of $f(X)$; in other words there exist $x_1, x_2 \in X$ such that

$$f(x_1) \leqslant f(x) \leqslant f(x_2) \qquad (x \in X).$$

This completes the proof.

This last result will be used frequently in later chapters. It also enables us to produce another standard example of a metric space. Given any compact metric space X let $\mathscr{C}_{\mathbf{R}}(X)$ denote the set of all continuous real functions on X. Since these functions are all bounded we may define d on $\mathscr{C}_{\mathbf{R}}(X) \times \mathscr{C}_{\mathbf{R}}(X)$ by

$$d(f, g) = \sup \{|f(x) - g(x)| : x \in X\}.$$

It is straightforward to verify that d is a metric on $\mathscr{C}_{\mathbf{R}}(X)$; we call d the *usual metric* on $\mathscr{C}_{\mathbf{R}}(X)$. It is also straightforward to verify that

$$\lim_{n \to \infty} d(f_n, f) = 0 \quad \Leftrightarrow \quad \lim_{n \to \infty} f_n = f \quad \text{(uniformly on } X\text{)}.$$

In other words, metric convergence in $\mathscr{C}_{\mathbf{R}}(X)$ is simply uniform convergence on X. The metric space $\mathscr{C}_{\mathbf{R}}(X)$ has been the subject of a great deal of research; related metric spaces are discussed in §8.5.

We conclude this section by discussing the concept of distance between sets of a metric space. The concept is very simple and yet turns out to be astonishingly useful. Let (X, d) be any metric space. Given a non-empty subset A of X and given $x \in X$ we define the *distance* from x to A by

$$\operatorname{dist}(x, A) = \inf \{d(x, y) : y \in A\}.$$

Given another non-empty subset B of X we define the *distance* from B to A by

$$\text{dist}(B, A) = \inf\{\text{dist}(x, A): x \in B\}.$$

We define the *diameter* of A by

$$\text{diam}(A) = \sup\{d(x, y): x, y \in A\}.$$

(If the set $\{d(x, y): x, y \in A\}$ is not bounded above we say that A has *infinite diameter* and we then write diam $(A) = +\infty$.)

Proposition 1.21. *Let (X, d) be a metric space and let A be a non-empty subset of X.*

(i) *The function $x \to \text{dist}(x, A)$ is uniformly continuous on X.*
(ii) *$\text{dist}(x, A) = 0$ if and only if $x \in A^-$.*
(iii) *If B is a non-empty compact subset of X and if A is closed, then $\text{dist}(B, A) = 0$ if and only if $B \cap A \neq \varnothing$.*

Proof. (i) Given $x, y \in X$ we have

$$d(x, z) \leqslant d(x, y) + d(y, z) \qquad (z \in A).$$

and so

$$\text{dist}(x, A) \leqslant d(x, y) + d(y, z) \qquad (z \in A).$$

This gives

$$\text{dist}(x, A) \leqslant d(x, y) + \text{dist}(y, A).$$

Similarly we obtain

$$\text{dist}(y, A) \leqslant d(y, x) + \text{dist}(x, A)$$

and therefore

$$|\text{dist}(x, A) - \text{dist}(y, A)| \leqslant d(x, y).$$

Given $\epsilon > 0$ choose $\delta = \epsilon$ and then

$$d(x, y) < \delta \quad \Rightarrow \quad |\text{dist}(x, A) - \text{dist}(y, A)| < \epsilon.$$

This proves that $x \to \text{dist}(x, A)$ is uniformly continuous on X.

(ii) Let $Z = \{x: \text{dist}(x, A) = 0\}$. It is clear from the definition of *dist* that $A \subset Z$. Since the mapping $x \to \text{dist}(x, A)$ is continuous and $\{0\}$ is closed in $\mathbf{R}$, it follows from Proposition 1.11 that Z is closed. It now follows from Proposition 1.9 (ii) that $A^- \subset Z$. Suppose conversely that $x \in Z$, so that $\inf\{d(x, y): y \in A\} = 0$. Then there is $\{x_n\} \subset A$ such that

$\lim_{n\to\infty} d(x, x_n) = 0$. Since $\{x_n\} \subset A^-$ it follows from Proposition 1.5 that $x \in A^-$. Thus $Z \subset A^-$ and so $Z = A^-$ as required.

(iii) Since $x \to \text{dist}(x, A)$ is a continuous real function on the compact metric space B, it follows from Proposition 1.20 that there is some $x \in B$ such that $\text{dist}(B, A) = \text{dist}(x, A)$. Since A is closed it follows from part (ii) above that

$$\text{dist}(x, A) = 0 \quad \Leftrightarrow \quad x \in A.$$

This completes the proof.

PROBLEMS 1

21. Which subsets of a discrete metric space are compact?

22. Let X be a non-empty set. A collection of subsets of X has the *finite intersection property* if every finite subcollection has non-empty intersection. Show that a metric space X is compact iff every collection of closed subsets of X with the finite intersection property has non-empty intersection.

23. Show that a subset K of $\mathbf{R}^n$ is compact iff every sequence in K contains a convergent subsequence. Does this result extend to any metric space?

24. Let X be the metric space $\{x: x \in \mathbf{Q}, 0 \leqslant x \leqslant 1\}$. It follows from Proposition 1.15 and Theorem 1.17 that X is not a compact metric space. Produce an open cover of X which does not have a finite subcover. (Hint. Choose irrational t such that $0 < t < 1$ and split up the interval X with the points $t + \dfrac{1-t}{n}$ $(n \in \mathbf{P})$.)

25. Let X be the metric space $\mathscr{B}_{\mathbf{R}}(\mathbf{R})$ and let $F = \{f: f \in X, |f| \leqslant 1\}$. Show that F is closed and bounded but not compact. (Hint. Consider the open cover $\{B(f, \tfrac{1}{2}): f \in F\}$.)

26. Let $a, b \in \mathbf{R}$ with $a < b$ and let $I = \{x: x \in \mathbf{R}, a < x < b\}$. Give an example of a continuous function $f: I \to \mathbf{R}$ which is not bounded.

27. Let A, B be non-empty subsets of a metric space (X, d). Show that

$$\begin{aligned}\text{dist}(B, A) &= \inf\{d(x, y): x \in B, y \in A\}\\ &= \text{dist}(A, B).\end{aligned}$$

28. Give an example of non-empty closed subsets A, B of $\mathbf{R}^2$ such that $B \cap A = \varnothing$ and $\text{dist}(B, A) = 0$.

29. Let $X = \{K_\lambda : \lambda \in \Lambda\}$ be a family of non-empty compact subsets of $\mathbf{R}^2$ such that $K_\lambda \cap K_\mu = \varnothing$ $(\lambda \neq \mu)$. Is *dist* a metric on X?

30. Show that a compact metric space has finite diameter.

31. Let (X, d) be a metric space with infinite diameter, and let e be the metric on X as defined in Problem 1.19. Show that the metric space (X, e) had diameter 1.

1.5. Completeness

In this section we shall discuss briefly another important metric space concept, namely that of completeness. Let (X, d) be a metric space and let $\{x_n\}$ be a sequence in X. We say that $\{x_n\}$ is a *Cauchy sequence* if for each $\epsilon > 0$ there is $N(\epsilon) \in \mathbf{P}$ such that

$$d(x_m, x_n) < \epsilon \qquad (m, n > N(\epsilon)).$$

It is easy to see that every convergent sequence is a Cauchy sequence. In fact, suppose $\{x_n\}$ converges to x. Given $\epsilon > 0$ there is $N(\epsilon) \in \mathbf{P}$ such that

$$d(x, x_n) < \tfrac{1}{2}\epsilon \qquad (n > N(\epsilon)).$$

Thus if $m, n > N(\epsilon)$ we have

$$\begin{aligned} d(x_m, x_n) &\leqslant d(x_m, x) + d(x, x_n) \\ &< \tfrac{1}{2}\epsilon + \tfrac{1}{2}\epsilon = \epsilon \end{aligned}$$

and hence $\{x_n\}$ is a Cauchy sequence. On the other hand, a sequence may be Cauchy and yet fail to converge. For example, let X be the metric space $\{x : x \in \mathbf{R}, 0 < x \leqslant 1\}$ and let $x_n = \dfrac{1}{n}$ $(n \in \mathbf{P})$. It is easy to verify that $\{x_n\}$ is a Cauchy sequence in X which fails to converge. We say that a metric space X is *complete* if every Cauchy sequence converges.

We assume that the reader is acquainted with the well known fact that every Cauchy sequence of real numbers converges, i.e. the metric space $\mathbf{R}$ is complete. This result generalizes to $\mathbf{R}^n$ and we shall show below in particular that $\mathbf{R}^2$ is complete. For the purposes of this book we really need no other results on completeness. The student should not now be deceived into thinking that completeness is not an important property. Indeed most theorems of real and complex analysis hinge strongly on the completeness of $\mathbf{R}$ and $\mathbf{R}^2$. Moreover completeness also plays a crucial role in many advanced topics in analysis. On the other hand it is true to say that the completeness of many standard metric spaces is essentially derived from the completeness of **R**. This will be illustrated in the proof below and in some of the problems that follow.

Theorem 1.22. $\mathbf{R}^n$ *is complete for* $n \in \mathbf{P}, n \geqslant 2$.

Proof. We shall give the proof for the case $n = 2$; it will be clear that our proof can be readily adapted to deal with the general case. Let $\{(x_n, y_n)\}$ be a Cauchy sequence in $\mathbf{R}^2$. Given $m, n \in \mathbf{P}$ it follows from the definition of the usual metric d on $\mathbf{R}^2$ that

$$|x_m - x_n| \leqslant d((x_m, y_m), (x_n, y_n))$$
$$|y_m - y_n| \leqslant d((x_m, y_m), (x_n, y_n)).$$

It is now clear that $\{x_n\}$, $\{y_n\}$ are Cauchy sequences in **R**. Since **R** is complete there exist $x, y \in \mathbf{R}$ such that

$$\lim_{n\to\infty} x_n = x, \qquad \lim_{n\to\infty} y_n = y.$$

It follows that

$$\lim_{n\to\infty} (x_n, y_n) = (x, y)$$

i.e. the sequence $\{(x_n, y_n)\}$ converges. Therefore $\mathbf{R}^2$ is complete as required.

PROBLEMS 1

32. Show that every discrete metric space is complete.

33. Let (X, d) be a complete metric space. Show that a non-empty subset E of X is closed iff E is a complete metric space (with the relative metric).

34. Let X, Y be metric spaces and let f be a continuous mapping from X onto Y. If X is complete does it follow that Y is complete? What if f is actually a homeomorphism?

35. Let $(X_1, d_1), \ldots, (X_n, d_n)$ be complete metric spaces and let $X = X_1 \times X_2 \times \cdots \times X_n$. Let d be a metric on X as in Problem 1.2. Show that (X, d) is complete.

36. Show that the metric space $\mathscr{B}_{\mathbf{R}}(S)$ of Problem 1.3 is complete.

37. Given any compact metric space X show that $\mathscr{C}_{\mathbf{R}}(X)$ is a complete metric space.

38. Given any metric space X let $\mathscr{BC}_{\mathbf{R}}(X)$ denote the set of bounded continuous real functions on X. Show that $\mathscr{BC}_{\mathbf{R}}(X)$ is a closed subset of $\mathscr{B}_{\mathbf{R}}(X)$ and hence is a complete metric space with the relative metric (the *usual metric* for $\mathscr{BC}_{\mathbf{R}}(X)$).

1.6 Connectedness

We still require one more basic topological concept, namely that of connectedness. Roughly speaking, a metric space is connected if it consists of a single 'piece'. For most of this book we shall be working with connected metric spaces.

Let (X, d) be a metric space. We say that X is *disconnected* if there exist non-empty open sets O_1, O_2 such that $X = O_1 \cup O_2$ and $O_1 \cap O_2 = \varnothing$. We say that X is *connected* if it is not disconnected. In any metric space X the subsets $\varnothing$ and X are both open and closed. It is easy to see from Proposition 1.3 that a metric space X is connected if and only if $\varnothing$ and X are the only subsets of X that are both open and closed. We say that a non-empty subset E of X is *connected* if E is a connected metric space (with the relative metric). In view of Proposition 1.6 we see that E is connected iff whenever O_1, O_2 are open sets in X with $E \subset O_1 \cup O_2$, $E \cap O_1 \cap O_2 = \varnothing$ then either $E \cap O_1 = \varnothing$ or $E \cap O_2 = \varnothing$. If E is itself an open subset of X then we see that E is connected iff whenever O_1, O_2 are open sets in X with $E = O_1 \cup O_2$, $O_1 \cap O_2 = \varnothing$ then either $O_1 = \varnothing$ or $O_2 = \varnothing$.

As a trivial example observe that any singleton is a connected subset in a metric space. The first result below shows how we may 'paste' together connected sets to form a larger connected set.

Proposition 1.23. *Let $\{E_\lambda : \lambda \in \Lambda\}$ be a family of connected subsets of a metric space X. Suppose there is $\lambda_0 \in \Lambda$ such that $E_{\lambda_0} \cap E_\lambda \neq \varnothing$ $(\lambda \in \Lambda)$. Then $E = \bigcup \{E_\lambda : \lambda \in \Lambda\}$ is connected.*

Proof. Let O_1, O_2 be open subsets of X such that $E \cap O_1 \cap O_2 = \varnothing$, $E \subset O_1 \cup O_2$. We shall show that either $E \cap O_1 = \varnothing$ or $E \cap O_2 = \varnothing$. Given $\lambda \in \Lambda$ we must have $E_\lambda \subset O_1$ or $E_\lambda \subset O_2$. Otherwise $E_\lambda \cap O_1 \neq \varnothing$, $E_\lambda \cap O_2 \neq \varnothing$, $E_\lambda \subset E \subset O_1 \cup O_2$, and $E_\lambda \cap O_1 \cap O_2 \subset E \cap O_1 \cap O_2 = \varnothing$, i.e. E_λ is not connected. Suppose that $E_{\lambda_0} \subset O_1$. We then have $E_\lambda \subset O_1$ for each $\lambda \in \Lambda$, for otherwise $E_\lambda \subset O_2$ and so $E_\lambda \cap E_{\lambda_0} \subset E \cap O_1 \cap O_2 = \varnothing$, which is a contradiction. It follows that $E \subset O_1$. Thus $E \cap O_1 = E$ and so $E \cap O_2 = \varnothing$. Similarly if $E_{\lambda_0} \subset O_2$ we may show $E \subset O_2$ and so $E \cap O_1 = \varnothing$. We conclude that E must be connected.

This last result leads us to look for the largest possible connected subsets of a metric space. We define a *component* of a metric space to be a maximal connected subset, i.e. a connected subset that is not properly contained in any other connected subset. In intuitive terms the components of a metric space are its separate 'pieces'. We thus expect that every metric space can be split up into its separate 'pieces'. This is indeed the case as we now show.

Proposition 1.24. *Every non-empty connected subset of a metric space is contained in exactly one component; every metric space admits a unique decomposition into components.*

Proof. Let E be a connected subset of a metric space X. Let A be the union of all those connected subsets of X that contain E. It follows from Proposition 1.23 that A is connected. Suppose now that B is a connected subset of X and $A \subset B$. Then B contains E and so by the definition of A we have $B \subset A$. This shows that A is a maximal connected subset, i.e. A is a component. Suppose now that C is any component of X that contains E. Then C is a connected set containing E and so $C \subset A$. Since A is connected and C is a component we must have $C = A$. This shows that E is contained in exactly one component.

For each $x \in X$ we have that $\{x\}$ is connected. By the above we thus have that each x belongs to exactly one component of X. This means that X is the union of its components. Further if A_1, A_2 are two components of X then $A_1 = A_2$ or $A_1 \cap A_2 = \varnothing$. This completes the proof.

Thus far our only specific examples of connected sets have been singletons. We now produce some non-trivial examples of connected sets.

Proposition 1.25. *Every interval of* $\mathbf{R}$ *is connected.*

Proof. Let I be any interval of $\mathbf{R}$. (I may be bounded or unbounded and may or may not include its finite end points.) Suppose that I is not connected. Then there exist open subsets O_1, O_2 of $\mathbf{R}$ such that $I \cap O_1 \neq \varnothing$, $I \cap O_2 \neq \varnothing$, $I \subset O_1 \cup O_2$, $I \cap O_1 \cap O_2 = \varnothing$. Choose $a \in I \cap O_1$, $b \in I \cap O_2$. We have $a \neq b$. Suppose that $a < b$ and let

$$c = \sup \{x: x \in O_1, x < b\}.$$

It is clear that $a \leqslant c \leqslant b$. Since I is an interval and $a, b \in I$ we must have $c \in I \subset O_1 \cup O_2$.

Suppose that $c \in O_1$. Since O_1 is open there is $\epsilon > 0$ such that $B(c, \epsilon) \subset O_1$. In particular there is $x \in O_1$ such that $c < x$. If $x \geqslant b$ we have $c \leqslant b \leqslant x$ and so $b \in B(c, \epsilon) \subset O_1$, which is impossible. On the other hand if $x < b$ we have $c < x < b$. Since $x \in O_1$ this contradicts the definition of c.

Suppose now that $c \in O_2$. Since O_2 is open there is $\epsilon > 0$ such that $B(c, \epsilon) \subset O_2$. By the definition of c we may choose $x \in O_1$ such that $c - \epsilon < x \leqslant c$. Thus $x \in B(c, \epsilon) \subset O_2$. This is again a contradiction. We may derive similar contradictions if $a > b$. We therefore conclude that I is connected.

The student may observe that in the above proof we used only once the fact that I is an interval. We used it to obtain the property that if $a, b \in I$ and $a \leqslant c \leqslant b$ then $c \in I$. We show next that any connected subset of $\mathbf{R}$ has this property. This property is in fact equivalent to being an interval; but we shall not need this result and accordingly we shall not stop to prove it.

Proposition 1.26. *Let E be a connected subset of $\mathbf{R}$. If $a, b \in E$, $a < b$ and $a \leqslant c \leqslant b$ then $c \in E$.*

Proof. Suppose that $a < c < b$ and $c \notin E$. Let $O_1 = \{x: x \in \mathbf{R}, x < c\}$ and $O_2 = \{x: x \in \mathbf{R}, x > c\}$. Then O_1, O_2 are open subsets of $\mathbf{R}$ with $a \in E \cap O_1$, $b \in E \cap O_2$, $E \cap O_1 \cap O_2 = \varnothing$. Since $c \notin E$ we also have $E \subset O_1 \cup O_2$. We now have that E is not connected. This contradiction shows that $c \in E$ as required.

We have now characterized all the connected subsets of $\mathbf{R}$. We postpone till the next chapter a detailed discussion of the connected subsets of $\mathbf{R}^2$. We show next that the property of connectedness is preserved under continuous mappings.

Proposition 1.27. *Let X, Y be metric spaces with X connected and let $f: X \to Y$ be continuous. Then $f(X)$ is a connected subset of Y.*

Proof. Suppose that $f(X)$ is not connected. Then there exist open sets O_1, O_2 of Y such that $f(X) \cap O_1 \neq \varnothing, f(X) \cap O_2 \neq \varnothing, f(X) \subset O_1 \cup O_2$, $f(X) \cap O_1 \cap O_2 = \varnothing$. Since f is continuous it follows from Proposition 1.11 that $f^{-1}(O_1), f^{-1}(O_2)$ are open in X. Moreover, we deduce from above that $f^{-1}(O_1) \neq \varnothing$, $f^{-1}(O_2) \neq \varnothing$, $X = f^{-1}(O_1) \cup f^{-1}(O_2)$, $f^{-1}(O_1) \cap f^{-1}(O_2) = \varnothing$. This says that X is disconnected. We conclude that $f(X)$ must therefore be connected.

This last result corresponds to the intuitive notion that a continuous mapping cannot tear a single 'piece' into several 'pieces'. To illustrate the result let $Y = \mathbf{R}$, and then $f(X)$ is a connected subset of $\mathbf{R}$. Suppose that $f(x_1) = p, f(x_2) = q$ with $p < q$. Given $p \leqslant r \leqslant q$ it follows from Proposition 1.26 that $r \in f(X)$, i.e. there exists $x_3 \in X$ such that $f(x_3) = r$. In particular if we take X to be the metric space $\{x: x \in \mathbf{R}, a \leqslant x \leqslant b\}$ we obtain the classical intermediate value theorem.

We conclude with a technical result that we shall wish to use in later chapters. Let X, Y be metric spaces and let $f: X \to Y$. We say that f is *locally constant* if for each $x \in X$ there is $\epsilon(x) > 0$ such that f is constant on $B(x, \epsilon(x))$.

Proposition 1.28. *Let X, Y be metric spaces with X connected, and let $f: X \to Y$ be continuous and locally constant. Then f is constant.*

Proof. Let $p \in X$, let $y = f(p)$, and let $E = \{x: x \in X, f(x) = y\}$. We shall show that E is both open and closed. Since X is connected it will follow that $E = \varnothing$ or $E = X$. Since $p \in E$ we shall have $E = X$, so that f is constant.

Since $\{y\}$ is closed in Y and f is continuous it follows from Proposition 1.11 that $E = f^{-1}(\{y\})$ is closed. Let $x \in E$. Since f is locally constant there is $\epsilon > 0$ such that

$$f(z) = f(x) = y \qquad (z \in B(x, \epsilon)).$$

Thus $B(x, \epsilon) \subset E$ and so E is open. The proof is now complete.

PROBLEMS 1

39. Which subsets of a discrete metric space are connected?

40. Show that a closed subset F of a metric space X is connected iff there exist non-empty closed sets F_1, F_2 of X such that $F = F_1 \cup F_2$, $F_1 \cap F_2 = \varnothing$.

41. A non-empty subset E of a metric space X is said to have property (C) if there exist non-empty open subsets O_1, O_2 of X such that $O_1 \cap O_2 = \varnothing$, $A \subset O_1 \cup O_2$. Is property (C) equivalent to being connected?

42. Let $\{E_n\}$ be a sequence of connected subsets of a metric space such that $E_n \cap E_{n+1} \neq \varnothing$ $(n \in \mathbf{P})$. Show that $\bigcup \{E_n: n \in \mathbf{P}\}$ is connected.

43. Given connected subsets A, B of $\mathbf{R}$ show that $A \times B$ is a connected subset of $\mathbf{R}^2$. State and prove some generalizations of this result.

44. Given a connected subset E of a metric space show that E^- is connected. Deduce that the components of a metric space are closed. Is E^0 connected?

45. A metric space is said to be *locally connected* if each point has a connected neighbourhood. Show that the components of a locally connected metric space are open. Give an example of a metric space that is not locally connected.

46. Show that a non-empty subset of $\mathbf{R}$ is an interval iff it has the property stated in Proposition 1.26.

47. Describe a general open subset of **R** by decomposing it into its components.

48. Can a compact metric space have an infinite number of components?

49. What are the components of the metric space **Q**?

50. Show that any continuous function from a connected metric space to a discrete metric space is constant.

2

THE COMPLEX NUMBERS

2.1 Definitions and notation

In this chapter we shall discuss some of the fundamental algebraic and topological properties of the complex numbers. We assume, as in Chapter 1, that the student is acquainted with the basic properties of the real numbers. In particular we recall that the real numbers **R** form a field under the operations of addition and multiplication.

Let us now consider $\mathbf{R}^2$. The algebraic structure on $\mathbf{R}^2$ that is usually encountered first by the student is the vector space (or linear space) structure defined by the operations of addition and scalar multiplication, namely,

$$\begin{aligned} (a, b) + (c, d) &= (a + c, b + d) \qquad (a, b, c, d \in \mathbf{R}) \\ \lambda(a, b) &= (\lambda a, \lambda b) \qquad (\lambda, a, b \in \mathbf{R}). \end{aligned}$$

Under these operations $\mathbf{R}^2$ becomes a two-dimensional real vector space. We now define a multiplication $*$ on $\mathbf{R}^2$ in such a way that $\mathbf{R}^2$ becomes a field under the operations of $+$ and $*$. We define

$$(a, b) * (c, d) = (ac - bd, ad + bc) \qquad (a, b, c, d \in \mathbf{R}).$$

Proposition 2.1. *$\mathbf{R}^2$ is a field under the operations of $+$ and $*$.*

Proof. We already know from the vector space structure of $\mathbf{R}^2$ that $\mathbf{R}^2$ is a commutative group under $+$, in which the zero element is $(0, 0)$. We show next that $\mathbf{R}^2 \setminus \{(0, 0)\}$ is a commutative group under $*$. Using the algebraic properties of **R** we obtain the following relations for all elements of $\mathbf{R}^2 \setminus \{(0, 0)\}$.

$$\begin{aligned} (a, b) * (c, d) &= (ac - bd, ad + bc) \\ &= (ca - db, da + cb) \\ &= (c, d) * (a, b) \end{aligned}$$

$$
\begin{aligned}
[(a, b) * (c, d)] * (e, f) &= (ac - bd, ad + bc) * (e, f) \\
&= (ace - bde - adf - bcf, acf - bdf + ade + bce) \\
&= (a, b) * (ce - df, cf + de) \\
&= (a, b) * [(c, d) * (e, f)] \\
(a, b) * (1, 0) &= (a, b) \\
(1, 0) &= (a, b) * \left(\frac{a}{a^2 + b^2}, \frac{-b}{a^2 + b^2}\right).
\end{aligned}
$$

We thus see that $\mathbf{R}^2 \setminus \{(0, 0)\}$ is a commutative group under $*$. It remains to show that $*$ is distributive over $+$. For all elements of $\mathbf{R}^2$ we have

$$
\begin{aligned}
(a, b) * [(c, d) + (e, f)] &= (a, b) * (c + e, d + f) \\
&= (ac + ae - bd - bf, ad + af + bc + be) \\
&= [(a, b) * (c, d)] + [(a, b) * (e, f)]
\end{aligned}
$$

and so the proof is complete.

The above field is called the *complex field* and we shall denote it by **C**. We shall also refer to **C** as the *complex numbers* or the *complex plane*. The student should realize that $\mathbf{R}^2$ and **C** are precisely the same set, namely, all ordered pairs of real numbers. When we wish to emphasize the vector space structure on this set, we use $\mathbf{R}^2$; when we wish to emphasize the field structure we use **C**. Since the subject of this book is *complex* analysis we shall almost always use **C** rather than $\mathbf{R}^2$.

Our next step is to show how the real field **R** can be *embedded* in the complex field **C**. In other words we shall show that the field **C** contains an isomorphic copy of the field **R**. We can then follow the usual practice of *identifying* **R** with its isomorphic copy in **C**. We define $h: \mathbf{R} \to \mathbf{C}$ by

$$h(a) = (a, 0) \qquad (a \in \mathbf{R}).$$

It is clear that h maps **R** one-to-one into **C** and that

$$
\begin{aligned}
h(a + b) &= (a + b, 0) = (a, 0) + (b, 0) = h(a) + h(b) \\
h(ab) &= (ab, 0) = (a, 0) * (b, 0) = h(a) * h(b).
\end{aligned}
$$

This shows that the set $\{(a, 0) : a \in \mathbf{R}\}$ with the operations $+$ and $*$ is an isomorphic copy of the real field. From now on we shall feel free to regard **R** as a subset of **C** by identifying any real number a with the ordered pair $(a, 0)$. Furthermore we shall now discard the symbol $*$ for multiplication and follow the standard practice of writing simply $(a, b)(c, d)$ for $(a, b) * (c, d)$.

The complex field has three special elements, namely $(0, 0)$, $(1, 0)$ and $(0, 1)$. The element $(0, 0)$ is the zero element of the additive structure of **C** and the element $(1, 0)$ is the unit element of the multiplicative structure of

C. Under the above identification these elements are denoted by 0 and 1 respectively. The element (0, 1) is denoted by i and has the property

$$i^2 = (0, 1)(0, 1) = (-1, 0) = -1.$$

By extending the real field to the complex field we have thus produced a concrete 'square root of minus one'. We also have that

$$(a, b) = (a, 0) + (0, b) = (a, 0) + (0, 1)(b, 0) = a + ib.$$

From now on we may denote a complex number by (a, b) or $a + ib$ according to which seems more appropriate to a given context. More often than not we shall use the form $a + ib$.

It is usually most convenient to denote a complex number by a single letter, say $\alpha = a + ib$. We then say that a is the *real part* of α and b the *imaginary part* of α. We write

$$a = \operatorname{Re} \alpha, \qquad b = \operatorname{Im} \alpha.$$

We define the *complex conjugate* of α by

$$\bar{\alpha} = a - ib.$$

Observe that

$$\operatorname{Re} \alpha = \frac{1}{2}(\alpha + \bar{\alpha}), \qquad \operatorname{Im} \alpha = \frac{1}{2i}(\alpha - \bar{\alpha}).$$

The mapping $\alpha \to \bar{\alpha}$ ($\alpha \in \mathbf{C}$) is easily seen to have the following properties.

$$\overline{\alpha + \beta} = \bar{\alpha} + \bar{\beta}$$
$$\overline{\alpha\beta} = \bar{\alpha}\bar{\beta}$$
$$\bar{\bar{\alpha}} = \alpha.$$

Geometrically the mapping $\alpha \to \bar{\alpha}$ is simply reflection in the real axis $\{(a, 0): a \in \mathbf{R}\}$. The perpendicular axis $\{(0, b): b \in \mathbf{R}\}$ is called the *imaginary* axis. Complex numbers of the form $(0, b)$ are called *pure imaginary*. Note that α is real if and only if $\alpha = \bar{\alpha}$, and α is pure imaginary if and only if $\alpha = -\bar{\alpha}$.

Given a complex number $\alpha = a + ib$ we define the *modulus* of α by

$$|\alpha| = (a^2 + b^2)^{\frac{1}{2}}.$$

Geometrically the modulus of a complex number is its distance from the origin. It is obvious from the definition that

$$|\bar{\alpha}| = |\alpha|, \qquad \alpha\bar{\alpha} = |\alpha|^2.$$

It is also obvious that

$$|\operatorname{Re} \alpha| \leqslant |\alpha|, \qquad |\operatorname{Im} \alpha| \leqslant |\alpha|.$$

On the other hand we easily verify the inequality

$$|\alpha| \leqslant |\text{Re}\,\alpha| + |\text{Im}\,\alpha|.$$

We give below three further simple properties of the modulus function. We single out these three properties since they are significant for generalizations of **R** and **C** (see Problem 2.6).

Proposition 2.2 *The modulus function has the following properties.*

(i) $|\alpha| = 0$ *if and only if* $\alpha = 0$.
(ii) $|\alpha\beta| = |\alpha|\,|\beta| \qquad (\alpha, \beta \in \mathbf{C})$.
(iii) $|\alpha + \beta| \leqslant |\alpha| + |\beta| \qquad (\alpha, \beta \in \mathbf{C})$.

Proof. (i) Let $\alpha = a + ib$. We have $|\alpha| = 0$ iff $a^2 + b^2 = 0$. Since a and b are real, this occurs iff $a = b = 0$, i.e. $\alpha = 0$.

(ii) $|\alpha\beta|^2 = (\alpha\beta)(\overline{\alpha\beta}) = \alpha\bar{\alpha}\beta\bar{\beta} = |\alpha|^2|\beta|^2$. Since the modulus is non-negative we conclude that $|\alpha\beta| = |\alpha||\beta|$.

(iii) Making use of several earlier properties we obtain

$$\begin{aligned} |\alpha + \beta|^2 &= (\alpha + \beta)(\bar{\alpha} + \bar{\beta}) \\ &= \alpha\bar{\alpha} + \beta\bar{\beta} + \alpha\bar{\beta} + \beta\bar{\alpha} \\ &= |\alpha|^2 + |\beta|^2 + 2\,\text{Re}\,(\alpha\bar{\beta}) \\ &\leqslant |\alpha|^2 + |\beta|^2 + 2|\alpha\bar{\beta}| \\ &= |\alpha|^2 + |\beta|^2 + 2|\alpha|\,|\beta|. \end{aligned}$$

Since the modulus is non-negative we conclude that

$$|\alpha + \beta| \leqslant |\alpha| + |\beta|.$$

Roughly speaking the *argument* of a non-zero complex number α is the angle t between the positive real axis and the line segment joining 0 to α. This gives the equations

$$\cos t = \frac{\text{Re}\,\alpha}{|\alpha|}, \ \sin t = \frac{\text{Im}\,\alpha}{|\alpha|}.$$

It is well known that there are an infinite number of real numbers t which satisfy the above equations and that any two such numbers differ by a multiple of 2π. We shall therefore find it convenient to define a whole family of argument functions. In order to do this we recall the following property of the cos and sin functions. Given $\lambda \in \mathbf{R}$ and given $a, b \in \mathbf{R}$ with $a^2 + b^2 = 1$ there is exactly one real number t such that $\cos t = a$, $\sin t = b$, and $\lambda < t \leqslant \lambda + 2\pi$.

Given $\theta \in \mathbf{R}$ we define the function $\arg_\theta : \mathbf{C} \setminus \{0\} \to \mathbf{R}$ by

$$\arg_\theta (\alpha) = t$$

where

$$\cos t = \frac{\operatorname{Re} \alpha}{|\alpha|}, \qquad \sin t = \frac{\operatorname{Im} \alpha}{|\alpha|}, \qquad \theta - \pi < t \leqslant \theta + \pi.$$

Observe that the range of $\arg_\theta$ is $\{t: \theta - \pi < t \leqslant \theta + \pi\}$. We call $\arg_0$ the *principal value* of the argument and denote it more simply by arg.

For any $\theta \in \mathbf{R}$ and any $\alpha \in \mathbf{C} \setminus \{0\}$ we have

$$\alpha = |\alpha|\{\cos \arg_\theta (\alpha) + i \sin \arg_\theta (\alpha)\}.$$

Further, $\arg_\theta$ is constant on the ray $\{r\alpha: r > 0\}$ determined by α. This is clear since

$$\frac{\operatorname{Re} (r\alpha)}{|r\alpha|} = \frac{\operatorname{Re} \alpha}{|\alpha|}, \qquad \frac{\operatorname{Im} (r\alpha)}{|r\alpha|} = \frac{\operatorname{Im} \alpha}{|\alpha|}.$$

Conversely suppose that $\arg_\theta (\alpha) = \arg_\theta (\beta)$. Then $\dfrac{\operatorname{Re} \alpha}{|\alpha|} = \dfrac{\operatorname{Re} \beta}{|\beta|}$ and so $\operatorname{Re} \beta = \dfrac{|\beta|}{|\alpha|} \operatorname{Re} \alpha$, and similarly $\operatorname{Im} \beta = \dfrac{|\beta|}{|\alpha|} \operatorname{Im} \alpha$. It follows that $\beta = \dfrac{|\beta|}{|\alpha|}\, \alpha$. We have thus shown that $\arg_\theta (\alpha) = \arg_\theta (\beta)$ if and only if $\beta = r\alpha$ for some $r > 0$. We define

$$N_\theta = \{0\} \cup \{\alpha: \arg_\theta (\alpha) = \theta + \pi\}.$$

It follows from the above that if $\alpha \in N_\theta$, $\alpha \neq 0$ then

$$N_\theta = \{r\alpha: r \geqslant 0\}.$$

We write N_0 more simply as N. Observe that N is simply the negative real axis (including the origin)*.

We turn now to the metric structure on **C**. Since the set **C** is the same as the set $\mathbf{R}^2$ it follows that we have a metric on **C** given by

$$\rho(a + ib, c + id) = [(a - c)^2 + (b - d)^2]^{\frac{1}{2}}.$$

If $z = a + ib$, $w = c + id$ then the metric is given more succinctly by

$$\rho(z, w) = |z - w|.$$

The metric space **C** is complete (Theorem 1.22). A subset of **C** is compact if and only if it is closed and bounded (Theorem 1.17). For obvious geometrical reasons we shall refer to balls in **C** as *discs*.

The following important notation will be observed throughout this book. We make no claims as to the general acceptance or otherwise of our notation, but it will be essential for the student of this book to familiarize

* Note that we also use N for a positive integer. The context will make clear which meaning is intended.

himself with our notation. In the list that follows, α and β denote complex numbers, r and R positive real numbers with $r < R$.

$$\begin{aligned}
\Delta(\alpha, r) &= \{z : z \in \mathbf{C}, |z - \alpha| < r\} \\
\Delta'(\alpha, r) &= \Delta(\alpha, r) \setminus \{\alpha\} \\
\bar{\Delta}(\alpha, r) &= \{z : z \in \mathbf{C}, |z - \alpha| \leqslant r\} \\
C(\alpha, r) &= \{z : z \in \mathbf{C}, |z - \alpha| = r\} \\
\nabla(\alpha, r) &= \{z : z \in \mathbf{C}, |z - \alpha| > r\} \\
\bar{\nabla}(\alpha, r) &= \{z : z \in \mathbf{C}, |z - \alpha| \geqslant r\} \\
A(\alpha; R, r) &= \{z : z \in \mathbf{C}, r < |z - \alpha| < R\} \\
\bar{A}(\alpha; R, r) &= \{z : z \in \mathbf{C}, r \leqslant |z - \alpha| \leqslant R\} \\
(\alpha, \beta) &= \{(1 - t)\alpha + t\beta : t \in \mathbf{R}, 0 < t < 1\} \\
(\alpha, \beta] &= \{(1 - t)\alpha + t\beta : t \in \mathbf{R}, 0 < t \leqslant 1\} \\
[\alpha, \beta) &= \{(1 - t)\alpha + t\beta : t \in \mathbf{R}, 0 \leqslant t < 1\} \\
[\alpha, \beta] &= \{(1 - t)\alpha + t\beta : t \in \mathbf{R}, 0 \leqslant t \leqslant 1\}.
\end{aligned}$$

We conclude this section with a technical result on compact sets that we shall need in Chapter 6.

Proposition 2.3. *Let* $\{K_n\}$ *be a sequence of non-empty compact subsets of* $\mathbf{C}$ *such that* $K_{n+1} \subset K_n$ $(n \in \mathbf{P})$. *Then* $K = \bigcap \{K_n : n \in \mathbf{P}\}$ *is non-empty. If, further,* $\lim_{n\to\infty} \operatorname{diam} K_n = 0$, *then there is* $\alpha \in \mathbf{C}$ *such that*

(i) $K = \{\alpha\}$
(ii) *given* $\epsilon > 0$ *there is* $N \in \mathbf{P}$ *such that*

$$K_n \subset \Delta(\alpha, \epsilon) \qquad (n > N).$$

Proof. For each $n \in \mathbf{P}$, K_n is compact and so is closed by Proposition 1.16. It follows that $\mathbf{C} \setminus K_n$ is open in $\mathbf{C}$. We now have from Proposition 1.6 that $K_1 \setminus K_n = K_1 \cap (\mathbf{C} \setminus K_n)$ is open in the metric space K_1. We note from Proposition 1.15 that K_1 is a compact metric space. Suppose that K is empty. Then

$$\bigcup \{K_1 \setminus K_n : n \in \mathbf{P}\} = K_1 \setminus K = K_1.$$

This means that $\{K_1 \setminus K_n : n \in \mathbf{P}\}$ is an open cover of the compact metric space K_1 and so has a finite subcover, say $\{K_1 \setminus K_{n_r} : r = 1, 2, \ldots, m\}$. Let $p = \max \{n_r : r = 1, 2, \ldots, m\}$. Since $K_{n+1} \subset K_n$ $(n \in \mathbf{P})$ it follows simply that

$$K_1 = \bigcup \{K_1 \setminus K_{n_r} : r = 1, 2, \ldots, m\} = K_1 \setminus K_p.$$

This gives the contradiction that $K_p = \varnothing$. We conclude that $K \neq \varnothing$.

Suppose now that $\lim_{n \to \infty} \operatorname{diam}(K_n) = 0$. Since K is non-empty we may choose some $\alpha \in K$. Given $\beta \in K$ we have $\alpha, \beta \in K_n$ $(n \in \mathbf{P})$ and so

$$0 \leqslant |\alpha - \beta| \leqslant \operatorname{diam}(K_n) \qquad (n \in \mathbf{P}).$$

It follows that $|\alpha - \beta| = 0$ and so $\beta = \alpha$. This proves that $K = \{\alpha\}$. Given $\epsilon > 0$ there is $N \in \mathbf{P}$ such that

$$\operatorname{diam}(K_n) < \epsilon \qquad (n > N).$$

Given $n > N$, $z \in K_n$ we then have

$$|z - \alpha| \leqslant \operatorname{diam}(K_n) < \epsilon.$$

This shows that $K_n \subset \Delta(\alpha, \epsilon)$ $(n > N)$ as required.

PROBLEMS 2

1. Show that the natural ordering $\leqslant$ on $\mathbf{R}$ cannot be extended to $\mathbf{C}$ in such a way that $\mathbf{C}$ becomes an ordered field. (Hint: consider the possibilities $i \leqslant 1$, $i \geqslant 1$.)

2. (i) Use the multiplication table

	1	i
1	1	i
i	i	-1

to show that the complex field is isomorphic with the field (under matrix addition and multiplication) of all 2×2 real matrices of the form $\begin{pmatrix} a & b \\ -b & a \end{pmatrix}$.

(ii) Show that there is no multiplication $*$ for which the vector space $\mathbf{R}^3$ becomes a real division algebra. (Hint: suppose there is such a multiplication $*$. Show that there is an isomorphic copy of the division algebra consisting of certain 3×3 real matrices. Use the fact that every such matrix has a real eigenvalue to obtain a contradiction.)

(iii) Let $*$ be defined on $\mathbf{R}^4$ by

$$\begin{aligned}(a_1, a_2, a_3, a_4) * (b_1, b_2, b_3, b_4) \\ = (a_1b_1 - a_2b_2 - a_3b_3 - a_4b_4,\ a_1b_2 + a_2b_1 + a_3b_4 - a_4b_3, \\ a_1b_3 - a_2b_4 + a_3b_1 + a_4b_2,\ a_1b_4 + a_2b_3 - a_3b_2 + a_4b_1).\end{aligned}$$

Show that $(\mathbf{R}^4, +, *)$ is a skew field, i.e. has all the properties of a field except commutativity of multiplication.

3. Let $\{\alpha_n\}$, $\{\beta_n\}$ be complex sequences with $\lim_{n\to\infty} \alpha_n = \alpha$, $\lim_{n\to\infty} \beta_n = \beta$. Show that

(i) $\lim_{n\to\infty} (\alpha_n + \beta_n) = \alpha + \beta$
(ii) $\lim_{n\to\infty} (\lambda\alpha_n) = \lambda\alpha \ (\lambda \in \mathbf{C})$
(iii) $\lim_{n\to\infty} (\alpha_n\beta_n) = \alpha\beta$
(iv) $\lim_{n\to\infty} \dfrac{\alpha_n}{\beta_n} = \dfrac{\alpha}{\beta} \quad (\beta_n, \beta \in \mathbf{C} \setminus \{0\})$

using (a) Proposition 2.2 (b) the real and imaginary part of the sequences and the analogous results for real sequences.

4. Given $z_r \in \mathbf{C} \setminus \{0\}$ $(r = 1, 2, \ldots, n)$ show that

$$\left|\sum_{r=1}^{n} z_r\right| \leqslant \sum_{r=1}^{n} |z_r|$$

with equality if and only if $\arg(z_r)$ is constant. Show also that

$$|z_n - z_1| \leqslant \sum_{r=1}^{n-1} |z_{r+1} - z_r|$$

and find necessary and sufficient conditions for equality.

5. Given $z = x + iy$ let $\|z\| = |x| + |y|$. Show that

(i) $\|z_1 + z_2\| \leqslant \|z_1\| + \|z_2\|$, $\|z_1z_2\| \leqslant \|z_1\| \, \|z_2\| \ (z_1, z_2 \in \mathbf{C})$
(ii) $|z| \leqslant \|z\| \leqslant \sqrt{2}|z| \ (z \in \mathbf{C})$.

6. Let **F** denote either **R** or **C**, and let X be a vector space over **F**. We say that a real function $\|\cdot\|$ on X is a *norm* if

(i) $\|x\| \geqslant 0 \ (x \in X)$, $\|x\| = 0$ iff $x = 0$
(ii) $\|\lambda x\| = |\lambda| \, \|x\| \ (x \in X, \lambda \in \mathbf{F})$
(iii) $\|x + y\| \leqslant \|x\| + \|y\| \ (x, y \in X)$.

If $d(x, y) = \|x - y\| \ (x, y \in X)$ show that d is a metric on X. If (X, d) is complete we say that $(X, \|\cdot\|)$ is a *Banach space* over **F**. If further X is a linear algebra over **F** and

(iv) $\|xy\| \leqslant \|x\| \, \|y\| \ (x, y \in X)$

we then say that X is a *Banach algebra* over **F**. If X is a Banach algebra and $\{x_n\}$, $\{y_n\}$ are sequences in X with $\lim_{n\to\infty} x_n = x$, $\lim_{n\to\infty} y_n = y$, show that

(a) $\lim_{n\to\infty} (x_n + y_n) = x + y$
(b) $\lim_{n\to\infty} (\lambda x_n) = \lambda x \ (\lambda \in \mathbf{F})$
(c) $\lim_{n\to\infty} (x_n y_n) = xy$.

7. (i) Let $\|\cdot\|$ be defined on $\mathbf{R}^n$ by

$$\|(x_1, x_2, \ldots, x_n)\| = \left(\sum_{r=1}^{n} |x_r|^2\right)^{\frac{1}{2}}.$$

Show that $(\mathbf{R}^n, \|\cdot\|)$ is a Banach space over $\mathbf{C}$ (with the usual vector space operations).

(ii) Let $\|\cdot\|$ be defined on $\mathbf{C}^n$ by

$$\|(z_1, z_2, \ldots, z_n)\| = \max\{|z_r|: r = 1, 2, \ldots, n\}.$$

Show that $(\mathbf{C}^n, \|\cdot\|)$ is a Banach algebra over $\mathbf{C}$ (with multiplication defined coordinate-wise).

8. Show that $C(0, 1) = \{\cos t + i \sin t: t \in [0, 2\pi]\}$.

9. Let $\theta \in \mathbf{R}$. For which complex numbers z_1, z_2 is it true that

(i) $\arg_\theta (z_1 z_2) = \arg_\theta (z_1) + \arg_\theta (z_2)$

(ii) $\arg_\theta (z_1 + z_2) = \arg_\theta (z_1) + \arg_\theta (z_2)$.

10. Sketch the sets of complex numbers specified by each of the following conditions.

(i) $|z| > 1, |\arg (z)| < \dfrac{\pi}{4}$

(ii) $|z| \arg (z) = \dfrac{\pi}{2}$

(iii) $|z| + \arg (z) \geqslant 1$

(iv) $|z + 1| = |2z - 1|$

(v) $\arg (z + 1) = \arg (2z - 1)$.

11. Find a sequence $\{I_n\}$ of non-empty closed intervals in $\mathbf{Q}$ such that $I_{n+1} \subset I_n$ $(n \in \mathbf{P})$ and $\bigcap\{I_n: n \in \mathbf{P}\} = \varnothing$.

2.2 Domains in the complex plane

A *domain* D in the complex plane is an open connected subset of $\mathbf{C}$. Much of our subsequent analysis will be performed on domains in the complex plane. We shall therefore attempt in this section to give the student some idea as to what a domain looks like. We shall also show that the components of an open subset of $\mathbf{C}$ are domains. In other words any open subset of $\mathbf{C}$ can be decomposed into domains. This fact will be important when we work with functions defined on open subsets of $\mathbf{C}$, for we can then study the behaviour of the functions on each of the components of the open set.

We have previously remarked (without complete proof) that the only open connected subsets of $\mathbf{R}$ are the open intervals. The situation for open connected subsets of **C**, i.e. domains, is much more complicated, but we can give a very useful geometrical characterization of when an open subset of **C** is connected. This will lead us to discuss special classes of domains in **C**. It will also enable us to prove very simply that certain commonly occurring open subsets of **C** are domains.

A *polygonal line* in **C** is a set of the form

$$\bigcup \{[z_j, z_{j+1}]: j = 1, 2, \ldots, n\}$$

where $n \geqslant 1$ and $z_j \in \mathbf{C}$ ($j = 1, 2, \ldots, n + 1$). We say that the above polygonal line *joins* z_1 and z_{n+1}, and that each $[z_j, z_{j+1}]$ is a *segment* of the polygonal line. We say that a non-empty subset E of **C** is *polygonally connected* if any two distinct points of E can be joined by a polygonal line contained in E.

Theorem 2.4. *A non-empty open subset D of* **C** *is connected if and only if it is polygonally connected.*

Proof. Suppose first that D is connected and let $w \in D$. Let O_1 consist of w and those points of D that can be joined to w by a polygonal line contained in D. Let $O_2 = D \setminus O_1$. We clearly have $D = O_1 \cup O_2$, $O_1 \cap O_2 = \varnothing$, $O_1 \neq \varnothing$. We shall now show that O_1 and O_2 are open. Let $z_1 \in O_1$. Since D is open, there is $r_1 > 0$ such that $\Delta(z_1, r_1) \subset D$. Since $z_1 \in O_1$ there is a polygonal line L in D joining w to z_1. Given $z \in \Delta(z_1, r_1)$, $L \cup [z_1, z]$ is then a polygonal line in D joining w to z. Thus $z \in O_1$ and so $\Delta(z_1, r_1) \subset O_1$. This shows that O_1 is open. If O_2 is empty it is certainly open. If $O_2 \neq \varnothing$ let $z_2 \in O_2$. Since D is open there is $r_2 > 0$ such that $\Delta(z_2, r_2) \subset D$. Let $z \in \Delta(z_2, r_2)$ and suppose $z \in O_1$. Then there is a polygonal line L in D joining w to z. It follows that $L \cup [z, z_2]$ is a polygonal line in D joining w to z_2, so that $z_2 \in O_1$. This contradiction shows that $\Delta(z_2, r_2) \subset O_2$ and so O_2 is open. Since D is connected we now conclude that $O_2 = \varnothing$ and $O_1 = D$. Since w was arbitrary we have now shown that any two distinct points of D can be joined by a polygonal line contained in D, i.e. D is polygonally connected.

Suppose now that D is polygonally connected. Suppose that D is disconnected so that there exist non-empty open subsets O_1, O_2 of **C** such that $D = O_1 \cup O_2$, $O_1 \cap O_2 = \varnothing$. Choose $z_1 \in O_1$, $z_2 \in O_2$. Since D is polygonally connected there is a polygonal line L in D joining z_1 and z_2. A straightforward (finite) induction argument shows that there must be a segment $[w_1, w_2]$ of L such that $w_1 \in O_1$, $w_2 \in O_2$. We now define $\varphi: [0, 1] \to \mathbf{C}$ by

$$\varphi(t) = (1 - t)w_1 + tw_2.$$

Clearly the range of φ is $[w_1, w_2] \subset O_1 \cup O_2$. Since

$$|\varphi(s) - \varphi(t)| = |w_1 - w_2|\,|s - t| \qquad (s, t \in [0, 1])$$

it is clear that φ is continuous. Let $L_1 = \varphi^{-1}(O_1)$, $L_2 = \varphi^{-1}(O_2)$ so that L_1, L_2 are open subsets of $[0, 1]$ by Proposition 1.11. Moreover $0 \in L_1$, $1 \in L_2$, $L_1 \cap L_2 = \varnothing$ and $[0, 1] = L_1 \cup L_2$. This means that the metric space $[0, 1]$ is disconnected. This contradicts Proposition 1.25 and we therefore conclude that D must be connected.

The student should realize that the above result does not quite enable us to visualize the most general possible domain in **C**. Indeed domains can be so complicated that no adequate visualization is possible. We can, however, now produce several classes of domains. It is clear for example that **C** itself is a domain since any two points $\alpha, \beta \in \mathbf{C}$ can be joined by the line segment $[\alpha, \beta]$. This immediately suggests the following class of domains.

We say that a subset E of **C** is *convex* if $\alpha, \beta \in E$ implies $[\alpha, \beta] \subset E$. It is thus immediate that any convex open subset of **C** is a domain. In particular any open disc is a convex domain. To see this, suppose that $w, z \in \Delta(\alpha, r)$ and $t \in [0, 1]$. Then

$$\begin{aligned} |(1 - t)w + tz - \alpha| &= |(1 - t)(w - \alpha) + t(z - \alpha)| \\ &\leqslant (1 - t)|w - \alpha| + t|z - \alpha| \\ &< r. \end{aligned}$$

In the same way we may easily verify that any open half-plane, e.g. $\{z: \operatorname{Re} z > 1)$ or $\{z: \operatorname{Im} z < 2\}$ is a convex domain.

We may now construct further examples of convex domains by observing that the intersection of any family of convex sets is convex. To see this let $\{E_\lambda: \lambda \in \Lambda\}$ be any family of convex subsets of **C** and suppose that $E = \bigcap\{E_\lambda: \lambda \in \Lambda\} \neq \varnothing$. Given $\alpha, \beta \in E$ we have $\alpha, \beta \in E_\lambda$ and so $[\alpha, \beta] \subset E_\lambda$ for each $\lambda \in \Lambda$. Thus $[\alpha, \beta] \subset E$ and so E is convex. Since the intersection of any finite number of open sets is open it follows that the intersection of any finite number of convex domains is a convex domain. In particular any open infinite strip e.g. $\{z: 0 < \operatorname{Re} z < 1\}$ is a convex domain. So also is any open rectangle or open triangle. Similarly we may see that any open segment or open sector of a disc is a convex domain.

There is a simple generalization of convex domains which we shall find very useful. We say that an open subset D of **C** is *starlike* if there is some point α of D such that $\beta \in D$ implies $[\alpha, \beta] \subset D$. We then say that α is a *star centre* of D. It is clear that any convex open set is starlike and that each point is then a star centre. Moreover, if D is starlike with star centre α, then given $\beta, \gamma \in D$ we have that $[\beta, \alpha] \cup [\alpha, \gamma]$ is a polygonal line in D

joining β and γ. It thus follows from Theorem 2.3 that any star-like open set is a domain.

Example 2.5. *The 'cut plane'* $\mathbf{C} \setminus N$ *is a non-convex starlike domain.*

Proof. Recall that N is just the negative real axis. To see that N is closed we need only observe that if $\{z_n\} \subset N$ and $\lim_{n\to\infty} z_n = z$ then $z_n \leqslant 0$ $(n \in \mathbf{P})$ and so $z \leqslant 0$. It follows from Proposition 1.5 that N is closed, and hence $\mathbf{C} \setminus N$ is open. We show now that 1 is a star centre for $\mathbf{C} \setminus N$. If not there is $z \in \mathbf{C} \setminus N$ and $t \in (0, 1)$ such that $(1 - t)1 + tz \in N$, and hence $1 - t + tz \leqslant 0$. It follows that $z \in \mathbf{R}$. Since $1 - t > 0$ it also follows that $z \leqslant 0$. This contradiction shows that 1 is a star centre and so $\mathbf{C} \setminus N$ is a starlike domain. Clearly, $i, -i \in \mathbf{C} \setminus N$ and $0 \in [i, -i]$. Since $0 \in N$, this shows that $\mathbf{C} \setminus N$ is not convex.

Further examples of non-convex starlike domains are illustrated in Figure 2.1. Not all domains are starlike. For example it is easy to see that

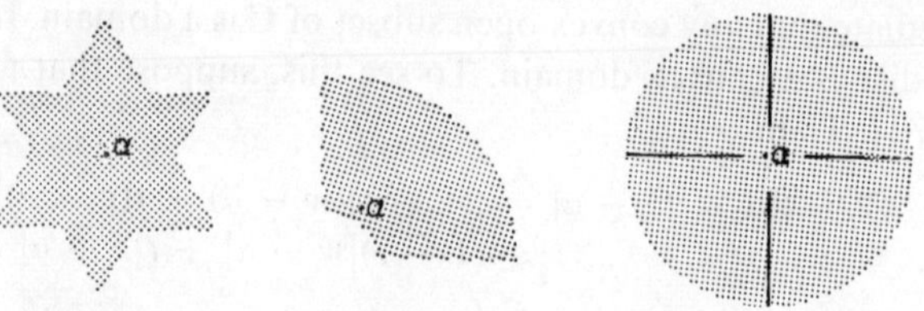

Figure 2.1

the 'punctured' disc $\Delta'(\alpha, r)$ is open. It is also straightforward to verify that $\Delta'(\alpha, r)$ is polygonally connected. Thus $\Delta'(\alpha, r)$ is a domain. Given $\beta \in \Delta'(\alpha, r)$ let $\gamma = \beta - \alpha$. Then $\alpha + \gamma$, $\alpha - \gamma \in \Delta'(\alpha, r)$ and evidently $\alpha \in [\alpha + \gamma, \alpha - \gamma]$. This shows that no point of $\Delta'(\alpha, r)$ is a star centre and so $\Delta'(\alpha, r)$ is not starlike. In the same way we may verify that $\mathbf{C} \setminus \{\alpha\}$, $\nabla(\alpha, r)$, and $A(\alpha; R, r)$ are non-starlike domains. These examples of non-starlike domains will occur frequently in subsequent chapters.

The student should now have some idea of how to produce increasingly complicated examples of domains. One such construction is given in Problem 2.20.

The problem of describing a general domain in $\mathbf{C}$ can be simplified slightly in that it is enough to study bounded domains. To see this we need first a simple lemma.

Lemma 2.6. *Let* $f\colon \mathbf{C} \to \mathbf{C}$ *be defined by*

$$f(z) = \frac{z}{1 + |z|}.$$

Then f *is a homeomorphism of* $\mathbf{C}$ *with* $\Delta(0, 1)$.

Proof. We have

$$|f(z)| = \frac{|z|}{1 + |z|} < 1 \qquad (z \in \mathbf{C})$$

so that f maps $\mathbf{C}$ into $\Delta(0, 1)$. If $f(z) = f(w)$, then $|f(z)| = |f(w)|$ and so

$$\frac{|z|}{1 + |z|} = \frac{|w|}{1 + |w|}.$$

It follows simply that $|z| = |w|$ and so

$$\frac{z}{1 + |z|} = \frac{w}{1 + |z|}.$$

Therefore $z = w$ and so f is one-to-one. Given $\alpha \in \Delta(0, 1)$, let $z = \dfrac{\alpha}{1 - |\alpha|}$. We may easily verify that $f(z) = \alpha$ and thus f maps $\mathbf{C}$ onto $\Delta(0, 1)$. Now let $\{z_n\} \subset \mathbf{C}$, with $\lim_{n \to \infty} z_n = z$. Since

$$\big||z_n| - |z|\big| \leqslant |z_n - z|$$

it follows that $\lim_{n \to \infty} |z_n| = |z|$. It then follows that

$$\lim_{n \to \infty} \frac{z_n}{1 + |z_n|} = \frac{z}{1 + |z|}.$$

We conclude from Proposition 1.10 that f is continuous. Since

$$f^{-1}(z) = \frac{z}{1 - |z|} \qquad (z \in \Delta(0, 1))$$

a similar argument shows that f^{-1} is continuous. This completes the proof.

Suppose now that D is any domain in $\mathbf{C}$ and that f is as above. Since f is a homeomorphism it follows from Propositions 1.11 and 1.27 that $f(D)$ is open and connected. Moreover we have $f(D) \subset \Delta(0, 1)$. This shows that D is homeomorphic with the bounded domain $f(D)$. Suppose further that D is starlike with star centre α. It is trivial to verify that the mapping $z \to z - \alpha$ is a homeomorphism of D with a domain D' that has star centre 0. It now follows that D' is homeomorphic with the bounded domain $f(D')$. Moreover, it is easily seen that 0 is a star centre for $f(D')$. We have thus shown that any starlike domain is homeomorphic with some bounded domain that has star centre 0. It is in fact the case that any starlike domain is homeomorphic with $\Delta(0, 1)$. We shall not prove this general result but we shall prove a special case of it in Chapter 5.

We can give one more general piece of information about an arbitrary domain D in $\mathbf{C}$. If $D \neq \mathbf{C}$, D cannot be closed. Otherwise D would be a non-empty subset of $\mathbf{C}$ that is both open and closed, and this contradicts the fact that $\mathbf{C}$ is connected. It follows that a domain can never be compact. We can however approximate a domain by compact sets in the following sense. We say that a sequence $\{K_n\}$ of compact subsets of D is a *compact exhaustion* of D if

(i) $K_n \subset K_{n+1}$ $(n \in \mathbf{P})$,
(ii) $D = \bigcup \{K_n : n \in \mathbf{P}\}$,
(iii) for each compact subset K of D there is $m \in \mathbf{P}$ such that $K \subset K_m$.

Proposition 2.7. *Every domain D in $\mathbf{C}$ has a compact exhaustion.*

Proof. If $D = \mathbf{C}$ we easily verify that $\{\bar{\Delta}(0, n)\}$ is a compact exhaustion of D. If $D \neq \mathbf{C}$ we take $K_n = \bar{\Delta}(0, n) \cap F_n$ where

$$F_n = \left\{z : \operatorname{dist}(z, \mathbf{C} \setminus D) \geqslant \frac{1}{n}\right\}.$$

It is clear that $F_n \subset D$ and so $K_n \subset D$ for each $n \in \mathbf{P}$. By Proposition 1.21 the mapping $z \to \operatorname{dist}(z, \mathbf{C} \setminus D)$ is a continuous mapping. Since $\left\{x : x \in \mathbf{R}, x \geqslant \frac{1}{n}\right\}$ is a closed set it now follows from Proposition 1.11 that each F_n is closed. Therefore each K_n is closed and bounded, and so is compact. It is clear from the definition that $K_n \subset K_{n+1}$ $(n \in \mathbf{P})$. Now let K be any compact set such that $K \subset D$. It follows from Proposition 1.21 that $\operatorname{dist}(K, \mathbf{C} \setminus D) > 0$. Hence we may choose $n_0 \in \mathbf{P}$ such that

$$\operatorname{dist}(z, \mathbf{C} \setminus D) \geqslant \frac{1}{n_0} \qquad (z \in K)$$

and we then have $K \subset F_n$ $(n \geqslant n_0)$. Since K is bounded we may now choose $m \geqslant n_0$ such that $K \subset \bar{\Delta}(0, m)$. We now have $K \subset K_m$. It remains to show that $D \subset \bigcup \{K_n : n \in \mathbf{P}\}$. Given $z \in D$, $\{z\}$ is a compact subset of D and so there is some $m \in \mathbf{P}$ such that $\{z\} \subset K_m$. This completes the proof.

There is still one piece of unfinished business in this section, namely to show that the components of an open subset of $\mathbf{C}$ are open.

Proposition 2.8. *The components of a non-empty open subset of* $\mathbf{C}$ *are domains.*

Proof. Let D be a component of a non-empty open subset O of $\mathbf{C}$. Let $z \in D$. Since O is open there is $r > 0$ such that $\Delta(z, r) \subset O$. We have seen earlier in this section that $\Delta(z, r)$ is connected. Thus $\Delta(z, r)$ is a connected subset of O containing z. It follows that $\Delta(z, r)$ is contained in that component of O which contains z, i.e. $\Delta(z, r) \subset D$. This proves that D is open. Since components are connected, D is thus a domain.

PROBLEMS 2

12. Show that any two points of a domain may be joined by a polygonal line in the domain whose segments are parallel to the coordinate axes.

13. A non-empty subset E of $\mathbf{C}$ is said to be *arcwise connected* if for each pair of points α, β of E there is a continuous function h: $[0, 1] \to \mathbf{C}$ such that $h([0, 1]) \subset E$ and $h(0) = \alpha$, $h(1) = \beta$. Show that an arcwise connected subset of $\mathbf{C}$ is connected and that an open connected subset of $\mathbf{C}$ is arcwise connected. Let E be the set

$$\left\{x + i \sin\left(\frac{1}{x}\right): x \in \left(0, \frac{\pi}{2}\right]\right\} \cup \{iy: y \in [-1, 1]\}.$$

Show that E is a connected closed subset of $\mathbf{C}$ that is not arcwise connected.

14. Every non-empty subset E of $\mathbf{C}$ is contained in a convex set, namely $\mathbf{C}$. We may thus define the *convex hull* of E, co (E), to be the intersection of all the convex sets that contain E. Show that co (E) is the smallest convex set that contains E. What is co (E) if E consists of three points z_1, z_2, z_3 not all on the same line? In this case show that

$$\text{co}\,(E) = \{t_1 z_1 + t_2 z_2 + t_3 z_3: t_1 \geqslant 0, t_2 \geqslant 0, t_3 \geqslant 0, t_1 + t_2 + t_3 = 1\}$$
$$\text{co}\,(E)^0 = \{t_1 z_1 + t_2 z_2 + t_3 z_3: t_1 > 0, t_2 > 0, t_3 > 0, t_1 + t_2 + t_3 = 1\}.$$

What is co (E) if the three points lie on a line? What is co (E) if E consists of 4 points, n points ($n \geqslant 4$)?

15. Give an example of a starlike domain with exactly one star centre.

16. Let D be a starlike domain. Show that the set of star centres for D is a convex closed subset of the metric space D.

17. Give examples of starlike domains D_1, D_2 such that

(i) $D_1 \cap D_2$ is connected but not starlike,
(ii) $D_1 \cap D_2$ is disconnected.

18. Let E be a non-empty finite subset of a domain D. Show that $D \setminus E$ is a domain and is never starlike.

19. Prove that any open annulus $A(\alpha; R, r)$ is a domain.

20. Show that $D = \mathbf{C} \setminus \bigcup \{\bar{\Delta}(3n, 1) : n \in \mathbf{P}\}$ is a domain. Note that D has an infinite number of 'holes'. Sketch the image of D under the homeomorphism f of Lemma 2.6.

21. Let D be a domain with $D \neq \mathbf{C}$. Show that $b(D) \neq \varnothing$. Give examples in which $b(D)$ is (i) connected, (ii) disconnected. Can $b(D)$ be (i) a finite set, (ii) a line segment, (iii) a closed disc?

22. Show that an open subset of $\mathbf{C}$ has at most a countable infinity of components. Give an example of an open set that has an infinite number of components.

2.3 The extended complex plane

In this section we shall formalize the notion of 'the point at infinity'. The ideas of this section will provide a very convenient language for discussing the behaviour of functions for 'large' values of z, i.e. values for which $|z|$ is large. They will also provide us with a basic model for our study in Chapter 10 of certain generalizations of domains in the complex plane.

The complex plane is of course not a compact metric space since it is not bounded. But we have seen in Lemma 2.6 that $\mathbf{C}$ is homeomorphic with $\Delta(0, 1)$. We know that the metric space $\bar{\Delta}(0, 1)$ is compact since it is closed and bounded. This suggests that we might try to extend $\mathbf{C}$ to obtain a compact metric space by adding on additional points corresponding to the points of the circle $C(0, 1)$. We could then 'transfer' the metric on $\bar{\Delta}(0, 1)$ to this larger set and so obtain a compact metric space. This would certainly be a reasonable course to pursue in view of the analogy with the real numbers where one adds on two 'points at infinity', namely $+\infty$ and $-\infty$. Nonetheless it turns out to be more useful for our problem to extend $\mathbf{C}$ simply by adding on one 'point at infinity'. We begin by obtaining another interesting homeomorphic image of $\mathbf{C}$.

Let S be the boundary of the closed sphere in $\mathbf{R}^3$ with centre $(0, 0, \frac{1}{2})$ and radius $\frac{1}{2}$, i.e.

$$S = \{(r, s, t) : r, s, t \in \mathbf{R}, r^2 + s^2 + (t - \tfrac{1}{2})^2 = \tfrac{1}{4}\}.$$

The line joining the 'north pole' $(0, 0, 1)$ to the point $(a, b, 0)$ is given by $\{(\lambda a, \lambda b, 1 - \lambda) : \lambda \in \mathbf{R}\}$. It is readily verified that this line meets S again in exactly one point and that this point has coordinates

$$\left(\frac{a}{1 + a^2 + b^2}, \frac{b}{1 + a^2 + b^2}, \frac{a^2 + b^2}{1 + a^2 + b^2}\right).$$

Let h be the mapping which takes $(a, b, 0)$ into the above point. Then h maps $\mathbf{R}^2 \setminus \{0\}$ one-to-one onto $S \setminus \{(0, 0, 1)\}$. It is readily verified that the inverse mapping is given by

$$h^{-1}(r, s, t) = \left(\frac{r}{1-t}, \frac{s}{1-t}, 0\right).$$

Lemma 2.9. *The mapping h defined above is a homeomorphism.*

Proof. Suppose that $\lim_{n\to\infty} (a_n, b_n, 0) = (a, b, 0)$. We then have $\lim_{n\to\infty} a_n = a$, $\lim_{n\to\infty} b_n = b$. Since $1 + a^2 + b^2 \neq 0$ we now obtain

$$\lim_{n\to\infty} \frac{a_n}{1 + a_n^2 + b_n^2} = \frac{a}{1 + a^2 + b^2}, \quad \lim_{n\to\infty} \frac{b_n}{1 + a_n^2 + b_n^2} = \frac{b}{1 + a^2 + b^2}$$

$$\lim_{n\to\infty} \frac{a_n^2 + b_n^2}{1 + a_2^2 + b_n^2} = \frac{a^2 + b^2}{1 + a^2 + b^2}.$$

This gives $\lim_{n\to\infty} h(a_n, b_n, 0) = h(a, b, 0)$. We conclude from Proposition 1.10 that h is continuous. Now let $(r_n, s_n, t_n), (r, s, t) \in S \setminus \{(0, 0, 1)\}$ with $\lim_{n\to\infty} (r_n, s_n, t_n) = (r, s, t)$. Then $\lim_{n\to\infty} r_n = r$, $\lim_{n\to\infty} s_n = s$, $\lim_{n\to\infty} t_n = t$ with $1 - t_n \neq 0$ $(n \in \mathbf{P})$, $1 - t \neq 0$. Arguing as above we now obtain $\lim_{n\to\infty} h^{-1}(r_n, s_n, t_n) = h^{-1}(r, s, t)$ so that h^{-1} is also continuous. This completes the proof.

Let $k: \mathbf{C} \to \mathbf{R}^3$ be defined by

$$k(a + ib) = (a, b, 0).$$

It is a trivial exercise to verify that k is a homeomorphism of $\mathbf{C}$ with $\mathbf{R}^2 \times \{0\}$. By the corollary to Proposition 1.12 we now have that $h \circ k$ is a homeomorphism of $\mathbf{C}$ with $S \setminus \{(0, 0, 1)\}$. We now wish to add another point to $\mathbf{C}$ to correspond with the 'north pole' $(0, 0, 1)$. Any symbol will do for this additional point provided it is not a complex number. We could for example use the symbol $\mathbf{C}$ for this additional point since $\mathbf{C} \notin \mathbf{C}$. It is customary to use the symbol ∞ with the agreement that ∞ is not a complex number. We now define

$$\mathbf{C}^\infty = \mathbf{C} \cup \{\infty\}$$

and $g: \mathbf{C}^\infty \to S$ by

$$\begin{cases} g(z) = h \circ k(z) & (z \in \mathbf{C}) \\ g(\infty) = (0, 0, 1). \end{cases}$$

It is now clear that g maps $\mathbf{C}^\infty$ one-to-one onto S. We define d^* on $\mathbf{C}^\infty \times \mathbf{C}^\infty$ by

$$d^*(z, w) = d(g(z), g(w)) \qquad (z, w \in \mathbf{C}^\infty)$$

where d is the usual metric on S. It is thus trivial to verify that d^* is a metric on $\mathbf{C}^\infty$. We say that d^* is the *chordal metric* on $\mathbf{C}^\infty$ and that $(\mathbf{C}^\infty, d^*)$ is the *extended complex plane*. The extended complex plane is of course isometric with (S, d). This latter metric space is called the *Riemann sphere*.

We leave the student the simple details of verifying that the 'formula' for d^* is given by

$$d^*(z, w) = \frac{|z - w|}{\{(1 + |z|^2)(1 + |w|^2)\}^{\frac{1}{2}}} \qquad (z, w \in \mathbf{C})$$

$$d^*(z, \infty) = d^*(\infty, z) = \frac{1}{(1 + |z|^2)^{\frac{1}{2}}} \qquad (z \in \mathbf{C})$$

$$d^*(\infty, \infty) = 0.$$

Note that the chordal distance between two points is at most one. The student may wish to know what open balls look like that are centred on the point of infinity. If $\epsilon \geqslant 1$ then $B_{d^*}(\infty, \epsilon) = \mathbf{C}^\infty$. If $0 < \epsilon < 1$, let $r = (\epsilon^{-2} - 1)^{\frac{1}{2}}$. Given $z \in \mathbf{C}$ we now have $d^*(\infty, z) < \epsilon$ if and only if $|z| > r$. This means that

$$B_{d^*}(\infty, \epsilon) = \{\infty\} \cup \nabla(0, r).$$

Observe that $B_{d^*}(\infty, \epsilon) \cap \mathbf{C}$ is always a very large open subset of $\mathbf{C}$.

Proposition 2.10. $(\mathbf{C}^\infty, d^*)$ *is a compact metric space and the identity mapping is a homeomorphism of* $(\mathbf{C}, \rho)$ *with* $(\mathbf{C}, d^*)$.

Proof. The Riemann sphere S is a bounded closed subset of $\mathbf{R}^3$ and so is compact. Therefore $\mathbf{C}^\infty$ is compact since it is isometric with S. We have seen that $h \circ k$ is a homeomorphism of $(\mathbf{C}, \rho)$ with $S \setminus \{(0, 0, 1)\}$. By the definition of d^*, $(h \circ k)^{-1}$ is an isometry and so homeomorphism of $S \setminus \{(0, 0, 1)\}$ with $(\mathbf{C}, d^*)$. It follows from the Corollary to Proposition 1.12 that the identity mapping is a homeomorphism of $(\mathbf{C}, \rho^*)$ with $(\mathbf{C}, d^*)$.

The above result gives us the basic facts that we shall need about the extended complex plane. The latter part of the result tells us that a subset of $\mathbf{C}$ is open with respect to the metric d^* if an only if it is open with respect to the usual metric. Now let E be any non-empty subset of $\mathbf{C}$. It follows from the Corollary to Proposition 1.6 that E has the same open subsets for the metrics induced by d^* and ρ. This means that (E, ρ) and (E, d^*) are essentially indistinguishable in terms of any properties that are defined by open sets.

PROBLEMS 2

23. Show that the metrics ρ, d^* on $\mathbf{C}$ are such that (i) $d^* \leqslant \rho$, (ii) there is no $K > 0$ with $\rho \leqslant Kd^*$.

24. Show that $\mathbf{C}$ is a dense subset of the metric space $(\mathbf{C}^\infty, d^*)$. Which sequences in $\mathbf{C}$ converge to ∞?

25. Let U be an open subset of $(\mathbf{C}^\infty, d^*)$ with $\infty \in U$. Show that $\mathbf{C}^\infty \setminus U$ is a compact subset of $\mathbf{C}$. Give a direct proof of the compactness of $\mathbf{C}^\infty$ using open coverings.

26. The extended complex plane is a connected metric space. Fill in the details of the following three proofs.

(i) $\mathbf{C}$ is connected. Use Proposition 1.27 and Problem 1.44.

(ii) Let U, V be non-empty open subsets of $\mathbf{C}^\infty$ with $\mathbf{C}^\infty = U \cup V$, $U \cap V = \varnothing$. Suppose that $\infty \in U$. Show that $U \setminus \{\infty\}$, V are open subsets of $\mathbf{C}$ contradicting the fact that $\mathbf{C}$ is connected.

(iii) Define arcwise connectedness for a subset of $\mathbf{R}^3$ and show that arcwise connectedness implies connectedness. Show that S is arcwise connected.

27. Let D be an open connected subset of $\mathbf{C}^\infty$. We say that D is *simply connected* if $\mathbf{C}^\infty \setminus D$ is connected. If $D \subset \mathbf{C}$ and D is starlike show that D is simply connected. Give an example of a domain in $\mathbf{C}$ that is not simply connected.

28. Show that $\mathbf{C}^\infty$ is a complete metric space.

29. Show that $(\mathbf{R} \cup \{\infty\}, d^*)$ is a compact connected metric space. Give another extension of $\mathbf{R}$ to a compact connected metric space by adding on two points, $+\infty$ and $-\infty$. Which is more appropriate for real analysis and why?

30. For each $z \in C(0, 1)$ let ∞_z be a distinct symbol that is not a complex number. Let $\hat{\mathbf{C}} = \mathbf{C} \cup \{\infty_z : z \in C(0, 1)\}$. Use Lemma 2.6 and the methods of the above section to produce a metric $d^\dagger$ on $\hat{\mathbf{C}}$ such that $(\hat{\mathbf{C}}, d^\dagger)$ is isometric with $\bar{\Delta}(0, 1)$, and $(\mathbf{C}, d^\dagger)$ is homeomorphic with $(\mathbf{C}, \rho)$. Give an example of a sequence $\{z_n\}$ in $\mathbf{C}$ such that $\{z_n\}$ converges in $(\mathbf{C}^\infty, d^*)$ but fails to converge in $(\hat{\mathbf{C}}, d^\dagger)$.

3

CONTINUOUS AND DIFFERENTIABLE COMPLEX FUNCTIONS

3.1 Continuous complex functions

We have already obtained in Chapter 1 several properties of continuous functions from one metric space X to another metric space Y. In this section we shall specialize to the case in which $Y = \mathbf{C}$ and obtain some results which reflect the algebraic structure of $\mathbf{C}$. We shall then specialize to the case in which X is a subset of $\mathbf{C}$ (with the relative metric). In subsequent work these subsets of $\mathbf{C}$ will usually be open sets, domains, or compact sets. We shall also study continuity in relation to the point at infinity.

Let (E, d) be a metric space and let f be a complex function on E, i.e. $f\colon E \to \mathbf{C}$. Associated with f are two real functions u, v defined on E by

$$u(z) = \operatorname{Re} f(z) \qquad (z \in E)$$
$$v(z) = \operatorname{Im} f(z) \qquad (z \in E).$$

We write $u = \operatorname{Re} f$, $v = \operatorname{Im} f$, and $f = u + iv$. In the case in which E is a non-empty subset of $\mathbf{C}$ it is sometimes convenient to think of u and v as functions of two real variables. We then write

$$u(x, y) = \operatorname{Re} f(x + iy) \qquad (x + iy \in E)$$
$$v(x, y) = \operatorname{Im} f(x + iy) \qquad (x + iy \in E).$$

This will be relevant for §3 of this chapter when we shall make use of the partial derivatives of u and v with respect to x and y.

When we write $f = u + iv$ in this chapter we shall always mean that u, v are real valued, $u = \operatorname{Re} f$, $v = \operatorname{Im} f$.

Proposition 3.1. *Given a metric space (E, d), a function $f\colon E \to \mathbf{C}$ is continuous if and only if both $\operatorname{Re} f$ and $\operatorname{Im} f$ are continuous.*

Proof. Let f be continuous, let $z \in E$ and let $\epsilon > 0$. Since f is continuous at z there is $\delta > 0$ such that $d(z, w) < \delta$ implies $|f(z) - f(w)| < \epsilon$. Since

$$|\text{Re} f(z) - \text{Re} f(w)| \leqslant |f(z) - f(w)|$$

it follows that $d(z, w) < \delta$ implies $|\text{Re} f(z) - \text{Re} f(w)| < \epsilon$. This shows that $\text{Re} f$ is continuous at z. Since z was arbitrary in E, $\text{Re} f$ is thus continuous. A similar argument shows that $\text{Im} f$ is continuous.

Suppose now that $\text{Re} f$ and $\text{Im} f$ are both continuous. Given $z \in E$, $\epsilon > 0$ we may choose $\delta > 0$ such that $d(z, w) < \delta$ implies

$$|\text{Re} f(z) - \text{Re} f(w)| < \tfrac{1}{2}\epsilon, \qquad |\text{Im} f(z) - \text{Im} f(w)| < \tfrac{1}{2}\epsilon.$$

Since

$$|f(z) - f(w)| \leqslant |\text{Re} f(z) - \text{Re} f(w)| + |\text{Im} f(z) - \text{Im} f(w)|$$

it follows that $d(z, w) < \delta$ implies $|f(z) - f(w)| < \epsilon$. This shows that f is continuous at z. Since z was arbitrary in E, f is thus continuous as required.

We shall denote the set of all continuous complex functions on E by $\mathscr{C}(E)$ and the set of all continuous real functions on E by $\mathscr{C}_{\mathbf{R}}(E)$. The above result then states that $\mathscr{C}(E) = \mathscr{C}_{\mathbf{R}}(E) + i\mathscr{C}_{\mathbf{R}}(E)$.

Given two complex functions f, g on any set E we define the associated functions λf, $(\lambda \in \mathbf{C})$, $f + g$, fg by

$$\begin{aligned} (\lambda f)(z) &= \lambda f(z) && (z \in E) \\ (f + g)(z) &= f(z) + g(z) && (z \in E) \\ (fg)(z) &= f(z)g(z) && (z \in E). \end{aligned}$$

It is trivial to verify that the set of all complex functions on E forms a complex linear algebra under these operations. To verify that a subset of this algebra is a subalgebra it is sufficient to show that the subset is closed under the above operations. The next result thus shows that for any metric space E, $\mathscr{C}(E)$ is a complex linear algebra.

Proposition 3.2. *Let E be a metric space and let $f, g \in \mathscr{C}(E)$. Then λf $(\lambda \in \mathbf{C}), f + g, fg \in \mathscr{C}(E)$.*

Proof. Let z_n, $z \in E$ with $\lim_{n\to\infty} z_n = z$. Then by Proposition 1.10

$$\lim_{n\to\infty} f(z_n) = f(z), \qquad \lim_{n\to\infty} g(z_n) = g(z).$$

It follows from Problem 2.3 that

$$\lim_{n\to\infty} \lambda f_n(z) = \lambda f(z) \qquad (\lambda \in \mathbf{C})$$

$$\lim_{n\to\infty} (f_n(z) + g_n(z)) = f(z) + g(z)$$

$$\lim_{n\to\infty} f_n(z)g_n(z) = f(z)g(z).$$

The proof is now complete in view of Proposition 1.10.

Let us now suppose that E is a non-empty subset of $\mathbf{C}$, with the relative metric. To avoid trivial considerations we shall suppose that E contains an infinite number of points. We show first that $\mathscr{C}(E)$ is always well supplied with elements. Let the functions $\mathbf{1}$, $\mathbf{u}$ be defined on $\mathbf{C}$ by

$$\mathbf{1}(z) = 1, \qquad \mathbf{u}(z) = z \qquad (z \in \mathbf{C}).$$

It is trivial to verify that $\mathbf{1}$ and $\mathbf{u}$ are continuous on E. For notational convenience we sometimes write $\mathbf{u}^0$ for $\mathbf{1}$ and $\mathbf{u}^1$ for $\mathbf{u}$. A *polynomial function* is a function p of the form

$$p = \sum_{r=0}^{n} \alpha_r \mathbf{u}^r$$

where $\alpha_r \in \mathbf{C}$ $(r = 0, 1, \ldots, n)$ and $\alpha_n \neq 0$. It is a simple result of linear algebra that if we also have

$$p = \sum_{r=0}^{m} \beta_r \mathbf{u}^r$$

with $\beta_r \in \mathbf{C}$ $(r = 0, 1, \ldots, m)$ and $\beta_m \neq 0$, then $m = n$ and $\beta_r = \alpha_r$ for each r. This result also holds when we restrict the functions to E. We say that p (or $p|_E$) has *degree n* and we write

$$\deg(p) = n.$$

It is clear from Proposition 3.2 that $\mathscr{C}(E)$ contains all polynomial functions (restricted to E).

Thus far we have specialized to complex functions on non-empty subsets of $\mathbf{C}$. We also wish to study functions defined on a non-empty subset $\mathbf{E}$ of $\mathbf{C}^\infty$ and taking values in $\mathbf{C}$ or in $\mathbf{C}^\infty$. Since the main interest centres round the point at infinity it is convenient to have certain algebraic conventions with respect to this point. We make the following definitions.

$$\infty \pm z = z \pm \infty = \infty \qquad (z \in \mathbf{C})$$

$$\infty . z = z . \infty = \infty \qquad (z \in \mathbf{C}^\infty, z \neq 0)$$

$$\frac{z}{0} = \infty \qquad (z \in \mathbf{C}^\infty, z \neq 0)$$

$$\frac{z}{\infty} = 0 \qquad (z \in \mathbf{C}).$$

There is a difficulty in trying to assign a convention for the meaning of $\infty + \infty$ or $\infty . 0$, and we shall return to this point shortly when we consider the rational functions. The main purpose in making the above definitions is to obtain notational brevity. For example we may now define the function $\mathbf{j}: \mathbf{C}^\infty \to \mathbf{C}^\infty$ by

$$\mathbf{j}(z) = \frac{1}{z} \qquad (z \in \mathbf{C}^\infty).$$

Our conventions then imply that $\mathbf{j}(0) = \infty$, $\mathbf{j}(\infty) = 0$.

Lemma 3.3. $\mathbf{j}$ *is an isometry of* $\mathbf{C}^\infty$ *with itself.*

Proof. If $z \in \mathbf{C}$, $z \neq 0$ then

$$d^*(\mathbf{j}(0), \mathbf{j}(z)) = \frac{1}{\left(1 + \left|\frac{1}{z}\right|^2\right)^{\frac{1}{2}}} = \frac{|z|}{(1 + |z|^2)^{\frac{1}{2}}} = d^*(0, z)$$

and $d^*(\mathbf{j}(\infty), \mathbf{j}(z)) = d^*(0, \mathbf{j}(z)) = d^*(\mathbf{j}(0), \mathbf{j}(\mathbf{j}(z))) = d^*(\infty, z)$. If $z, w \in \mathbf{C}$, $z \neq 0$, $w \neq 0$ then

$$d^*(\mathbf{j}(z), \mathbf{j}(w)) = \frac{\left|\frac{1}{z} - \frac{1}{w}\right|}{\left(1 + \left|\frac{1}{z}\right|^2\right)^{\frac{1}{2}}\left(1 + \left|\frac{1}{w}\right|^2\right)^{\frac{1}{2}}} = \frac{|z - w|}{(1 + |z|^2)^{\frac{1}{2}}(1 + |w|^2)^{\frac{1}{2}}}$$
$$= d^*(z, w).$$

Finally, $d^*(\mathbf{j}(0), \mathbf{j}(\infty)) = 1 = d^*(0, \infty)$. The proof is complete.

Proposition 3.4. *Let E be a non-empty subset of $\mathbf{C}^\infty$ and let f be a function from E into $\mathbf{C}$ or $\mathbf{C}^\infty$. If $z \in E \cap \mathbf{C}$ then f is continuous at z with respect to d^* iff it is continuous with respect to the usual metric. If $\infty \in E$, f is continuous at ∞ (with respect to d^*) iff $f \circ \mathbf{j}$ is continuous at 0 with respect to the usual metric.*

Proof. Recall from the final paragraph on page 52 that the metric spaces $(E \cap \mathbf{C}, d^*)$, $(E \cap \mathbf{C}, \rho)$ have precisely the same open sets. It follows from Proposition 1.11 that the identity mapping is a homeomorphism of $(E \cap \mathbf{C}, d^*)$ with $(E \cap \mathbf{C}, \rho)$. The first part of the proposition is now immediate from Proposition 1.12.

Suppose now that f is continuous at ∞. By Lemma 3.3 and Proposition 1.12 it follows that $f \circ \mathbf{j}$ is continuous at 0 with respect to d^* and so, by the above, with respect to the usual metric. Conversely, if $f \circ \mathbf{j}$ is continuous with respect to the usual metric then it is continuous with respect to d^*. Arguing as above we see that $f = f \circ \mathbf{j} \circ \mathbf{j}$ is continuous at ∞ with respect to d^*.

The above result indicates that continuous functions on subsets of $\mathbf{C}^\infty$ can be effectively described in terms of the usual metric on $\mathbf{C}$. The result will also serve as a motivation for subsequent definitions involving the point at infinity. Note in passing that a continuous *complex* valued function on $\mathbf{C}^\infty$ is automatically bounded by Proposition 1.20.

A *rational function* is a function $q: \mathbf{C} \to \mathbf{C}^\infty$ of the form $q = p_1/p_2$ where p_1, p_2 are polynomial functions such that

$$\{z: p_1(z) = 0\} \cap \{z: p_2(z) = 0\} = \varnothing.$$

If p_3, p_4, are other polynomial functions such that $q = p_3/p_4$ then we have $p_1 p_4 = p_2 p_3$ and so

$$\begin{aligned}\deg(p_1) + \deg(p_4) = \deg(p_1 p_4) &= \deg(p_2 p_3)\\ &= \deg(p_2) + \deg(p_3).\end{aligned}$$

Therefore

$$\deg(p_1) - \deg(p_2) = \deg(p_3) - \deg(p_4)$$

and so we may now define the *degree* of q by

$$\deg(q) = \deg(p_1) - \deg(p_2)$$

whenever $q = p_1/p_2$. We now extend q to $\mathbf{C}^\infty$ as follows: if

$$p_1(z) = \sum_{r=0}^{n} \alpha_r z^r, \quad (\alpha_n \neq 0), \qquad p_2(z) = \sum_{r=0}^{m} \beta_r z^r, \quad (\beta_m \neq 0),$$

we define $q(\infty) = \dfrac{\alpha_n}{\beta_m}$. (This is again independent of the choice of polynomials p_1, p_2 with $q = p_1/p_2$.) In particular $q(\infty) = 0$ if $\deg(q) < 0$ and $q(\infty) =$ if $\deg(q) > 0$. Since $\mathbf{j}$ is continuous on $\mathbf{C} \setminus \{0\}$ the function $1/p_2$ is a continuous complex function whenever $p_2(z) \neq 0$ and hence q is continuous at those points of $\mathbf{C}$ where $p_2(z) \neq 0$. If $p_2(z) = 0$, then $p_1(z) \neq 0$ and we easily show that

$$\lim_{w \to z} \frac{p_1(w)}{p_2(w)} = \infty = q(z).$$

Finally it is easy to see that q is continuous at ∞. We have thus shown that every rational function is a continuous function from $\mathbf{C}^\infty$ into $\mathbf{C}^\infty$.

We conclude this section by discussing the continuity of the modulus and argument functions. The continuity of the argument functions will be of particular significance for our discussion of logarithmic functions in the next chapter.

Proposition 3.5. (i) *The function* $z \to |z|$ *is continuous on* $\mathbf{C}$.

(ii) *Given* $\theta \in \mathbf{R}$, $\arg_\theta$ *is continuous on* $\mathbf{C} \setminus N_\theta$ *and is not continuous at each point of* $N_\theta \setminus \{0\}$.

Proof. (i) This follows immediately from the inequality

$$||z| - |w|| \leqslant |z - w| \qquad (z, w \in \mathbf{C}).$$

(ii) Let $z \in \mathbf{C} \setminus N_\theta$ and let $\arg_\theta(z) = t$. Choose any $\epsilon > 0$ such that $(t - \epsilon, t + \epsilon) \subset (\theta - \pi, \theta + \pi)$. It follows from the properties of cos and sin that there is $\eta > 0$ such that $|s - t| < \epsilon$ whenever $s \in (\theta - \pi, \theta + \pi)$ and $|\cos s - \cos t| < \eta$, $|\sin s - \sin t| < \eta$. Since $\mathbf{C} \setminus N_\theta$ is an open set (see Example 2.4) there is $\delta_1 > 0$ such that $\Delta(z, \delta_1) \subset \mathbf{C} \setminus N_\theta$. Let $\delta = \min(\delta_1, \frac{1}{2}|z|\eta)$ so that $\delta > 0$. Given $w \in \Delta(z, \delta)$ let $\arg_\theta(w) = s$. Then

$$\begin{aligned}
|\cos s - \cos t| &= \left| \frac{\operatorname{Re} w}{|w|} - \frac{\operatorname{Re} z}{|z|} \right| \\
&= \left| \frac{\operatorname{Re} w\,(|z| - |w|) + |w|\,(\operatorname{Re} w - \operatorname{Re} z)}{|w||z|} \right| \\
&\leqslant \frac{2\,|w - z|}{|z|} \\
&< \eta.
\end{aligned}$$

Similarly, we may show that $|\sin s - \sin t| < \eta$. It then follows that $|\arg_\theta(w) - \arg_\theta(z)| < \epsilon$. This shows that $\arg_\theta$ is continuous at each point of $\mathbf{C} \setminus N_\theta$.

Given $\beta \in N_\theta \setminus \{0\}$, let

$$\beta_n = |\beta| \left\{ \cos\left(\theta + \pi - \frac{1}{n}\right) + i \sin\left(\theta + \pi - \frac{1}{n}\right) \right\}.$$

We then have

$$\lim_{n \to \infty} \beta_n = \beta, \quad \lim_{n \to \infty} \arg_\theta(\beta_n) = \lim_{n \to \infty} \left(\theta + \pi - \frac{1}{n}\right) = \theta + \pi.$$

Now let

$$\gamma_n = |\beta| \left\{ \cos\left(\theta - \pi + \frac{1}{n}\right) + i \sin\left(\theta - \pi + \frac{1}{n}\right) \right\}.$$

Then

$$\lim_{n \to \infty} \gamma_n = \beta, \quad \lim_{n \to \infty} \arg_\theta(\gamma_n) = \lim_{n \to \infty} \left(\theta - \pi + \frac{1}{n}\right) = \theta - \pi.$$

It follows from Proposition 1.10 that $\arg_\theta$ is not continuous at β. The proof is complete.

PROBLEMS 3

1. (i) Let E be a compact metric space. Define $\|.\|$ on $\mathscr{C}(E)$ by

$$\|f\| = \sup\{|f(z)| : z \in E\}.$$

Show that $(\mathscr{C}(E), \|.\|)$ is a Banach algebra (see Problem 2.6).

(ii) Let E be an arbitrary metric space and let $\mathscr{BC}(E)$ denote the set of bounded continuous complex functions on E. Define $\|.\|$ on $\mathscr{BC}(E)$ by

$$\|f\| = \sup\{|f(z)| : z \in E\}.$$

Show that $(\mathscr{BC}(E), \|.\|)$ is a Banach algebra.

2. Prove Proposition 3.2 by $\epsilon - \delta$ arguments.

3. If p, q are polynomial functions show that

$$\deg(p + q) = \max(\deg(p), \deg(q))$$
$$\deg(pq) = \deg(p)\deg(q).$$

Do the same equations hold if p, q are rational functions?

4. Let $\alpha, \beta, \gamma, \delta \in \mathbf{C}$ with $\alpha\delta - \beta\gamma \neq 0$. Let f be defined on $\mathbf{C}^\infty$ by

$$f(z) = \frac{\alpha z + \beta}{\gamma z + \delta}$$

(f is called a *homography* or *fractional linear transformation*). Show that $f \in Hom(\mathbf{C}^\infty, \mathbf{C}^\infty)$.

5. Complete the details of the following sketch proof of the continuity of $\arg_\theta$ on $\mathbf{C} \setminus N_\theta$. Suppose $\arg_\theta$ is not continuous at $z \in \mathbf{C} \setminus N_\theta$. Choose a sequence $\{z_n\}$ in $\mathbf{C} \setminus N_\theta$ such that $\lim_{n\to\infty} z_n = z$ and no subsequence of $\{\arg_\theta(z_n)\}$ converges to $\arg_\theta(z)$. Since $\{\arg_\theta(z_n)\} \subset (\theta - \pi, \theta + \pi]$ some subsequence converges to a point $t \in [\theta - \pi, \theta + \pi]$. Consider cos and sin of this subsequence and show that

$$\cos t = \cos \arg_\theta(z), \qquad \sin t = \sin \arg_\theta(z).$$

Hence deduce a contradiction.

3.2 Differentiable complex functions

In this section we define differentiability for complex functions and deduce some elementary facts about differentiable functions. Let E be a non-empty subset of $\mathbf{C}$ and let $f: E \to \mathbf{C}$. As far as the text is concerned we shall define the concept of complex differentiability only at interior points of E. (The history of complex analysis has shown that this is the most fruitful definition to consider; a more general definition is given in

Problem 3.8.) Suppose then that $\alpha \in E$ and that $\Delta(\alpha, r) \subset E$ for some $r > 0$. We say that f is *differentiable at* α if there exists $\gamma \in \mathbf{C}$ such that the following condition holds.

(1) Given $\epsilon > 0$ there is δ such that $0 < \delta < r$ and

$$\left|\frac{f(\alpha + h) - f(\alpha)}{h} - \gamma\right| < \epsilon \qquad (h \in \Delta'(0, \delta)).$$

If f is differentiable at α, there is at most one such number γ. For if $\gamma_1 \in \mathbf{C}$ is such that given $\epsilon > 0$ there is δ_1 such that $0 < \delta_1 < r$ and

$$\left|\frac{f(\alpha + h) - f(\alpha)}{h} - \gamma_1\right| < \epsilon \qquad (h \in \Delta'(0, \delta_1))$$

then an application of the triangle inequality with any h in $\Delta'(0, \delta) \cap \Delta'(0, \delta_1)$ gives $|\gamma - \gamma_1| < 2\epsilon$. Since ϵ is an arbitrary positive number we conclude that $\gamma_1 = \gamma$. Following standard notation we denote γ by $f'(\alpha)$ and we write

$$f'(\alpha) = \lim_{h \to 0} \frac{f(\alpha + h) - f(\alpha)}{h}.$$

It is easy to see that f is continuous at α if f is differentiable at α. In fact if we choose $\epsilon = 1$ in condition **(1)** we obtain

$$|f(\alpha + h) - f(\alpha)| < (1 + |\gamma|)|h| \qquad (h \in \Delta'(0, \delta))$$

and this clearly implies that f is continuous at α. We shall now show that condition **(1)** is equivalent to each of the following conditions, both of which prove to be useful.

(2) There is $\eta: \Delta(\alpha, r) \to \mathbf{C}$ such that

$$f(\alpha + h) = f(\alpha) + h\{\gamma + \eta(\alpha + h)\} \qquad (h \in \Delta(0, r))$$

where η is continuous at α and $\eta(\alpha) = 0$.

(3) Given $\{h_n\} \subset \Delta'(0, r)$ with $\lim_{n \to \infty} h_n = 0$ we have

$$\lim_{n \to \infty} \frac{f(\alpha + h_n) - f(\alpha)}{h_n} = \gamma.$$

Proposition 3.6. *The above conditions* **(1)**, **(2)**, **(3)** *are equivalent.*

Proof. Let condition **(1)** hold and then define $\eta: \Delta(\alpha, r) \to \mathbf{C}$ by $\eta(\alpha) = 0$ and

$$\eta(\alpha + h) = \frac{f(\alpha + h) - f(\alpha)}{h} - \gamma \qquad (h \in \Delta'(0, r)).$$

It is then immediate that

$$f(\alpha + h) = f(\alpha) + h\{\gamma + \eta(a + h)\} \qquad (h \in \Delta(0, r)).$$

Moreover given $\epsilon > 0$ there is δ such that $0 < \delta < r$ and

$$\left|\eta(\alpha + h) - \eta(\alpha)\right| = \left|\frac{f(\alpha + h) - f(\alpha)}{h} - \gamma\right| < \epsilon \qquad (h \in \Delta'(0, r)).$$

Thus η is continuous at α and so condition **(2)** holds.

Let condition (2) hold and let $\{h_n\} \subset \Delta'(0, r)$ with $\lim_{n\to\infty} h_n = 0$. We have

$$\frac{f(\alpha + h_n) - f(\alpha)}{h_n} = \gamma + \eta(\alpha + h_n).$$

Since $\lim_{n\to\infty} (\alpha + h_n) = \alpha$, and η is continuous at α with $\eta(\alpha) = 0$, it follows from Proposition 1.10 that

$$\lim_{n\to\infty} \frac{f(\alpha + h_n) - f(\alpha)}{h_n} = \gamma,$$

i.e. condition **(3)** holds.

Finally, let condition (3) hold and suppose that condition **(1)** does not hold. Then there is some $\epsilon_0 > 0$ such that whenever $0 < \delta < r$ there is $h \in \Delta'(0, \delta)$ with

$$\left|\frac{f(\alpha + h) - f(\alpha)}{h} - \gamma\right| \geqslant \epsilon_0.$$

In particular we may take $\delta = \dfrac{r}{n + 1}$ and so obtain $\{h_n\} \subset \Delta'(0, r)$ with $\lim_{n\to\infty} h_n = 0$ and

$$\left|\frac{f(\alpha + h_n) - f(\alpha)}{h_n} - \gamma\right| \geqslant \epsilon_0 \qquad (n \in \mathbf{P}).$$

This contradicts condition **(3)** and so the proof is complete.

Observe that in the definition of differentiability it is immaterial which value of r we choose provided only that $\Delta(\alpha, r) \subset E$. We now wish to define differentiability at the point of infinity. There is clearly no hope of a sensible definition in terms of any of the above three conditions. We use Proposition 3.4 to motivate the definition. Let V be a neighbourhood of ∞ in $\mathbf{C}^\infty$, and let f: $V \to \mathbf{C}$. We say that f is *differentiable at* ∞ if $f \circ \mathbf{j}$ is differentiable at 0. Observe from Proposition 3.4 that f is continuous at ∞ if it is differentiable at ∞.

Given $f: E \to \mathbf{C}$ and $\alpha \in E$ we say that f is *regular at* α if there is $r > 0$ such that $\Delta(\alpha, r) \subset E$ and f is differentiable at each point of $\Delta(\alpha, r)$. This is a much stronger condition than simply asking that f be differentiable at α; it also turns out to be a much more interesting condition. Given a non-empty open set O of $\mathbf{C}$ (or of $\mathbf{C}^\infty$) we say that $f: O \to \mathbf{C}$ is *analytic on* O if it is differentiable at each point of O. Thus f is analytic on O if and only if it is regular at each point of O. Since differentiability is a local property of a function we may just as well study f separately on each component of O. In other words we are reduced to studying functions that are analytic on a domain D of $\mathbf{C}$ (or of $\mathbf{C}^\infty$). We shall denote by $\mathscr{A}(D)$ the set of all functions that are analytic on D. The main purpose of this book is to perform a detailed study of $\mathscr{A}(D)$. If f is analytic on $\mathbf{C}$ we say that f is an *entire* function.

Given $f \in \mathscr{A}(D)$ the *derivative* of f is the function $z \to f'(z)$ $(z \in D)$, denoted of course by f'. As usual the higher derivatives of f are defined by induction. Thus if $f' \in \mathscr{A}(D)$ than f'' is defined to be the derivative of f'. We say that f is *infinitely differentiable* on D if all the higher derivatives of f exist on D. One of the astonishing facts that we shall prove about functions in $\mathscr{A}(D)$ is that they are automatically infinitely differentiable on D.

Given $f: D \to \mathbf{C}$ we say that $\alpha \in D$ is a *singularity* of f if f is not regular at α. We say that $\alpha \in D$ is an *isolated singularity* of f if f is differentiable at all points of some neighbourhood of α except at the point α itself. Observe that an isolated singularity is indeed a special kind of singularity.

Several remarks are in order here on the matter of terminology. Some text books use the word *holomorphic* instead of analytic, and some use the word *integral* instead of entire. The phrase 'an analytic function' appears to have many different meanings in the literature on complex analysis. We shall avoid this phrase whenever possible and use the more precise terminology 'f is analytic on D' or '$f \in \mathscr{A}(D)$'. The above definition of a singularity may seem harsh in view of the fact (Problem 3.11) that α can be a singularity of a function f even if f is differentiable at α. On the other hand it is natural in that it contradicts the more important condition of regularity at a point, and as such it will avoid unnecessary pathological situations in subsequent chapters.

Proposition 3.7. *Given any domain D of $\mathbf{C}$, $\mathscr{A}(D)$ is a subalgebra of $\mathscr{C}(D)$. Given $f, g \in \mathscr{A}(D)$, $\lambda \in \mathbf{C}$ we have*

$$(\lambda f)' = \lambda f', \qquad (f + g)' = f' + g' \qquad (fg)' = f'g + fg'.$$

Proof. This is a straightforward exercise on any of the conditions **(1)**, **(2)** or **(3)**.

Proposition 3.8. *Let D_1, D_2 be domains of* $\mathbf{C}$, *let* $f: D_1 \to \mathbf{C}$, $g: D_2 \to \mathbf{C}$ *and let* $f(D_1) \subset D_2$.

(i) *If* $f \in \mathscr{A}(D_1)$, $g \in \mathscr{A}(D_2)$ *then* $g \circ f \in \mathscr{A}(D_1)$ *and* $(g \circ f)' = (g' \circ f)f'$.
(ii) *If* $f \in \mathscr{C}(D_1)$, $g \in \mathscr{A}(D_2)$, *with* $g'(z) \neq 0$ $(z \in D_2)$ *and*

$$g \circ f(z) = z \ (z \in D_1)$$

then $f \in \mathscr{A}(D_1)$ *and* $f' = 1/g' \circ f$.

Proof. (i) Let $\alpha \in D_1$. Then $f(\alpha) \in D_2$ and so there is $r > 0$ with $\Delta(f(\alpha), r) \subset D_2$. Since g is differentiable at $f(\alpha)$ it follows from condition **(2)** that there is $\xi: \Delta(f(\alpha), r) \to \mathbf{C}$ such that

$$g(f(\alpha) + k) = g(f(\alpha)) + k\{g'(f(\alpha)) + \xi(f(\alpha) + k)\} \qquad (k \in \Delta(0, r))$$

where ξ is continuous at $f(\alpha)$ and $\xi(f(\alpha)) = 0$. Since f is continuous at α there is $\delta > 0$ such that $\Delta(\alpha, \delta) \subset D_1$ and $f(\Delta(\alpha, \delta)) \subset \Delta(f(\alpha), r)$. It follows that for each $h \in \Delta(0, \delta)$ we have

$$g(f(\alpha + h)) = g(f(\alpha)) + \{f(\alpha + h) - f(\alpha)\}\{g'(f(\alpha)) + \xi(f(\alpha + h))\}.$$

Since f is differentiable at α there is $\eta: \Delta(\alpha, \delta) \to \mathbf{C}$ such that

$$f(\alpha + h) = f(\alpha) + h\{f'(\alpha) + \eta(\alpha + h)\} \qquad (h \in \Delta(0, \delta))$$

where η is continuous at α and $\eta(\alpha) = 0$. We now have

$$g \circ f(\alpha + h) = g \circ f(\alpha) + h\{g' \circ f(\alpha) . f'(\alpha) + \zeta(\alpha + h)\} \qquad (h \in \Delta(0, \delta))$$

where

$$\zeta(a + h) = f'(\alpha) \cdot \xi(f(\alpha + h)) + g' \circ f(\alpha) . \eta(\alpha + h) \\ + \xi(f(\alpha + h)) . \eta(\alpha + h) \qquad (h \in \Delta(0, \delta)).$$

It is clear that ζ is continuous at α and $\zeta(\alpha) = 0$. By condition **(2)** we conclude that $g \circ f$ is differentiable at α with $(g \circ f)'(\alpha) = (g' \circ f(\alpha) f')(\alpha)$. Since α was any point of D_1 this completes the proof of (i).

(ii) Observe that f is one-to-one on D_1 since $f(z_1) = f(z_2)$ implies $z_1 = g(f(z_1)) = g(f(z_2)) = z_2$. Let $\alpha \in D_1$. Then $f(\alpha) \in D_2$ and there is $r > 0$ with $\Delta(f(\alpha), r) \subset D_2$. Since f is continuous at α there is $\delta > 0$ such that $\Delta(\alpha, \delta) \subset D_1$ and $f(\Delta(\alpha, \delta)) \subset \Delta(f(\alpha), r)$. Let $\{h_n\} \subset \Delta'(0, \delta)$ with $\lim_{n \to \infty} h_n = 0$. Then $f(\alpha + h_n) \in \Delta'(f(\alpha), r)$ and $\lim_{n \to \infty} f(\alpha + h_n) = f(\alpha)$ by Proposition 1.10. Since g is differentiable at $f(\alpha)$, condition **(3)** gives

$$\lim_{n \to \infty} \frac{g(f(\alpha + h_n)) - g(f(\alpha))}{f(\alpha + h_n) - f(\alpha)} = g'(f(\alpha)).$$

Since $g'(f(\alpha)) \neq 0$ we have

$$\lim_{n\to\infty} \frac{f(\alpha + h_n) - f(\alpha)}{h_n} = \frac{1}{g'(f(\alpha))}.$$

By condition (3) f is thus differentiable at α with

$$f'(\alpha) = \frac{1}{g' \circ f(\alpha)}.$$

Since α was any point of D_1, the proof is complete.

We can now see that $\mathscr{A}(D)$ is well supplied with elements for any domain D of $\mathbf{C}$. Indeed it is trivial to verify that $\mathbf{1}$ and $\mathbf{u}$ are entire functions with $\mathbf{1}' = 0$, $\mathbf{u}' = \mathbf{1}$. It follows from Proposition 3.7 that any polynomial function is an entire function. Thus $\mathscr{A}(D)$ contains all polynomial functions (restricted to D). It is equally trivial to verify that $\mathbf{j}$ is analytic on $\mathbf{C} \setminus \{0\}$ with

$$\mathbf{j}'(z) = -\frac{1}{z^2} \quad (z \in \mathbf{C} \setminus \{0\}).$$

Now let q be a rational function, say $q = p_1 . (\mathbf{j} \circ p_2)$. If D is any domain in $\mathbf{C}$ such that

$$\{z: p_2(z) = 0\} \cap D = \varnothing$$

then $q|_D \in \mathscr{A}(D)$ by Propositions 3.8 and 3.7. In the next chapter we shall produce many more examples of functions in $\mathscr{A}(D)$.

PROBLEMS 3

6. Let I be an open interval in $\mathbf{R}$ and let $f: I \to \mathbf{C}$. We say that f is (real) *differentiable* at $a \in I$ if there is $\gamma \in \mathbf{C}$ such that for every $\epsilon > 0$ there is $\delta > 0$ with $(a - \delta, a + \delta) \subset I$ and

$$\left| \frac{f(a+h) - f(a)}{h} - \gamma \right| < \epsilon \quad (h \in \mathbf{R}, 0 < |h| < \delta).$$

If γ_1 also satisfies the above condition show that $\gamma_1 = \gamma$. As usual we denote γ by $f'(a)$. Show that f is differentiable at a iff $\operatorname{Re} f$ and $\operatorname{Im} f$ are differentiable at a, in which case $f'(a) = (\operatorname{Re} f)'(a) + i(\operatorname{Im} f)'(a)$.

7. Let $a, b \in \mathbf{R}$ with $a < b$. We say that $f: [a, b] \to \mathbf{C}$ is of *bounded variation on* $[a, b]$ if there is $M > 0$ such that

$$\sum_{r=0}^{n} |f(t_{r+1}) - f(t_r)| \leqslant M$$

whenever $a = t_0 < t_1 < \cdots < t_{n+1} = b$. We denote the set of all such functions by $\mathscr{BV}([a, b])$.

(i) Show that $f \in \mathscr{BV}([a, b])$ iff $\operatorname{Re} f$, $\operatorname{Im} f \in \mathscr{BV}([a, b])$.
(ii) Show that $\mathscr{BV}([a, b])$ is a complex linear algebra.
(iii) Let $f \in \mathscr{C}([a, b])$ be differentiable at each point of (a, b) with f' bounded and continuous on (a, b). Show that $f \in \mathscr{BV}([a, b])$.

(Further properties of $\mathscr{BV}([a, b])$ are considered in the Appendix.)

8. Let E be a non-empty subset of $\mathbf{C}$, let $f: E \to \mathbf{C}$ and let $\alpha \in E$. We say that f is *E-differentiable at* α with *E-derivative* γ if condition **(1)** holds for those h such that $0 < |h| < \delta$ and $\alpha + h \in E$.

(i) What does E-differentiability mean if $E = \mathbf{P}$, $\mathbf{R}$, $[0, 1]$, $\Delta(0, 1)$, $\bar{\Delta}(0, 1)$?
(ii) If f is E-differentiable at α, show that the E-derivative is unique iff α is a cluster point of E.

For the next two parts suppose that E is a dense subset of a domain D of $\mathbf{C}$.

(iii) Give an example of a function $f: D \to \mathbf{C}$ which is E-differentiable at each point of E but is not differentiable at any point of D.
(iv) Let $f \in \mathscr{C}(D)$ be E-differentiable at each point of E. Show that f is differentiable at each point of E.

9. Which polynomial functions are differentiable at ∞? Which rational functions are differentiable at ∞?

10. Differentiability has been defined only for complex valued functions. Can you formulate a meaningful definition for differentiability at points α where $f(\alpha) = \infty$? (see §10.3.)

11. Let $f: \mathbf{C} \to \mathbf{C}$ be defined by

$$f(x) = \begin{cases} x^2 & \text{if } x \in \mathbf{R} \\ 0 & \text{if } x \in \mathbf{C} \setminus \mathbf{R}. \end{cases}$$

Show that f is differentiable at 0 while 0 is a singularity of f.

12. Let $f: D \to \mathbf{C}$ and let $\{\alpha_n\} \subset D$ with $\lim_{n \to \infty} \alpha_n = \alpha \in D$. If each α_n is a singularity of f show that α is a singularity of f. Give an example in which f is differentiable at each α_n and at α.

13. Use each of conditions **(1)**, **(2)**, **(3)** to prove Proposition 3.7. Which proof is most elegant?

3.3 The Cauchy-Riemann equations

In this section we shall consider what the statement '$f \in \mathscr{A}(D)$' means in terms of the associated functions $u = \operatorname{Re} f$, $v = \operatorname{Im} f$. We have already shown that f is continuous if and only if u and v are continuous. In other words a continuous complex function can be obtained by piecing together arbitrary continuous real functions u and v. On the other hand we shall see that the functions u and v must be very intimately related if we are to obtain a differentiable function. We begin by recalling some real analysis.

Let U be an open subset of $\mathbf{R}^2$ and let $g: U \to \mathbf{R}$. We denote by $\dfrac{\partial g}{\partial x}, \dfrac{\partial g}{\partial y}$ the partial derivatives of g (if they exist) with respect to x, y respectively. Thus for example if g has a partial derivative with respect to x at (a, b) then

$$\frac{\partial g}{\partial x}(a, b) = \lim_{h \to 0} \frac{g(a + h, b) - g(a, b)}{h}.$$

Recall that g is said to be (real) *differentiable at* $(a, b) \in U$ if there exist $A, B \in \mathbf{R}$ such that for (x, y) in some neighbourhood of (a, b) we have

$$g(x, y) = g(a, b) + (x - a)\{A + \epsilon_1(x, y)\} + (y - b)\{B + \epsilon_2(x, y)\}$$

where ϵ_1, ϵ_2 are continuous at (a, b) with

$$\epsilon_1(a, b) = \epsilon_2(a, b) = 0.$$

It then follows that g has partial derivatives with respect to x and y at (a, b) and

$$\frac{\partial g}{\partial x}(a, b) = A, \qquad \frac{\partial g}{\partial y}(a, b) = B.$$

Theorem 3.9. *Let $f: D \to \mathbf{C}$ with $f = u + iv$ and let $\alpha \in D$. Then f is differentiable at α if and only if u, v are differentiable at α and satisfy the Cauchy-Riemann equations at α, i.e.*

$$\frac{\partial u}{\partial x}(\alpha) = \frac{\partial v}{\partial y}(\alpha), \qquad \frac{\partial u}{\partial y}(\alpha) = -\frac{\partial v}{\partial x}(\alpha).$$

Thus $f \in \mathscr{A}(D)$ if and only if u, v are differentiable and satisfy the Cauchy-Riemann equations on D.

Proof. Let $\alpha = (a, b)$ and choose $r > 0$ such that $\Delta(\alpha, r) \subset D$. Suppose that f is differentiable at α. Then by condition **(2)** there is $\zeta: \Delta(a, r) \to \mathbf{C}$ such that

$$f(z) = f(\alpha) + (z - \alpha)\{f'(\alpha) + \zeta(z)\} \qquad (z \in \Delta(\alpha, r))$$

where ζ is continuous at α with $\zeta(\alpha) = 0$. Let $\zeta = \xi + i\eta$ and $z = (x, y)$. On taking real parts of the above equation we obtain

$$\begin{aligned} u(x, y) = u(a, b) &+ (x - a)\{\operatorname{Re} f'(\alpha) + \xi(x, y)\} \\ &+ (y - b)\{-\operatorname{Im} f'(\alpha) - \eta(x, y)\}. \end{aligned}$$

Since ξ, η are continuous at (a, b) with $\xi(a, b) = \eta(a, b) = 0$, it follows that u is differentiable at (a, b) with

$$\frac{\partial u}{\partial x}(a, b) = \operatorname{Re} f'(\alpha), \qquad \frac{\partial u}{\partial y}(a, b) = -\operatorname{Im} f'(\alpha).$$

Similarly, by taking imaginary parts we see that v is differentiable at (a, b) with

$$\frac{\partial v}{\partial x}(a, b) = \operatorname{Im} f'(\alpha), \qquad \frac{\partial v}{\partial y}(a, b) = \operatorname{Re} f'(\alpha).$$

It follows that u and v satisfy the Cauchy-Riemann equations at α.

Suppose conversely that u and v are differentiable at α and satisfy the Cauchy-Riemann equations at α. It follows that there exist real functions ϵ_r $(r = 1, 2, 3, 4)$ in some neighbourhood of (a, b) such that

$$\begin{aligned} u(x, y) - u(a, b) = (x - a)&\left\{\frac{\partial u}{\partial x}(a, b) + \epsilon_1(x, y)\right\} \\ &+ (y - b)\left\{\frac{\partial u}{\partial y}(a, b) + \epsilon_2(x, y)\right\} \\ v(x, y) - v(a, b) = (x - a)&\left\{\frac{\partial v}{\partial x}(a, b) + \epsilon_3(x, y)\right\} \\ &+ (y - b)\left\{\frac{\partial v}{\partial y}(a, b) + \epsilon_4(x, y)\right\} \end{aligned}$$

where for $r = 1, 2, 3, 4$, ϵ_r is continuous at (a, b) with $\epsilon_r(a, b) = 0$. Using the Cauchy-Riemann equations we now obtain in some neighbourhood of α that

$$f(z) - f(\alpha) = (z - \alpha)\left\{\frac{\partial u}{\partial x}(a, b) + i\frac{\partial v}{\partial x}(a, b) + w(z)\right\}$$

where for $z \neq \alpha$

$$w(z) = \frac{x - a}{z - \alpha}\{\epsilon_1(z) + i\epsilon_3(z)\} + \frac{y - b}{z - \alpha}\{\epsilon_2(z) + i\epsilon_4(z)\}.$$

It follows that $|w| \leqslant |\epsilon_1| + |\epsilon_2| + |\epsilon_3| + |\epsilon_4|$ and so $\lim_{z \to \alpha} w(z) = 0$. We conclude from condition **(2)** that f is differentiable at α.

It should be noted from the above theorem that if $f \in \mathscr{A}(D)$ then the derivative of f is given by

$$f' = \frac{\partial u}{\partial x} + i\frac{\partial v}{\partial x}.$$

We conclude this section by giving two simple applications of the use of the Cauchy-Riemann equations. A more significant application will be considered in the next section.

Proposition 3.10. *Given $f \in \mathscr{A}(D)$ with $f' = 0$, f is constant.*

Proof. Given $\alpha \in D$ there is $r > 0$ such that $\Delta(\alpha, r) \subset D$. Given $z \in \Delta(\alpha, r)$ we have $[\alpha, z] \subset D$. Let $f = u + iv$. Since $f' = 0$ we have $\frac{\partial u}{\partial x} = \frac{\partial u}{\partial y} = 0$. Define $g: [0, 1] \to \mathbf{R}$ by

$$g(t) = u((1 - t)\alpha + tz) \qquad (t \in [0, 1]).$$

Then $g \in \mathscr{C}_{\mathbf{R}}([0, 1])$ and

$$\begin{aligned} g'(t) &= \operatorname{Re}(z - \alpha)\frac{\partial u}{\partial x}((1 - t)\alpha + tz) + \operatorname{Im}(z - \alpha)\frac{\partial u}{\partial y}((1 - t)\alpha + tz) \\ &= 0 \qquad (t \in (0, 1)). \end{aligned}$$

It follows from real analysis that g is constant and so $u(z) = u(\alpha)$. Thus u is constant on $\Delta(\alpha, r)$. A similar argument shows that v is constant on $\Delta(\alpha, r)$ and hence f is constant on $\Delta(\alpha, r)$. We now have that f is locally constant on D. Since f is continuous on the connected space D it follows from Proposition 1.28 that f is constant.

Corollary. *Given $f, g \in \mathscr{A}(D)$ with $f' = g'$, $f - g$ is constant.*

Proposition 3.11. *Given $f \in \mathscr{A}(D)$ with $|f|$ constant, f is constant.*

Proof. Let $f \in \mathscr{A}(D)$ with $|f| = k$. If $k = 0$, then $f = 0$ and so f is constant. Suppose that $k \neq 0$ and let $f = u + iv$. Then $u^2 + v^2 = k^2$ and so

$$u\frac{\partial u}{\partial x} + v\frac{\partial v}{\partial x} = u\frac{\partial u}{\partial y} + v\frac{\partial v}{\partial y} = 0.$$

Using the Cauchy-Riemann equations we obtain

$$u^2\frac{\partial u}{\partial x} = uv\frac{\partial u}{\partial y} = -v^2\frac{\partial u}{\partial x}.$$

Therefore $k^2\frac{\partial u}{\partial x} = 0$ and so $\frac{\partial u}{\partial x} = 0$. A similar argument gives $\frac{\partial v}{\partial x} = 0$. We now have $f' = 0$ and so f is constant by Proposition 3.10.

PROBLEMS 3

14. Let $f\colon D \to \mathbf{C}$ be such that $f(D) \subset \mathbf{R}$ and f is differentiable at $\alpha \in D$. Show that $f'(\alpha) = 0$. Generalize the result.

15. Use the Cauchy-Riemann equations to prove Propositions 3.7 and 3.8.

16. Show that the modulus and argument functions are nowhere differentiable.

17. Let $\alpha \in \mathbf{C}$ and let

$$f(z) = (z - \alpha)|z - \alpha| \qquad (z \in \mathbf{C}).$$

Show that f is differentiable only at α.

18. Let $f\colon \mathbf{C} \to \mathbf{C}$ be defined by

$$f(z) = \begin{cases} \dfrac{x^3 + iy^3}{x^2 + y^2} & \text{if} \quad z = x + iy \neq 0 \\ 0 & \text{if} \quad z = 0. \end{cases}$$

(i) At which points are the Cauchy-Riemann equations satisfied?
(ii) At which points is f differentiable?

19. Let the formal operators $\dfrac{\partial}{\partial z}, \dfrac{\partial}{\partial \bar{z}}$ be defined by

$$\frac{\partial}{\partial z} = \frac{\partial}{\partial x} + i\frac{\partial}{\partial y}, \frac{\partial}{\partial \bar{z}} = \frac{\partial}{\partial x} - i\frac{\partial}{\partial y}.$$

(i) Show that the Cauchy-Riemann equations correspond to

$$\frac{\partial f}{\partial \bar{z}} = 0.$$

(ii) If $f \in \mathscr{A}(D)$ show that $\dfrac{\partial f}{\partial z} = f'$.

20. Let $f \in \mathscr{A}(D)$ with $f = u + iv$.

(i) If u (or v) is constant show that f is constant.
(ii) If there exist $\alpha, \beta, \gamma \in \mathbf{C}$, not all zero, such that $\alpha u + \beta v = \gamma$, show that f is constant.

21. Let $f \in \mathscr{A}(D)$ be such that $f^{(n+1)} = 0$. Show that f is a polynomial function and that $\deg(f) \leqslant n$. Given $f \in \mathscr{A}(\mathbf{C})$ with

$$f(z_1 + z_2) = f(z_1) + f(z_2) \qquad (z_1, z_2 \in \mathbf{C}) \quad (\dagger)$$

show that $f(z) = f(1)z$ ($z \in \mathbf{C}$). Give an example of a continuous nowhere differentiable function f satisfying (†). Find all continuous functions satisfying (†).

22. Let $f \in \mathscr{A}(D)$ where D is a convex domain. Given $\alpha, \beta \in D$ with $\alpha \neq \beta$ show that there exist $\lambda, \mu \in [\alpha, \beta]$ such that

$$f(\alpha) - f(\beta) = (\alpha - \beta)\{\operatorname{Re} f'(\lambda) + i \operatorname{Im} f'(\mu)\}.$$

23. Let $g: D \to \mathbf{R}$ have partial derivatives on D with respect to x and y. If $\dfrac{\partial g}{\partial x}$ $\left(\text{or } \dfrac{\partial g}{\partial y}\right)$ is continuous on D show that g is differentiable on D.

3.4 Harmonic functions of two real variables

There is an interesting relationship between functions that are analytic on a domain and real functions that are harmonic on the domain. This relationship has fruitful implications in various branches of applied mathematics. In this section we make an introductory study of this relationship.

Let D be a domain in $\mathbf{C}$ and let $u: D \to \mathbf{R}$. We say that u is *harmonic on* D if u has continuous partial derivatives on D up to and including second order, and satisfies the (Laplace) equation

$$\frac{\partial^2 u}{\partial x^2} + \frac{\partial^2 u}{\partial y^2} = 0.$$

We denote by $\mathscr{H}(D)$ the set of all functions harmonic on D.

Proposition 3.12. *Let $f \in \mathscr{A}(D)$ with $f = u + iv$. If u and v have continuous partial derivatives on D up to second order then $u, v \in \mathscr{H}(D)$.*

Proof. Since u and v have continuous partial derivatives up to second order we have

$$\frac{\partial^2 u}{\partial y\, \partial x} = \frac{\partial^2 u}{\partial x\, \partial y}, \qquad \frac{\partial^2 v}{\partial y\, \partial x} = \frac{\partial^2 v}{\partial x\, \partial y}.$$

Using the Cauchy-Riemann equations we now obtain

$$\frac{\partial^2 u}{\partial x^2} = \frac{\partial}{\partial x}\left(\frac{\partial v}{\partial y}\right) = \frac{\partial}{\partial y}\left(\frac{\partial v}{\partial x}\right) = -\frac{\partial^2 u}{\partial y^2}$$

so that

$$\frac{\partial^2 u}{\partial x^2} + \frac{\partial^2 u}{\partial y^2} = 0.$$

This shows that $u \in \mathscr{H}(D)$ and a similar argument shows that $v \in \mathscr{H}(D)$.

We have already pointed out that we shall later prove that functions in $\mathscr{A}(D)$ are infinitely differentiable. It follows easily that if $f = u + iv \in \mathscr{A}(D)$ then u and v automatically have continuous partial derivatives of all orders.

The above result produces harmonic functions as the real parts of functions analytic on D. It is equally significant to be able to proceed in the opposite direction. Given $u \in \mathscr{H}(D)$ we say that $v \in \mathscr{H}(D)$ is an *harmonic conjugate* of u if $f = u + iv \in \mathscr{A}(D)$. The question immediately arises as to the existence and uniqueness of such an harmonic conjugate. We shall show by an example below that some restriction must be placed on the nature of the domain D in order to obtain an harmonic conjugate for each $u \in \mathscr{H}(D)$. Roughly speaking the general condition required is that the domain should have 'no holes'. With our present limited machinery we shall prove a positive result only for the case in which D is a disc.

Proposition 3.13. *Each $u \in \mathscr{H}(\Delta(\alpha, r))$ has an harmonic conjugate v on $\Delta(\alpha, r)$. If v_1 is another such harmonic conjugate then $v - v_1$ is constant.*

Proof. Let $\alpha = (a, b)$ and define $v: \Delta(\alpha, r) \to \mathbf{R}$ by

$$v(x, y) = \int_a^x -\frac{\partial u}{\partial y}(t, b)\, dt + \int_b^y \frac{\partial u}{\partial x}(x, t)\, dt.$$

It follows from Riemann integration theory that $\dfrac{\partial v}{\partial y} = \dfrac{\partial u}{\partial x}$ and

$$\begin{aligned}\frac{\partial v}{\partial x}(x, y) &= -\frac{\partial u}{\partial y}(x, b) + \int_b^y \frac{\partial^2 u}{\partial x^2}(x, t)\, dt \\ &= -\frac{\partial u}{\partial y}(x, b) - \int_b^y \frac{\partial^2 u}{\partial y^2}(x, t)\, dt \\ &= -\frac{\partial u}{\partial y}(x, b) - \frac{\partial u}{\partial y}(x, y) + \frac{\partial u}{\partial y}(x, b)\end{aligned}$$

so that $\dfrac{\partial v}{\partial x} = -\dfrac{\partial u}{\partial y}$. Since u has continuous partial derivatives on D up to second order it follows that v also has this property. In particular (Problem 3.23) u and v are differentiable on D. It follows from Theorem 3.9 that $f = u + iv \in \mathscr{A}(\Delta(\alpha, r))$. Moreover $v \in \mathscr{H}(\Delta(\alpha, r))$ by Proposition 3.12 so that v is an harmonic conjugate of u.

Suppose now that v_1 is any harmonic conjugate of u. Then $f_1 = u + iv_1 \in \mathscr{A}(\Delta(\alpha, r))$ and so $f - f_1 \in \mathscr{A}(\Delta(\alpha, r))$. Since $f - f_1 = i(v - v_1)$ it follows from the Cauchy-Riemann equations that

$$\frac{\partial}{\partial x}(v - v_1) = \frac{\partial}{\partial y}(v - v_1) = 0$$

so that $v - v_1$ is constant on $\Delta(\alpha, r)$ as required.

Example 3.14. *Let* $D = \Delta'(0, 2)$ *and let*

$$u(x, y) = \log(x^2 + y^2) \qquad ((x, y) \in D).$$

Then $u \in \mathscr{H}(D)$ *and* u *has no harmonic conjugate on* D.

Proof. It is routine to verify that $u \in \mathscr{H}(D)$. Suppose that u has an harmonic conjugate v on D. Define $g\colon [0, 2\pi] \to \mathbf{R}$ by

$$g(t) = v(\cos t, \sin t) \qquad (t \in [0, 2\pi]).$$

It follows that $g \in \mathscr{C}_{\mathbf{R}}([0, 2\pi])$ with $g(2\pi) = g(0)$. Moreover

$$g'(t) = \frac{\partial v}{\partial x}(\cos t, \sin t)(-\sin t) + \frac{\partial v}{\partial y}(\cos t, \sin t)(\cos t).$$

Since u and v satisfy the Cauchy-Riemann equations we have

$$\begin{aligned} g'(t) &= -\frac{\partial u}{\partial y}(\cos t, \sin t)(-\sin t) + \frac{\partial u}{\partial x}(\cos t, \sin t)(\cos t) \\ &= \frac{2\sin^2 t}{\cos^2 t + \sin^2 t} + \frac{2\cos^2 t}{\cos^2 t + \sin^2 t} \\ &= 2 \qquad (t \in (0, 2\pi)). \end{aligned}$$

It follows that

$$g(t) = g(0) + 2t \qquad (t \in [0, 2\pi])$$

and so $g(2\pi) \neq g(0)$. This contradiction shows that u cannot have an harmonic conjugate on D.

PROBLEMS 3

24. Given that $f = u + iv \in \mathscr{A}(D)$ is infinitely differentiable on D show that u, v have partial derivatives of all orders on D.

25. Show that the following functions are harmonic on $\mathbf{C}$ and find harmonic conjugates for each.

(i) $u(x, y) = x^2 - y^2 + 2x$.
(ii) $u(x, y) = e^x \cos y$.
(iii) $u(x, y) = \sin x \cosh y$.

26. Show that the Laplace equation corresponds to

$$\frac{\partial^2 f}{\partial z\, \partial \bar{z}} = 0.$$

27. Let $f \in \mathscr{A}(D)$ with $f(z) \neq 0$ $(z \in D)$. Show that $\log|f| \in \mathscr{H}(D)$, assuming that f is suitably differentiable.

28. Let $D = \{z\colon \operatorname{Re} z > 0\}$ and let

$$u(x, y) = \log (x^2 + y^2) \qquad ((x, y) \in D).$$

Show that $u \in \mathscr{H}(D)$ and u has an harmonic conjugate on D.

29. Let $u_1, u_2, \ldots, u_n, u_1^2 + u_2^2 + \cdots + u_n^2 \in \mathscr{H}(D)$. Show that each u_r is constant on D.

30. Given $f\colon D \to \mathbf{C}$ with $f = u + iv$ we say that f is *complex harmonic on* D if both u and v are harmonic on D. Let $g\colon D \to \mathbf{C}$ be such that g and ug are complex harmonic on D. Show that $g \in \mathscr{A}(D)$.

4

POWER SERIES FUNCTIONS

The purpose of this chapter is twofold; first to introduce further classes of functions analytic on certain domains and then to study in detail certain special functions which occur repeatedly throughout complex analysis. We recall from the last chapter that any polynomial function is an entire function. It is now a natural step for the analyst to investigate 'infinite polynomial functions' or power series functions as we shall call them. This necessitates the development of some machinery to discuss infinite series of complex numbers. In subsequent chapters we shall wish to consider special cases of double sequences of complex numbers; the necessary machinery will also be developed in this chapter.

4.1 Infinite series of complex numbers

Given a complex sequence $\{\alpha_n\}$ we define the associated *infinite series* to be the sequence $\{S_n\}$ where

$$S_n = \alpha_1 + \alpha_2 + \cdots + \alpha_n \qquad (n \in \mathbf{P}).$$

If the infinite series *converges*, i.e. if the sequence $\{S_n\}$ converges, we call $\lim_{n\to\infty} S_n$ the *sum* of the infinite series. We say that the infinite series *diverges* if it does not converge. The classical notation denotes both the infinite series and its sum (if it exists) by $\sum_{n=1}^{\infty} \alpha_n$. No satisfactory rationalization yet seems to have been produced for this unfortunate terminology. Wherever possible we shall try to distinguish the series and its sum as follows. We shall denote the series by $\sum \alpha_n$ (so that $\sum$ may be thought of as the transformation which maps the sequence $\{\alpha_n\}$ into the sequence $\{S_n\}$). We shall then denote the sum, if it exists, by $\sum_{n=1}^{\infty} \alpha_n$. In places we shall wish to discuss the infinite series associated with 'sequences' of the form β_n $(n = 0, 1, 2, \ldots)$ or γ_n $(n = n_0, n_0 + 1, n_0 + 2, \ldots)$. As usual we shall denote the corresponding sums by

$$\sum_{n=0}^{\infty} \beta_n, \qquad \sum_{n=n_0}^{\infty} \gamma_n.$$

The modifications in the theory are so trivial (relabelling of sequences) that we shall not weary the reader with them.

We shall assume that the reader is acquainted with the rudiments of the theory of infinite series of real numbers. This enables us to give simple proofs of most of the results. Although the proofs are simple they are worth giving since they illustrate standard methods for extending results for real functions to corresponding results for complex functions.

Proposition 4.1. *Given* $\{\alpha_n\} \subset \mathbf{C}$, $\sum \alpha_n$ *converges if and only if both* $\sum \operatorname{Re} \alpha_n$ *and* $\sum \operatorname{Im} \alpha_n$ *converge, in which case*

$$\sum_{n=1}^{\infty} \alpha_n = \sum_{n=1}^{\infty} \operatorname{Re} \alpha_n + i \sum_{n=1}^{\infty} \operatorname{Im} \alpha_n.$$

Proof. This follows immediately from the now familiar statement that if $\{x_n\}, \{y_n\}$ are real sequences

$$\lim_{n \to \infty} (x_n + iy_n) = x + iy$$

if and only if

$$\lim_{n \to \infty} x_n = x, \qquad \lim_{n \to \infty} y_n = y.$$

We say that $\sum \alpha_n$ is *absolutely convergent* if $\sum |\alpha_n|$ is convergent. Given that $\varphi: \mathbf{P} \to \mathbf{P}$ is one-to-one and onto we say that $\sum \alpha_{\varphi(n)}$ is a *rearrangement* of $\sum \alpha_n$.

Proposition 4.2. *If* $\sum \alpha_n$ *is absolutely convergent then it is convergent. If* $\sum \alpha_{\varphi(n)}$ *is a rearrangement of* $\sum \alpha_n$ *then* $\sum \alpha_{\varphi(n)}$ *is absolutely convergent and*

$$\sum_{n=1}^{\infty} \alpha_{\varphi(n)} = \sum_{n=1}^{\infty} \alpha_n.$$

Proof. Let $T_n = |\alpha_1| + |\alpha_2| + \cdots + |\alpha_n|$ $(n \in \mathbf{P})$. Since $\sum \alpha_n$ is absolutely convergent $\{T_n\}$ is convergent and therefore Cauchy. Since

$$|S_m - S_n| \leqslant |T_m - T_n|$$

it follows that $\{S_n\}$ is Cauchy and therefore convergent since $\mathbf{C}$ is a complete metric space.

Since $|\text{Re } \alpha_n| \leqslant |\alpha_n|$, $|\text{Im } \alpha_n| \leqslant |\alpha_n|$ $(n \in \mathbf{P})$ it follows from the comparison test for real series that $\sum \text{Re } \alpha_n$ and $\sum \text{Im } \alpha_n$ are absolutely convergent. It now follows from the theory of real series that $\sum \text{Re } \alpha_{\varphi(n)}$ and $\sum \text{Im } \alpha_{\varphi(n)}$ are absolutely convergent with

$$\sum_{n=1}^{\infty} \text{Re } \alpha_{\varphi(n)} = \sum_{n=1}^{\infty} \text{Re } \alpha_n, \qquad \sum_{n=1}^{\infty} \text{Im } \alpha_{\varphi(n)} = \sum_{n=1}^{\infty} \text{Im } \alpha_n.$$

Since

$$|\alpha_{\varphi(n)}| \leqslant |\text{Re } \alpha_{\varphi(n)}| + |\text{Im } \alpha_{\varphi(n)}| \qquad (n \in \mathbf{P})$$

it follows from the comparison test for real series that $\sum \alpha_{\varphi(n)}$ is absolutely convergent, and so also convergent. Using Proposition 4.1 we now obtain

$$\begin{aligned}\sum_{n=1}^{\infty} \alpha_{\varphi(n)} &= \sum_{n=1}^{\infty} \text{Re } \alpha_{\varphi(n)} + i \sum_{n=1}^{\infty} \text{Im } \alpha_{\varphi(n)} \\ &= \sum_{n=1}^{\infty} \text{Re } \alpha_n + i \sum_{n=1}^{\infty} \text{Im } \alpha_n \\ &= \sum_{n=1}^{\infty} \alpha_n.\end{aligned}$$

In the theory of absolutely convergent real series the main tools are the comparison test and the nth root test (or the weaker ratio test). The same is true for complex series. We recall first the definition of the limit superior of a bounded sequence of real numbers. Given a bounded real sequence $\{a_n\}$ the *limit superior* of $\{a_n\}$, denoted by $\overline{\lim\limits_{n\to\infty}}\, a_n$, is that real number a with the following properties:

(i) Given $\epsilon > 0$ there is $N \in \mathbf{P}$ such that

$$a_n < a + \epsilon \qquad (n > N).$$

(ii) Given $\epsilon > 0$ there is a subsequence $\{a_{n\sigma(n)}\}$ of $\{a_n\}$ such that

$$a_{n\sigma(n)} > a - \epsilon \qquad (n \in \mathbf{P}).$$

If $\{a_n\}$ is not bounded above we write

$$\overline{\lim_{n\to\infty}}\, \alpha_n = +\infty.$$

Proposition 4.3. (i) *If $\sum \alpha_n$ is absolutely convergent and if there is $N \in \mathbf{P}$ such that $|\beta_n| \leqslant |\alpha_n|$ $(n > N)$ then $\sum \beta_n$ is absolutely convergent and so convergent.*

(ii) *Let $A = \overline{\lim_{n \to \infty}} |\alpha_n|^{1/n}$. Then $\sum \alpha_n$ is absolutely convergent if $A < 1$ and diverges if $A > 1$.*

Proof. (i) This follows from the comparison test for real series and Proposition 4.2.

(ii) Suppose $A < 1$ and choose $\epsilon > 0$ such that $t = A + \epsilon < 1$. It follows that there is $N \in \mathbf{P}$ such that

$$|\alpha_n|^{1/n} < t \qquad (n > N)$$

and so

$$|\alpha_n| < t^n \qquad (n > N).$$

It follows from the comparison test for real series that $\sum |\alpha_n|$ is convergent.

Suppose now that $A > 1$ and choose $\epsilon = A - 1$. Then there is a subsequence $\{\alpha_{\sigma(n)}\}$ of $\{\alpha_n\}$ such that

$$|\alpha_{\sigma(n)}|^{1/\sigma(n)} > A - \epsilon = 1 \qquad (n \in \mathbf{P})$$

and so

$$|\alpha_{\sigma(n)}| > 1 \quad (n \in \mathbf{P}).$$

It follows that $\{\alpha_n\}$ does not converge to zero and hence $\sum \alpha_n$ cannot converge.

PROBLEMS 4

1. Given $z \in \mathbf{C}$ discuss the convergence of the series

$$\sum \frac{1}{n^2 + z^2}, \qquad \sum \frac{\arg(z + n)}{1 + in}.$$

2. Given $0 < a < b$ show that $\sum \dfrac{1}{n^a + in^b}$ converges iff $2a - b > 1$. If

$$f(a) = \sum_{n=1}^{\infty} \frac{1}{n^a + in^3} \qquad (2 < a < 3)$$

show that f is continuous on $(2, 3)$.

3. Given $\alpha_n, \beta_n \in \mathbf{C}$ $(n \in \mathbf{P})$ let $A_n = \sum_{r=1}^{n} \alpha_r$. Show that

$$\sum_{r=1}^{N} \alpha_r\beta_r = \sum_{r=1}^{N-1} (\beta_r - \beta_{r+1})A_r + \beta_N A_N.$$

Discuss the convergence of $\sum_{n=1}^{\infty} \frac{1}{n}\{\cos nt + i \sin nt\}$.

4. Given $\alpha, \beta, \gamma, z \in \mathbf{C}$ where $\gamma \neq 0, -1, -2, \ldots$, discuss the convergence of

$$\sum \frac{\alpha(\alpha + 1)\ldots(\alpha + n - 1)\beta(\beta + 1)\ldots(\beta + n - 1)}{\gamma(\gamma + 1)\ldots(\gamma + n - 1)} \frac{z^n}{n!}.$$

5. Let B be a Banach space (see Problem 2.6). If $\{x_n\} \subset B$ and $\sum \|x_n\|$ converges show that $\sum x_n$ converges. Let X be a compact metric space, let $B = \mathscr{C}(X)$ and let $f \in B$ with $\|f\| < 1$. Show that $1 + \sum f^n$ converges. What is the sum? Suppose now that B is a Banach algebra and $y \in B$ is such that $\overline{\lim_{n\to\infty}} \|y^n\|^{1/n} < 1$. Show that $\sum y^n$ converges. What is the sum if B has a *unit element* e, i.e. an element e such that $ex = xe = x$ $(x \in B)$?

4.2 Double sequences of complex numbers

We define a complex *double sequence* to be a function α from $\mathbf{P} \times \mathbf{P}$ to $\mathbf{C}$. We denote the image of (m, n) under α by $\alpha_{m,n}$ and we usually denote the double sequence by $\{\alpha_{m,n}\}$. We say that the double sequence $\{\alpha_{m,n}\}$ *converges* to λ if for each $\epsilon > 0$ there is $N(\epsilon) \in \mathbf{P}$, such that

$$|\alpha_{m,n} - \lambda| < \epsilon \qquad (m > N(\epsilon), n > N(\epsilon)).$$

If $\{\alpha_{m,n}\}$ also converges to μ it is simple to show that $\mu = \lambda$. We write

$$\lim_{m,n\to\infty} \alpha_{m,n} = \lambda.$$

It is sometimes helpful to visualize a double sequence as a doubly infinite array as indicated below.

$$\begin{matrix} \alpha_{1,1} & \alpha_{1,2} & \alpha_{1,3} & \cdots \\ \alpha_{2,1} & \alpha_{2,2} & \alpha_{2,3} & \cdots \\ \alpha_{3,1} & \alpha_{3,2} & \alpha_{3,3} & \cdots \\ \vdots & \vdots & \vdots & \cdots \end{matrix}$$

It follows trivially from the definition that if $\{\alpha_{m,n}\}$ converges to λ then $\lim_{n\to\infty} \alpha_{n,n} = \lambda$. In pictorial terms this says that the dexter diagonal converges to the limit of the double sequence.

Our interest is essentially confined to double sequences which arise in special ways.

Proposition 4.4. *Given* $\{\alpha_n\}, \{\beta_n\} \subset \mathbf{C}$ *let*

$$S_{m,n} = \sum_{r=1}^{m} \alpha_r + \sum_{r=1}^{n} \beta_r.$$

Then $\{S_{m,n}\}$ *converges if and only if both* $\sum \alpha_r, \sum \beta_r$ *converge, in which case*

$$\lim_{m,n\to\infty} S_{m,n} = \sum_{r=1}^{\infty} \alpha_r + \sum_{r=1}^{\infty} \beta_r.$$

Proof. Suppose that $\sum \alpha_r, \sum \beta_r$ converge to λ, μ respectively. Given $\epsilon > 0$ there is $m_1(\epsilon) \in \mathbf{P}$ such that

$$\left|\sum_{r=1}^{m} \alpha_r - \lambda\right| < \tfrac{1}{2}\epsilon \qquad (m > m_1(\epsilon)).$$

Similarly, there is $n_1(\epsilon) \in \mathbf{P}$ such that

$$\left|\sum_{r=1}^{n} \beta_r - \mu\right| < \tfrac{1}{2}\epsilon \qquad (n > n_1(\epsilon)).$$

Choose $N(\epsilon) = \max(m_1(\epsilon), n_1(\epsilon))$ and we have

$$|S_{m,n} - (\lambda + \mu)| < \epsilon \qquad (m > N(\epsilon), n > N(\epsilon))$$

so that $\{S_{m,n}\}$ converges to $\lambda + \mu$.

Suppose now that $\{S_{m,n}\}$ converges to S. Given $\epsilon > 0$ there is $N(\epsilon) \in \mathbf{P}$ such that

$$|S_{m,n} - S| < \tfrac{1}{2}\epsilon \qquad (m > N(\epsilon), n > N(\epsilon)).$$

Fix $n > N(\epsilon)$. Then for $m_1 > m_2 > N(\epsilon)$ we have

$$\left|\sum_{r=m_2+1}^{m_1} \alpha_r\right| = |S_{m_1,n} - S_{m_2,n}| < \epsilon.$$

This shows that $\sum \alpha_r$ is Cauchy and thus converges, say to λ. A similar argument shows that $\sum \beta_r$ converges to μ, say. It follows from the first part that $\{S_{m,n}\}$ converges to $\lambda + \mu$, and thus $S = \lambda + \mu$ as required.

The above result enables us to discuss the convergence of series associated with 'sequences' defined on the set of all integers. More precisely, given $\alpha\colon \mathbf{Z} \to \mathbf{C}$ define

$$S_{m,n} = \sum_{r=-m}^{n} \alpha_r.$$

By the above proposition we have that $\{S_{m,n}\}$ converges if and only if both $\sum \alpha_r$, $\sum \alpha_{-r}$ converge, in which case we have

$$\lim_{m,n\to\infty} S_{m,n} = \sum_{r=0}^{\infty} \alpha_r + \sum_{r=1}^{\infty} \alpha_{-r}.$$

We adopt the usual convention of writing

$$\lim_{m,n\to\infty} S_{m,n} = \sum_{-\infty}^{\infty} \alpha_r.$$

Given a double sequence $\{\alpha_{m,n}\}$ there are several possible ways of 'summing' the double array. Probably the most natural method is to consider the double sequence defined by

$$S_{m,n} = \sum_{r=1}^{m} \sum_{s=1}^{n} \alpha_{r,s}.$$

Pictorially this corresponds to summing by rectangles. We might also sum the 'rows' separately and then sum the row sums; or sum the 'columns' separately and then sum the column sums. A further possibility is to sum by triangles, i.e. to consider the sequence defined by

$$T_n = \sum_{r+s\leqslant n} \alpha_{r,s}.$$

We give in Problem 4.7 a general result on these methods of summation. We shall here confine our attention to the simpler situation in which $\alpha_{r,s}$ is of the form $\beta_r\gamma_s$.

Proposition 4.5. *Given convergent series* $\sum \alpha_r$, $\sum \beta_r$ *with sums* λ, μ, *respectively, let*

$$S_{m,n} = \sum_{r=1}^{m} \alpha_r \sum_{r=1}^{n} \beta_r.$$

Then $\{S_{m,n}\}$ *converges to* $\lambda\mu$.

Proof. Let $A_m = \sum_{r=1}^{m} \alpha_r$, $B_n = \sum_{r=1}^{n} \beta_r$. Then

$$\begin{aligned} |S_{m,n} - \lambda\mu| &= |A_m B_n - \lambda\mu| \\ &= |A_m(B_n - \mu) + \mu(A_m - \lambda)| \\ &\leqslant |A_m|\,|B_n - \mu| + |\mu|\,|A_m - \lambda|. \end{aligned}$$

Since $\{A_m\}$ converges it is bounded and so there exists $M > 0$ such that $|A_m| \leqslant M$ ($m \in \mathbf{P}$). Given $\epsilon > 0$ we may choose $N(\epsilon) \in \mathbf{P}$ such that

$$|B_n - \mu| < \frac{\epsilon}{2M} \qquad (n > N(\epsilon))$$

$$|A_m - \lambda| < \frac{\epsilon}{2(1 + |\mu|)} \qquad (m > N(\epsilon)).$$

We now have

$$|S_{m,n} - \lambda\mu| < \epsilon \qquad (m > N(\epsilon), n > N(\epsilon)),$$

and so $\{S_{m,n}\}$ converges to $\lambda\mu$.

Observe that the above result sums the double array $\{\alpha_r\beta_s\}$ by rectangles. It is trivial to verify that the same sum is obtained by the row or column method. The situation is much more complicated with regard to summing by triangles. This latter method is very significant for the next section of this chapter and warrants special definition. Given complex sequences $\{\alpha_n\}$, $\{\beta_n\}$ ($n = 0, 1, 2, \ldots$) we define their *convolution* to be the sequence $\{\gamma_n\}$ defined by

$$\gamma_n = \sum_{r=0}^{n} \alpha_r\beta_{n-r} \qquad (n = 0, 1, 2, \ldots)$$

and we write

$$\{\gamma_n\} = \{\alpha_n\} * \{\beta_n\}.$$

Proposition 4.6. *Let $\sum \alpha_n$, $\sum \beta_n$ be absolutely convergent and let $\{\gamma_n\} = \{\alpha_n\} * \{\beta_n\}$. Then $\sum \gamma_n$ is absolutely convergent and*

$$\sum_{n=0}^{\infty} \gamma_n = \sum_{n=0}^{\infty} \alpha_n \sum_{n=0}^{\infty} \beta_n.$$

Proof. Since

$$\sum_{r=0}^{n} |\gamma_r| \leqslant \sum_{r=0}^{n} |\alpha_r| \sum_{r=0}^{n} |\beta_r| \leqslant \sum_{r=0}^{\infty} |\alpha_r| \sum_{r=0}^{\infty} |\beta_r|$$

it follows that $\left\{\sum_{r=0}^{n} |\gamma_r|\right\}$ is monotonic and bounded. Thus $\sum \gamma_n$ is absolutely convergent and so convergent. Let $A_n = \sum_{r=0}^{n} \alpha_r$, $B_n = \sum_{r=0}^{n} \beta_r$, $C_n = \sum_{r=0}^{n} \gamma_r$, $D_n = \sum_{r=0}^{n} |\alpha_r|$, $E_n = \sum_{r=0}^{n} |\beta_r|$, $\lambda = \sum_{r=0}^{\infty} \alpha_r$, $\mu = \sum_{r=0}^{\infty} \beta_r$. We then have

$$\begin{aligned} |C_{2n} - \lambda\mu| &\leqslant |C_{2n} - A_{2n}B_{2n}| + |A_{2n}B_{2n} - \lambda\mu| \\ &\leqslant |D_nE_n - D_{2n}E_{2n}| + |A_{2n}B_{2n} - \lambda\mu|. \end{aligned}$$

Since $\{A_{2n}B_{2n}\}$ converges to $\lambda\mu$ and since $\{D_nE_n\}$ converges it follows that

$$\lim_{n\to\infty} |C_{2n} - \lambda\mu| = 0.$$

Since any subsequence of a convergent sequence converges to the limit of the sequence we conclude that

$$\sum_{n=0}^{\infty} \gamma_n = \lim_{n\to\infty} C_{2n} = \lambda\mu.$$

PROBLEMS 4

6. (i) Let

$$\alpha_{m,n} = \frac{mn}{(m+in)^2} \qquad (m, n \in \mathbf{P}).$$

Show that

$$\lim_{m\to\infty} (\lim_{n\to\infty} \alpha_{m,n}) = 0 = \lim_{n\to\infty} (\lim_{m\to\infty} \alpha_{m,n})$$

but that $\{\alpha_{m,n}\}$ does not converge.

(ii) Let

$$\alpha_{m,n} = (-1)^{m+n}\left(\frac{1}{m} + \frac{i}{n}\right) \qquad (m, n \in \mathbf{P}).$$

Show that $\{\alpha_{m,n}\}$ converges to 0 but that for each fixed $m \in \mathbf{P}$ the sequence $\{\alpha_{m,n}\}$ fails to converge.

7. Let $\{\alpha_{r,s}\}$ be such that the double sequence $\left\{\sum_{r=1}^{m}\sum_{s=1}^{n} |\alpha_{r,s}|\right\}$ converges. Show that

$$\lim_{m,n\to\infty} \sum_{r=1}^{m}\sum_{s=1}^{n} \alpha_{r,s} = \sum_{r=1}^{\infty}\left(\sum_{s=1}^{\infty} \alpha_{r,s}\right) = \sum_{s=1}^{\infty}\left(\sum_{r=1}^{\infty} \alpha_{r,s}\right)$$
$$= \lim_{n\to\infty} \sum_{r+s\leqslant n} \alpha_{r,s}.$$

(This shows that the double array has the same sum by rectangles, rows, columns, or triangles.)

8. For which z do the following series converge?

$$\sum_{-\infty}^{\infty} z^n, \sum_{-\infty}^{\infty} \frac{\log(1+|n|)}{n^2+n+z}, \sum_{-\infty}^{\infty} \frac{2z}{z^2-n^2}.$$

9. Let

$$\alpha_n = \frac{(-1)^n}{(n+1)^{\frac{1}{2}}} \qquad (n = 0, 1, 2, \ldots)$$

and let $\{\gamma_n\} = \{\alpha_n\} * \{\alpha_n\}$. Show that $\sum \alpha_n$ converges while $\sum \gamma_n$ diverges.

10. (i) Let $L^1(\mathbf{Z}^+)$ denote the set of all complex sequences α_n ($n = 0, 1, 2, \ldots$) such that $\sum |\alpha_n|$ converges. Show that $L^1(\mathbf{Z}^+)$ is a complex linear algebra with the operations defined by

$$\lambda\{\alpha_n\} = \{\lambda\alpha_n\}$$
$$\{\alpha_n\} + \{\beta_n\} = \{\alpha_n + \beta_n\}$$
$$\{\alpha_n\}.\{\beta_n\} = \{\alpha_n\} * \{\beta_n\}.$$

If

$$\|\{\alpha_n\}\| = \sum_0^\infty |\alpha_n|$$

show that $(L^1(\mathbf{Z}^+), \|.\|)$ is a Banach algebra.

(ii) Let $L^1(\mathbf{Z})$ denote the set of all functions $\alpha: \mathbf{Z} \to \mathbf{C}$ such that $\sum |\alpha_n|$ converges. Given $\alpha, \beta \in L^1(\mathbf{Z})$ let

$$\alpha * \beta = \left\{\sum_{-\infty}^{\infty} \alpha_r \beta_{n-r}\right\}.$$

Show that $L^1(\mathbf{Z})$ is a complex linear algebra (with the above multiplication). If

$$\|\alpha\| = \sum_{-\infty}^{\infty} |\alpha_n|$$

show that $(L^1(\mathbf{Z}), \|.\|)$ is a Banach algebra.

4.3 Power series functions

A *power series function* is a function of the form

$$f(z) = \sum_{n=0}^{\infty} \alpha_n (z - \alpha)^n$$

where $\alpha \in \mathbf{C}$, $\alpha_n \in \mathbf{C}$ ($n = 0, 1, 2, \ldots$), being defined for those points z for which the series converges. The first problem is thus to determine the nature of the set on which f is defined.

Proposition 4.7. *Let f be a power series function as above. The set on which f is defined is one of the following:*

(i) *the single point α;*

(ii) *a disc $\Delta(\alpha, \rho)$ with part (possibly void) of $C(\alpha, \rho)$;*

(iii) *the whole of* $\mathbf{C}$.

Proof. Suppose the above series converges at some point $z_0 \neq \alpha$. It then follows from Proposition 4.3(ii) that

$$\overline{\lim_{n\to\infty}} |\alpha_n(z_0 - \alpha)^n|^{1/n} \leqslant 1.$$

This implies in particular that $\{|\alpha_n|^{1/n}\}$ is bounded above and hence $k = \overline{\lim_{n\to\infty}} |\alpha_n|^{1/n} < +\infty$. We now have for each $z \in \mathbf{C}$

$$\overline{\lim_{n\to\infty}} |\alpha_n(z - \alpha)^n|^{1/n} = k|z - \alpha|.$$

If $k = 0$ the series converges (absolutely) for all $z \in \mathbf{C}$ by Proposition 4.3(ii). If $k > 0$ it follows from Proposition 4.3(ii) that the series converges (absolutely) for $|z - \alpha| < \dfrac{1}{k}$ and does not converge for $|z - \alpha| > \dfrac{1}{k}$. Take $\rho = \dfrac{1}{k}$ and the proof is complete.

The above number ρ is called the *radius of convergence* of the power series $\sum \alpha_n(z - \alpha)^n$. If we take $\rho = 0$ for case (i) and $\rho = +\infty$ for case (iii) then we have the general formula

$$\rho = \frac{1}{\overline{\lim_{n\to\infty}} |\alpha_n|^{1/n}}$$

(with the convention $\dfrac{1}{+\infty} = 0$). Observe that we have also shown that the power series converges absolutely at each point of the (open) disc of convergence. There is a considerable range of possibility for the behaviour of the series on $C(\alpha, \rho)$ (in the case $0 < \rho < +\infty$). For example the series may converge at no point, at some of the points or all points of $C(\alpha, \rho)$ (see Problem 4.11). Our main interest is simply with the behaviour of the function f on the disc $\Delta(\alpha, \rho)$. More can be said about the nature of the convergence of the series on this disc.

Proposition 4.8. *Let ρ be the radius of convergence of the power series $\sum \alpha_n(z - \alpha)^n$. Then the series converges uniformly on every closed disc $\bar{\Delta}(a, r)$ $(0 < r < \rho)$.*

Proof. Given $0 < r < \rho$ choose $z_0 \in \Delta(\alpha, \rho)$ such that $|z_0 - \alpha| = r$. We thus have

$$|\alpha_n(z - \alpha)^n| \leqslant |\alpha_n| r^n \qquad (z \in \bar{\Delta}(a, r))$$

and therefore

$$\left| \sum_{n=N+1}^{\infty} \alpha_n(z - \alpha)^n \right| \leqslant \sum_{n=N+1}^{\infty} |\alpha_n| r^n \qquad (z \in \bar{\Delta}(\alpha, r),\ N \in \mathbf{P}).$$

Since $\sum |\alpha_n| r^n$ converges it follows readily that $\sum \alpha_n(z - \alpha)^n$ converges uniformly on $\bar{\Delta}(\alpha, r)$ as required.

As a simple application of the above result we shall show that the associated power series function f is continuous on $\Delta(\alpha, \rho)$. Observe that f is the pointwise limit of a sequence of polynomial functions. If the series converged uniformly on $\Delta(\alpha, \rho)$ it would follow immediately from Proposition 1.14 that $f \in \mathscr{C}(\Delta(\alpha, \rho))$. In general it is not true that the series converges uniformly on $\Delta(\alpha, \rho)$ (see Problem 4.12) and we need to employ a finer argument. Given $z \in \Delta(\alpha, \rho)$ choose r such that $|z - \alpha| < r < \rho$. The series then converges uniformly on $\bar{\Delta}(\alpha, r)$ and so on $\Delta(\alpha, r)$. It now follows from Proposition 1.14 that $f|_{\Delta(\alpha, r)} \in \mathscr{C}(\Delta(\alpha, r))$ and so by Proposition 1.13 f is continuous on $\Delta(\alpha, r)$. In particular f is continuous at z. Since z was arbitrary this shows that f is continuous on $\Delta(\alpha, \rho)$. The above technique might well be called the 'squeezing argument'.

Proposition 4.9. *Let the radius of convergence of the power series functions*

$$f(z) = \sum_{n=0}^{\infty} \alpha_n (z - \alpha)^n, \qquad g(z) = \sum_{n=0}^{\infty} \beta_n (z - \alpha)^n$$

be ρ_1, ρ_2 *respectively. Let* $\rho = \min(\rho_1, \rho_2)$ *and let* $\{\gamma_n\} = |\alpha_n\} * \{\beta_n\}$. *Then*

$$(fg)(z) = \sum_{n=0}^{\infty} \gamma_n (z - \alpha)^n \qquad (z \in \Delta(\alpha, \rho)).$$

Proof. This follows immediately from Proposition 4.6 since

$$\{\gamma_n (z - \alpha)^n\} = \{\alpha_n (z - \alpha)^n\} * \{\beta_n (z - \alpha)^n\}$$

and since power series converge absolutely on the (open) disc of convergence.

We come now to the main result of this chapter.

Theorem 4.10. *Let* $\rho\,(>0)$ *be the radius of convergence of the power series function*

$$f(z) = \sum_{n=0}^{\infty} \alpha_n (z - \alpha)^n.$$

Then f *is infinitely differentiable on* $\Delta(\alpha, \rho)$ *and we have*

$$f^{(k)}(z) = \sum_{n=k}^{\infty} n(n-1)\ldots(n-k+1)\alpha_n (z - \alpha)^{n-k} \qquad (z \in \Delta(\alpha, \rho), k \in \mathbf{P})$$

and

$$f(z) = f(\alpha) + \sum_{n=1}^{\infty} \frac{f^{(n)}(\alpha)}{n!}(z - \alpha)^n \qquad (z \in \Delta(\alpha, \rho)).$$

Proof. It is well known from real analysis that $\lim_{n\to\infty} n^{1/n} = 1$. It follows that

$$\overline{\lim_{n\to\infty}} |n\alpha_n|^{1/n} = \overline{\lim_{n\to\infty}} |\alpha_n|^{1/n}$$

so that ρ is also the radius of convergence for the power series function

$$g(z) = \sum_{n=1}^{\infty} n\alpha_n(z - \alpha)^{n-1}.$$

Let $z \in \Delta(\alpha, \rho)$ with $|z - \alpha| = r$ and choose R such that $r < R < \rho$. Given $h \in \mathbf{C}$ with $|h| = \delta$, $0 < \delta \leqslant \frac{1}{2}(R - r)$ we have $\alpha + h \in \Delta(\alpha, \rho)$. For each $n \in \mathbf{P}$ we have

$$\begin{aligned}\left|\frac{(z - \alpha - h)^n - (z - \alpha)^n}{h} - n(z - \alpha)^{n-1}\right| &= \left|\sum_{t=2}^{n} \binom{n}{t}(z - \alpha)^{n-t}h^{t-1}\right|\\ &\leqslant |h| \sum_{t=2}^{n} \binom{n}{t} r^{n-t}\delta^{t-2}\\ &\leqslant |h| \frac{n(n-1)}{2} \sum_{t=2}^{n} \binom{n-2}{t-2} r^{n-t}\delta^{t-2}\\ &= |h| \frac{n(n-1)}{2}(r + \delta)^{n-2}.\end{aligned}$$

Since $\sum \alpha_n(z - \alpha)^n$ converges on $C(\alpha, R)$ there is $M > 0$ such that $|\alpha_n R^n| \leqslant M$ $(n \in \mathbf{P})$. It now follows that

$$\begin{aligned}&\left|\frac{f(z + h) - f(z)}{h} - g(z)\right|\\ &= \left|\sum_{n=1}^{\infty} \alpha_n \left\{\frac{(z - \alpha - h)^n - (z - \alpha)^n}{h} - n(z - \alpha)^{n-1}\right\}\right|\\ &\leqslant \sum_{n=1}^{\infty} \frac{M}{R^n} |h| \frac{n(n-1)}{2}(r + \delta)^{n-2}\\ &= \frac{M|h|}{2R^2} \sum_{n=1}^{\infty} n(n-1)\left(\frac{r + \delta}{R}\right)^{n-2}.\end{aligned}$$

Since $r + \delta \leqslant r + \frac{1}{2}(R - r) < R$ the above series converges and we conclude that f is differentiable at z with $f'(z) = g(z)$. Since z was arbitrary we have thus shown that f is differentiable on $\Delta(\alpha, \rho)$ with

$$f'(z) = \sum_{n=1}^{\infty} n\alpha_n(z - \alpha)^{n-1} \qquad (z \in \Delta(\alpha, \rho)).$$

Suppose now that f is differentiable k times on $\Delta(\alpha, \rho)$ with

$$f^{(k)}(z) = \sum_{n=k}^{\infty} n(n-1)\ldots(n-k+1)\alpha_n(z - \alpha)^{n-k} \qquad (z \in \Delta(\alpha, \rho)).$$

By applying the above result to $f^{(k)}$ we deduce that $f^{(k)}$ is differentiable on $\Delta(\alpha, \rho)$ with

$$f^{(k+1)}(z) = \sum_{n=k+1}^{\infty} n(n-1)\ldots(n-k)\alpha_n(z-\alpha)^{n-k-1} \qquad (z \in \Delta(\alpha, \rho)).$$

It follows by induction that f is infinitely differentiable on $\Delta(\alpha, \rho)$ with $f^{(k)}(z)$ given as above. In particular we have

$$f^{(k)}(\alpha) = k!\alpha_k \qquad (k \in \mathbf{P}).$$

Since $f(\alpha) = \alpha_0$ it now follows that

$$f(z) = \sum_{n=0}^{\infty} \alpha_n(z-\alpha)^n = f(\alpha) + \sum_{n=1}^{\infty} \frac{f^{(n)}(\alpha)}{n!}(z-\alpha)^n \qquad (z \in \Delta(\alpha, \rho)).$$

The above theorem extends considerably our list of functions analytic on discs. (In fact we shall see later that any function in $\mathscr{A}(\Delta(\alpha, r))$ is a power series function.) We may proceed a little further as follows. Let us consider 'upside-down' power series functions, namely functions of the form

$$g(z) = \sum_{n=1}^{\infty} \beta_n(z-\alpha)^{-n}$$

being defined at those points for which the series converges. It follows from Proposition 4.7 that g is defined for no points, for some $\nabla(\alpha, \rho)$ together with part (possibly void) of $C(\alpha, \rho)$, or for $\mathbf{C} \setminus \{\alpha\}$. Such functions might well be thought of as power series functions about the point at infinity with disc of convergence $\nabla(\alpha, \rho)$. Indeed we have

$$g(z) = (f \circ \mathbf{j})(z - \alpha)$$

where f is the power series function

$$f(z) = \sum_{n=1}^{\infty} \beta_n z^n.$$

It now follows from Theorem 4.10 and Proposition 3.8 that the above function g is analytic on $\nabla(\alpha, \rho)$ with

$$g'(z) = \sum_{n=1}^{\infty} -n\beta_n(z-\alpha)^{-n-1}.$$

Combining the above two types of function we obtain functions of the form

$$h(z) = \sum_{-\infty}^{\infty} \alpha_n(z-\alpha)^n$$

being defined at those points for which the series converges. Such functions are either defined for no points, for an open annulus together with part (possibly void) of its boundary, or for $\mathbf{C} \setminus \{\alpha\}$. Moreover it follows easily from the work above that the series converges absolutely at each point of the open annulus and uniformly on any closed subannulus. Furthermore h is infinitely differentiable on the open annulus and the series may be differentiated 'term by term'. Functions of this type will appear frequently in Chapter 7. We shall call such functions *quasi power series functions*. In particular it will turn out that any function in $\mathscr{A}(\mathbf{C} \setminus \{\alpha\})$ is a quasi power series function.

PROBLEMS 4

11. Show that the following series have radius of convergence 1.

$$\sum z^n, \qquad \sum \frac{z^n}{n}, \qquad \sum \frac{z^n}{n^2}.$$

Show that the first converges nowhere on $C(0, 1)$, the second converges on $C(0, 1) \setminus \{1\}$, and the third converges on $C(0, 1)$.

12. (i) Show that $\sum z^n$ does not converge uniformly on $\Delta(0, 1)$.

(ii) Show that $\sum \dfrac{z^{n+1}}{n(n+1)}$ converges uniformly on $\Delta(0, 1)$ while the derived series $\sum \dfrac{z^n}{n}$ does not converge uniformly on $\Delta(0, 1)$.

13. If $\sum \alpha_n (z - \alpha)^n$ has radius of convergence ρ, what is the radius of convergence of the following power series?

$$\sum \alpha_n (z - \alpha)^{2n}, \qquad \sum \alpha_{2n} (z - \alpha)^n, \qquad \sum \alpha_n^2 (z - \alpha)^n.$$

14. Which power series functions are differentiable at ∞?

15. Suppose that $\sum \alpha_n z^n$ has radius of convergence 1 and that $\sum \alpha_n$ converges. If

$$f(z) = \sum_{n=0}^{\infty} \alpha_n z^n \qquad (z \in \Delta(0, 1))$$

show that

$$\lim_{x \to 1-} f(x) = \sum_{n=0}^{\infty} \alpha_n \qquad \text{(Abel)}.$$

(*Hint.* Show first that it is sufficient to consider the case in which $\sum_{n=0}^{\infty} \alpha_n = 0$. Let $S_n = \alpha_0 + \alpha_1 + \cdots + \alpha_n$ and show that

$$f(z) = (1 - z) \sum_{n=0}^{\infty} S_n z^n \qquad (z \in \Delta(0, 1)).$$

Given $\epsilon > 0$ there is $N \in \mathbf{P}$ such that $|S_n| < \epsilon$ $(n > N)$. Now split the series into two parts and deduce the result.)

4.4 The exponential function

We shall now examine the so-called standard or elementary functions. We shall assume that the reader has a working knowledge of the corresponding real functions.

Aside from the polynomial functions the most important standard function is the *exponential* function. We define $\exp: \mathbf{C} \to \mathbf{C}$ by

$$\exp(z) = 1 + \sum_{n=1}^{\infty} \frac{z^n}{n!} \qquad (z \in \mathbf{C}).$$

Proposition 4.11. *The function* exp *is an entire function with the following properties*:

(i) $\exp' = \exp$;
(ii) $\exp(z_1 + z_2) = \exp(z_1)\exp(z_2) \quad (z_1, z_2 \in \mathbf{C})$;
(iii) $\exp(z) = e^x\{\cos y + i \sin y\} \quad (z = x + iy \in \mathbf{C})$.

Proof. (i) It is well known from real analysis that $(n!)^{1/n} \to \infty$ as $n \to \infty$ so that the power series for exp has infinite radius of convergence. It thus follows from Theorem 4.10 that $\exp \in \mathscr{A}(\mathbf{C})$ and moreover

$$\exp'(z) = \sum_{n=1}^{\infty} \frac{nz^{n-1}}{n!} = \exp(z) \qquad (z \in \mathbf{C}).$$

(ii) Given $\alpha \in \mathbf{C}$ define f on $\mathbf{C}$ by

$$f(z) = \exp(z)\exp(\alpha - z) \qquad (z \in \mathbf{C}).$$

It follows from Propositions 3.7 and 3.8 that $f \in \mathscr{A}(\mathbf{C})$ and

$$f'(z) = \exp(z)\exp(\alpha - z) + \exp(z)\{-\exp(\alpha - z)\} \qquad (z \in \mathbf{C}).$$

We now have $f' = 0$ and so by Proposition 3.10

$$f(z) = f(0) = \exp(\alpha) \qquad (z \in \mathbf{C}),$$

i.e.

$$\exp(z)\exp(\alpha - z) = \exp(\alpha) \qquad (z \in \mathbf{C}).$$

Given $z_1, z_2 \in \mathbf{C}$ let $z = z_1$, $\alpha = z_1 + z_2$ and we obtain

$$\exp(z_1 + z_2) = \exp(z_1)\exp(z_2)$$

as required.

(iii) Using (ii) we now obtain for $z = x + iy$

$$\begin{aligned}\exp(z) &= \exp(x+iy) = \exp(x)\exp(iy)\\ &= e^x\left\{1 + \sum_{n=1}^{\infty}\frac{(iy)^n}{n!}\right\}\\ &= e^x\left\{1 + \sum_{m=1}^{\infty}(-1)^m\frac{y^{2m}}{(2m)!} + i\sum_{n=0}^{\infty}(-1)^n\frac{y^{2n+1}}{(2n+1)!}\right\}\\ &= e^x\{\cos y + i\sin y\}.\end{aligned}$$

Corollary. *The function* exp *has no zeros.*

Proof. It follows from (ii) that

$$\exp(z)\exp(-z) = \exp(0) = 1 \qquad (z \in \mathbf{C}).$$

and therefore exp cannot have any zeros.

Since the function exp is an extension of the usual exponential function on the real line we shall often write $\exp(z)$ more briefly as e^z. The occurrences of the exponential function in complex analysis are too numerous to mention. It is thus important to have as clear a geometrical picture as possible of how the exponential function behaves on $\mathbf{C}$.

Given $f: \mathbf{C} \to \mathbf{C}$ we say that γ is a *period* of f if

$$f(z+\gamma) = f(z) \qquad (z \in \mathbf{C}).$$

It is easy to see that such a function f repeats itself in strips of width $|\gamma|$. If γ is a period of f so is 2γ, for

$$f(z+2\gamma) = f(z+\gamma) = f(z) \qquad (z \in \mathbf{C}).$$

More generally we have that $n\gamma$ is a period of f for each $n \in \mathbf{Z}$. We say that γ is a *fundamental period* of f if γ is a period of f, $\gamma \neq 0$, and no proper submultiple of γ is a period of f. If γ is a fundamental period of f so also is $-\gamma$.

Proposition 4.12. *The function* exp *has fundamental periods* $2\pi i$, $-2\pi i$ *and these only. Given* $\theta \in \mathbf{R}$, exp *maps the strip* $\{z: \theta - \pi < \operatorname{Im} z \leqslant \theta + \pi\}$ *one-to-one onto* $\mathbf{C} \setminus \{0\}$, *and the domain* $\{z: \theta - \pi < \operatorname{Im} z < \theta + \pi\}$ *one-to-one onto* $\mathbf{C} \setminus N_\theta$.

Proof. We have by Proposition 4.11 (ii) and (iii)

$$\begin{aligned}\exp(z+2\pi i) = \exp(z)\exp(2\pi i) &= \exp(z)\{\cos 2\pi + i\sin 2\pi\}\\ &= \exp(z) \qquad (z \in \mathbf{C})\end{aligned}$$

so that $2\pi i$ is a period of exp. If γ is any period of exp we have

$$\exp(\gamma) = \exp(\gamma + 0) = \exp(0) = 1.$$

If $\gamma = a + ib$ we then have

$$e^a \{\cos b + i \sin b\} = 1.$$

On taking the modulus of both sides we obtain $e^a = 1$ and thus $a = 0$. We then have $\cos b = 1$, $\sin b = 0$ so that γ is an integral multiple of $2\pi i$. This proves the first statement.

Let $z_1 = x_1 + iy_1$, $z_2 = x_2 + iy_2$ be such that $\exp(z_1) = \exp(z_2)$. It follows that $|\exp(z_1)| = |\exp(z_2)|$, i.e. $e^{x_1} = e^{x_2}$ so that $x_1 = x_2$. We now have

$$\cos y_1 + i \sin y_1 = \cos y_2 + i \sin y_2.$$

It follows that exp is one-to-one on any strip $\{z: \theta - \pi < \text{Im}\, z \leqslant \theta + \pi\}$. We have already observed that exp maps $\mathbf{C}$ into $\mathbf{C} \setminus \{0\}$. Given $w \in \mathbf{C} \setminus \{0\}$ let $z = \log|w| + i \arg_\theta(w)$. Then $\theta - \pi < \text{Im}\, z \leqslant \theta + \pi$ and

$$\begin{aligned} \exp(z) &= \exp(\log|w|) \exp(i \arg_\theta(w)) \\ &= |w| \{\cos(\arg_\theta(w)) + i \sin(\arg_\theta(w))\} \\ &= w. \end{aligned}$$

Therefore exp maps the strip $\{z: \theta - \pi < \text{Im}\, z \leqslant \theta + \pi\}$ one-to-one onto $\mathbf{C} \setminus \{0\}$. For the final part it is sufficient to verify that exp maps the line $\{z: \text{Im}\, z = \theta + \pi\}$ onto $N_\theta \setminus \{0\}$.

As a further illustration of the exponential function the student should verify that exp maps the segment $[a + ic, b + ic]$ onto the segment $[e^a e^{ic}, e^b e^{ic}]$. We may now obtain the image of rectangles under the exponential mapping. The situation is described pictorially in Figure 4.1.

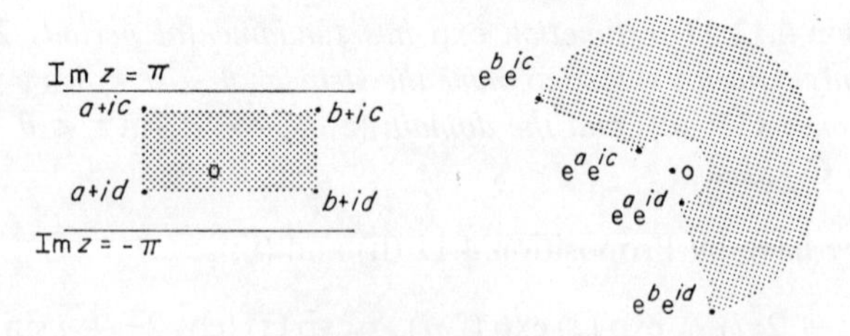

Figure 4.1

There are several standard functions related to the exponential function. We define the complex functions sin, cos, sinh, cosh by

$$\sin(z) = \frac{1}{2i}\{\exp(iz) - \exp(-iz)\} \qquad (z \in \mathbf{C})$$
$$\cos(z) = \tfrac{1}{2}\{\exp(iz) + \exp(-iz)\} \qquad (z \in \mathbf{C})$$
$$\sinh(z) = \tfrac{1}{2}\{\exp(z) - \exp(-z)\} \qquad (z \in \mathbf{C})$$
$$\cosh(z) = \tfrac{1}{2}\{\exp(z) + \exp(-z)\} \qquad (z \in \mathbf{C}).$$

It is clear from Propositions 3.7 and 3.8 that each of these functions is an entire function. By considering their power series expansions we may easily see that the above functions agree on the real line with the associated real functions. We set out below some of the simple properties of these functions leaving the proofs as exercises. The student should familiarize himself with these properties since they will be required for many subsequent problems.

(i) $\sin' = \cos$, $\cos' = -\sin$, $\sinh' = \cosh$, $\cosh' = \sinh$.
(ii) $\cos^2 + \sin^2 = \cosh^2 - \sinh^2 = \mathbf{1}$.
(iii) $\sin(z_1 + z_2) = \sin(z_1)\cos(z_2) + \cos(z_1)\sin(z_2)$
$\cos(z_1 + z_2) = \cos(z_1)\cos(z_2) - \sin(z_1)\sin(z_2)$.
(iv) $|\sin(z)|^2 = \sin^2 x + \sinh^2 y$, $|\cos(z)|^2 = \cos^2 x + \sinh^2 y$.
(v) sin, cos have fundamental period 2π; sinh, cosh have fundamental period $2\pi i$.
(vi) The zeros of sin, cos, sinh, cosh consist respectively of the sets $\{n\pi : n \in \mathbf{Z}\}$, $\{(n + \frac{1}{2})\pi : n \in \mathbf{Z}\}$, $\{in\pi : n \in \mathbf{Z}\}$, $\{i(n + \frac{1}{2})\pi : n \in \mathbf{Z}\}$.

We shall feel free to speak of other such complex functions as

$$\tan = \frac{\sin}{\cos}, \qquad \text{sech} = \frac{1}{\cosh}$$

and to state some of their obvious properties. For example it is clear that $\tan \in \mathscr{A}(D)$ where $D = \mathbf{C} \setminus \{(n + \frac{1}{2})\pi : n \in \mathbf{Z}\}$.

PROBLEMS 4

16. Show that exp is the only entire function f such that $f(\mathbf{R}) \subset \mathbf{R}$ and

$$f(1) = e, \qquad f(z_1 + z_2) = f(z_1)f(z_2) \qquad (z_1, z_2 \in \mathbf{C}). \qquad (*)$$

Find all entire functions satisfying (*). Find a continuous nowhere differentiable function f satisfying (*). (*Hint.* Functions of $\bar{z}$ are helpful for the latter part.)

17. Let $g\colon \mathbf{C} \to \mathbf{C}$ be defined by

$$g(z) = \begin{cases} \exp\left(-\dfrac{1}{z}\right) & \text{if} \quad |\arg(z)| \leqslant \dfrac{\pi}{4}, \quad |z| > 0, \\ 0 & \text{otherwise.} \end{cases}$$

Where is g differentiable? Where is g regular?

18. Use Proposition 4.9 to prove Proposition 4.11 (ii).

19. At which points do the following series converge? On which sets do they converge uniformly?

$$\sum_0^\infty e^{-nz}, \qquad \sum_0^\infty e^{-nz^2}, \qquad \sum_0^\infty 2nze^{-nz^2}, \qquad \sum_{-\infty}^\infty e^{-nz^2}, \qquad \sum_{-\infty}^\infty e^{-n^2z^2}.$$

20. Establish the following results.

(i) $\sin(iz) = i\sinh(z)$, $\sinh(iz) = i\sin(z)$,
$\cos(iz) = \cosh(z)$, $\cosh(iz) = \cos(z) \quad (z \in \mathbf{C})$.

(ii) $\overline{\sin(z)} = \sin(\bar{z})$, $\overline{\cos(z)} = \cos(\bar{z}) \quad (z \in \mathbf{C})$.

(iii) $\lim\limits_{y \to +\infty} e^{-y}\sin(x + iy) = \frac{1}{2}\{\sin x + i\cos x\}$.

(iv) $\lim\limits_{y \to +\infty} \tan(x + iy) = i$.

21. Let $D_1 = \left\{z\colon |\text{Re}\, z| < \dfrac{\pi}{2}\right\}$, $D_2 = \{z\colon 0 < \text{Re}\, z < \pi\}$, $D_3 = \{z\colon |\text{Im}\, z| < 1\}$, $D_4 = \mathbf{C} \setminus \{(-\infty, -1] \cup [1, \infty)\}$. Show that

(i) sin maps D_1 one-to-one onto D_4

(ii) cos maps D_2 one-to-one onto D_4

(iii) tan maps D_1 one-to-one onto D_3.

The associated inverse functions are denoted by $\sin^{-1}$, $\cos^{-1}$, $\tan^{-1}$; their properties will be given in later problems. Use Problem 4.20 to obtain analogous results for sinh, cosh, tanh.

22. Let D_1 be as above. Show that cosec maps $D_1 \setminus \{0\}$ one-to-one onto $\mathbf{C} \setminus [-1, 1]$. This map is different from those above in that its 'natural' domain of definition has a 'hole' in it. At a later stage the student will be able to give simpler proofs of the results of these last two problems (see Theorem 8.52).

23. Given $n \in \mathbf{Z}$ show that

$$\lim_{z \to n\pi} (z - n\pi)\,\text{cosec}\,(z) = (-1)^n.$$

What information does this give about the behaviour of cosec near the point $n\pi$ $(n \in \mathbf{Z})$? Show also that

$$\lim_{z \to \frac{1}{n\pi}} \left(z - \frac{1}{n\pi}\right) \operatorname{cosec}\left(\frac{1}{z}\right) = \frac{(-1)^{n+1}}{n^2\pi^2}$$

but that $z^k \operatorname{cosec}\left(\frac{1}{z}\right)$ fails to converge as $z \to 0$ for any $k \in \mathbf{P}$.

4.5 Branches-of-log

In the theory of real analysis one of the classical definitions of the logarithmic function is that it is the inverse of the exponential function. It is not possible to make such a definition in complex analysis since the complex exponential function is not one-to-one and therefore fails to have an inverse function. We therefore proceed with some care, and in fact define a whole class of logarithmic functions.

To be precise, let D be a domain of $\mathbf{C}$ with $0 \notin D$, and let $f \in \mathscr{C}(D)$. We say that f is a *branch-of-log* on D if

$$\exp \circ f(z) = z \qquad (z \in D).$$

It is immediate that such a branch-of-log is one-to-one on D. Given D with $0 \notin D$, there need not exist any branch-of-log on D. Indeed we shall show later in this section that there is no branch-of-log on the domain $\mathbf{C} \setminus \{0\}$. Given that there is a branch-of-log on a domain D it is easy to obtain all the branches-of-log on that domain.

Proposition 4.13. *If f is a branch-of-log on D so is $f + 2k\pi i\mathbf{1}$ for each $k \in \mathbf{Z}$. If g is any branch-of-log on D then there is an integer k such that $g = f + 2k\pi i\mathbf{1}$.*

Proof. Let f be a branch-of-log on D and let $k \in \mathbf{Z}$. Since $f \in \mathscr{C}(D)$ we have $f + 2k\pi i\mathbf{1} \in \mathscr{C}(D)$. Since

$$\exp \circ (f + 2k\pi i\mathbf{1})(z) = \exp(f(z)) \exp(2k\pi i) = z \qquad (z \in D)$$

it follows that $f + 2k\pi i\mathbf{1}$ is a branch-of-log on D.

Suppose now that g is any branch-of-log on D. Let $h = g - f$ and we have

$$\exp(h(z)) = \frac{\exp(g(z))}{\exp(f(z))} = \frac{z}{z} = 1 \qquad (z \in D).$$

Therefore $h(D) \subset \{2k\pi i\colon k \in \mathbf{Z}\}$. Since D is connected and h is continuous it follows from Proposition 1.26 that $h(D)$ is connected. It follows that

$h(D)$ must consist of a single point, i.e. there is some $k \in \mathbf{Z}$ such that $h(z) = 2k\pi i$ $(z \in D)$. The proof is complete.

We show next that any branch-of-log on D is analytic on D and has the derivative we would expect from real analysis.

Proposition 4.14. *If f is a branch-of-log on D then $f \in \mathscr{A}(D)$ and*

$$f'(z) = \frac{1}{z} \quad (z \in D).$$

Proof. Since $\exp' = \exp$, we have $\exp'(z) \neq 0$ $(z \in \mathbf{C})$. Since $f \in \mathscr{C}(D)$ and $\exp \circ f(z) = z$ $(z \in D)$ it follows from Proposition 3.8 (ii) that $f \in \mathscr{A}(D)$ and

$$f'(z) = \frac{1}{\exp' \circ f(z)} = \frac{1}{\exp(f(z))} = \frac{1}{z} \qquad (z \in D).$$

The above result has an interesting converse. We need first a general definition. Let D be any domain of $\mathbf{C}$ and let $g \in \mathscr{C}(D)$. We say that h is a *primitive* of g if $h \in \mathscr{A}(D)$ and $h' = g$. If h is a primitive of g then so is $h + \alpha\mathbf{1}$ for any $\alpha \in \mathbf{C}$. Conversely if h_1 is any primitive of g we then have $h_1 - h \in \mathscr{A}(D)$ and $(h_1 - h)' = g - g = 0$ so that $h_1 - h$ is constant on D.

Proposition 4.15. *Let D be a domain with $0 \notin D$ and let $\mathbf{j}|_D$ have a primitive f on D. Then there is a branch-of-log g on D such that $f - g$ is constant.*

Proof. Define

$$h(z) = \frac{\exp(f(z))}{z} \qquad (z \in D).$$

We then have

$$h'(z) = -\frac{1}{z^2}\exp(f(z)) + \frac{1}{z}\exp'(f(z))f'(z) = 0 \qquad (z \in D)$$

so that there is $\beta \in \mathbf{C}$ such that $h(z) = \beta$ $(z \in D)$. Since exp has no zeros, $\beta \neq 0$ and so by Proposition 4.12 there is $\alpha \in \mathbf{C}$ such that $\exp(\alpha) = \beta$. We now have

$$\exp(f(z)) = z\exp(\alpha) \qquad (z \in D)$$

and therefore

$$\exp \circ (f - \alpha\mathbf{1})(z) = z \qquad (z \in D).$$

Thus $f - \alpha\mathbf{1}$ is a branch-of-log on D and the proof is complete.

These last two results show that the existence of a branch-of-log on D is equivalent to the existence of a primitive on D for $\mathbf{j}|_D$. In this connection the student may recall that one of the definitions of the real logarithmic function is given by

$$\log x = \int_1^x \frac{dt}{t} \qquad (x > 0).$$

We shall now construct some specific branches-of-log. Given $\theta \in \mathbf{R}$ we define $\log_\theta : \mathbf{C} \setminus \{0\} \to \mathbf{C}$ by

$$\log_\theta (z) = \log |z| + i \arg_\theta (z) \qquad (z \in \mathbf{C} \setminus \{0\}).$$

We shall write $\log_0$ more simply as log, since it agrees with the real logarithmic function on the positive real numbers.

Proposition 4.16. *For each $\theta \in \mathbf{R}$, $\log_\theta$ is a branch-of-log on $\mathbf{C} \setminus N_\theta$ and each point of $N_\theta \setminus \{0\}$ is a singularity of $\log_\theta$.*

Proof. It follows from Propositions 3.1 and 3.5 that $\log_\theta$ is continuous on $\mathbf{C} \setminus N_\theta$ and not continuous at any point of $N_\theta \setminus \{0\}$. Since

$$\begin{aligned} \exp (\log_\theta (z)) &= \exp (\log |z|) \exp (i \arg_\theta (z)) \\ &= |z| \{\cos (\arg_\theta (z)) + i \sin (\arg_\theta (z))\} \\ &= z \qquad (z \in \mathbf{C} \setminus N_\theta) \end{aligned}$$

the result now follows.

Corollary. *If $0 \notin \Delta(\alpha, r)$ there is a branch-of-log on $\Delta(\alpha, r)$.*

Proof. Let $\theta = \arg (\alpha)$ and let f be the restriction of $\log_\theta$ to $\Delta(\alpha, r)$.

It should be noted that $\log_\theta$ is simply the inverse of the restriction of the exponential function to the strip $\{z: \theta - \pi < \operatorname{Im} z \leqslant \theta + \pi\}$. We might well have begun with this definition but it is more illuminating to consider the general situation and so obtain as many views as possible of the important logarithmic functions.

Example 4.17. *There is no branch-of-log on the domain $\mathbf{C} \setminus \{0\}$.*

Proof. Suppose that there is a branch-of-log f on $\mathbf{C} \setminus \{0\}$. Then the restriction of f to $\mathbf{C} \setminus N$ is clearly a branch-of-log on $\mathbf{C} \setminus N$. It follows from Propositions 4.13 and 4.16 that there is $k \in \mathbf{Z}$ such that

$$f(z) = \log (z) + 2k\pi i \qquad (z \in \mathbf{C} \setminus N).$$

Since arg is not continuous at -1 we now see that f is not continuous at -1. This contradiction completes the proof.

It is clear from Proposition 4.14 that any branch-of-log f on D is infinitely differentiable on D. The question then arises as to whether f can be represented by a power series on appropriate discs in D. The answer is in the affirmative (Problem 4.28) and we shall now illustrate the technique for the case of the function log.

Example 4.18.

$$\log(z) = \sum_{n=1}^{\infty} \frac{(-1)^{n+1}}{n}(z-1)^n \qquad (z \in \Delta(1, 1)).$$

Proof. Since $\lim_{n\to\infty} n^{1/n} = 1$, the power series $\sum \frac{(-1)^{n+1}}{n}(z-1)^n$ has radius of convergence 1. We may thus define

$$g(z) = \log(z) - \sum_{n=1}^{\infty} \frac{(-1)^{n+1}}{n}(z-1)^n \qquad (z \in \Delta(1, 1)).$$

We now have $g \in \mathscr{A}(\Delta(1, 1))$ and

$$\begin{aligned} g'(z) &= \frac{1}{z} - \sum_{n=1}^{\infty} (-1)^{n+1}(z-1)^{n-1} \\ &= \frac{1}{z} - \frac{1}{1+(z-1)} \\ &= 0 \qquad (z \in \Delta(1, 1)). \end{aligned}$$

Therefore g is constant on $\Delta(1, 1)$. Since $g(1) = 0$ the proof is now complete.

We conclude this chapter by considering the generalized 'power' functions. When n is an integer the properties of the function $z \to z^n$ are well known. We now ask what meaning can be given to z^n when n is an arbitrary real or complex number. Recall that if $t > 0$ and $\alpha \in \mathbf{R}$ then t^α is defined by $e^{\alpha \log t}$. This motivates the following definition. Given $\lambda \in \mathbf{C}$, $\theta \in \mathbf{R}$ we define $p_\theta^\lambda : \mathbf{C} \setminus \{0\} \to \mathbf{C}$ by

$$p_\theta^\lambda(z) = \exp(\lambda \log_\theta(z)) \qquad (z \neq 0).$$

If we take $\lambda = 1$ in the above definition we obtain for any θ that

$$p_\theta^1(z) = z \qquad (z \neq 0).$$

More generally, for any integer $n \in \mathbf{Z}$ we may easily show that

$$p_\theta^n(z) = z^n \qquad (z \neq 0).$$

For the case $\lambda = \frac{1}{2}$ we have

$$(p_\theta^{\frac{1}{2}}(z))^2 = \exp\{\tfrac{1}{2}\log_\theta(z) + \tfrac{1}{2}\log_\theta(z)\} = z \qquad (z \neq 0)$$

so that each function $p_\theta^{\frac{1}{2}}$ is a 'square root' function. Moreover for each $z \neq 0$

$$\begin{aligned} p_{\theta+2\pi}^{\frac{1}{2}}(z) &= \exp\{\tfrac{1}{2}\log_{\theta+2\pi}(z)\} \\ &= \exp\{\tfrac{1}{2}\log_\theta(z) + \pi i\} \\ &= -p_\theta^{\frac{1}{2}}(z). \end{aligned}$$

This coincides with the intuitive notion of 'positive' and, negative' square roots. On the other hand when we take λ to be complex our intuition breaks down in the face of formulae such as

$$p_0^i(i) = \exp(i\log(i)) = \exp\left(i\frac{\pi}{2}i\right) = e^{-\pi/2}.$$

In line with our earlier practice we write p_0^λ more simply as p^λ. If $\lambda = \mu + i\nu$ and $t > 0$ then

$$\begin{aligned} p^\lambda(t) &= \exp\{(\mu + i\nu)\log t\} \\ &= t^\mu\{\cos(\nu\log t) + i\sin(\nu\log t)\}. \end{aligned}$$

We shall sometimes write t^λ for $p^\lambda(t)$ $(t > 0)$ to show its close relation with the corresponding real function.

Proposition 4.19. *Given* $\lambda, \mu \in \mathbf{C}$, $\theta \in \mathbf{R}$ *we have*

(i) $p_\theta^{\lambda+\mu} = p_\theta^\lambda p_\theta^\mu$

(ii) $p_\theta^\lambda \in \mathscr{A}(\mathbf{C} \setminus N_\theta)$ *and* $(p_\theta^\lambda)' = \lambda p_\theta^{\lambda-1}$.

Proof. (i)
$$\begin{aligned} p_\theta^{\lambda+\mu}(z) &= \exp\{(\lambda+\mu)\log_\theta(z)\} \\ &= \exp\{\lambda\log_\theta(z)\}\exp\{\mu\log_\theta(z)\} \\ &= p_\theta^\lambda(z)\,p_\theta^\mu(z). \end{aligned}$$

(ii) It follows from Propositions 4.16 and 3.8 that p_θ^λ is analytic on $\mathbf{C} \setminus N_\theta$ and

$$\begin{aligned} (p_\theta^\lambda)'(z) &= \exp'(\lambda\log_\theta(z))\lambda\log_\theta'(z) \\ &= \frac{\lambda}{z}p_\theta^\lambda(z) \\ &= \lambda p_\theta^{\lambda-1}(z). \end{aligned}$$

If λ is an integer then p_θ^λ is in fact analytic on $\mathbf{C} \setminus \{0\}$. If λ is not an integer we may use the argument of Proposition 3.5 to show that p_θ^λ is not continuous at any point of $N_\theta \setminus \{0\}$, and hence in this case each point of $N_\theta \setminus \{0\}$ is a singularity of p_θ^λ. It may also be verified (see Problem 4.28) that given $\alpha \in \mathbf{C} \setminus N_\theta$ we may represent p_θ^λ by a power series on some disc about α. In particular using the technique of Example 4.18 we may verify that

$$p^\lambda(z) = 1 + \sum_{n=1}^{\infty} \lambda(\lambda-1)\ldots(\lambda-n+1)\frac{(z-1)^n}{n!} \qquad (z \in \Delta(1, 1)).$$

PROBLEMS 4

24. Given $n \in \mathbf{P}$ show that

$$n \log\left(1 + \frac{z}{n}\right) = z + f_n(z) \qquad (z \in \bar{\Delta}(0, \tfrac{1}{2}n))$$

where $|f_n(z)| \leqslant \frac{|z|^2}{n}$ $(z \in \bar{\Delta}(0, \frac{1}{2}n))$. Deduce that

$$\lim_{n \to \infty} \left(1 + \frac{z}{n}\right)^n = \exp(z) \qquad (z \in \mathbf{C})$$

and show that the convergence is uniform on any compact subset of $\mathbf{C}$.

25. Let D be a domain in $\mathbf{C}$ with $0 \notin D$. We say that $\chi: D \to \mathbf{R}$ is a *continuous argument* on D if χ is continuous and

$$\chi(z) \equiv \arg(z) \qquad (\text{mod } 2\pi).$$

Show that there is a branch-of-log on D iff there is a continuous argument on D.

26. Given $z_1, z_2 \in C \setminus N_\theta$ find necessary and sufficient conditions for

$$\log_\theta(z_1 z_2) = \log_\theta(z_1) + \log_\theta(z_2).$$

27. Let D be a domain in $\mathbf{C}$ such that for some $R > 0$

$$C(0, R) \subset D \subset \mathbf{C} \setminus \{0\}.$$

Show that there is no branch-of-log on D.

28. (i) Let f be a branch-of-log on D. Given $\Delta(\alpha, r) \subset D$ represent f on $\Delta(\alpha, r)$ by a power series.

(ii) Given $\alpha \in \mathbf{C} \setminus N_\theta$ represent p_θ^λ by a power series on some disc about α. What is the largest such disc?

29. Given $\lambda = \mu + i\nu$ show that

$$|p_\theta^\lambda(z)| = |z|^\mu \, e^{-\nu \arg_\theta(z)}.$$

30. Discuss the convergence of the following series.

$$\sum_{n=1}^{\infty} n^{-z}, \qquad \sum_{n=0}^{\infty} p_{2n\pi}^\lambda(z), \qquad \sum_{n=0}^{\infty} p^{n\lambda}(z), \qquad \sum_{n=1}^{\infty} \frac{\log(1 + n^2 z^2)}{1 + n^2 z^2}.$$

5

ARCS, CONTOURS, AND INTEGRATION

The main purpose of this chapter is to develop the machinery of integration along arcs and contours in the complex plane. Every mathematical theory of integration that is not trivial involves a considerable amount of technical detail, and this chapter will prove no exception. We have attempted some streamlining of the treatment of integration by relegating to the Appendix the technical details of the Riemann-Stieltjes integral. Consequently, the larger part of the chapter is concerned with the topological aspects of arcs and contours in the complex plane. This study is of interest in itself apart from applications to complex analysis. We include a simple proof of the Jordan curve theorem for a special class of simple closed curves.

5.1 Arcs

An intuitive idea of an arc in the complex plane is given by a picture that looks like a piece of bent wire. As a first step in formalizing this picture we might regard an arc, Γ say, as the range of a continuous complex function defined on the interval [0, 1]. If we do not wish the arc to intersect itself we must take the continuous function to be one-to-one. (Indeed without this restriction the curve* could fill out a square!) Unless we wish to label Γ with exactly one such function we are obliged to consider all the continuous functions on [0, 1] that are one-to-one and have range Γ. Since [0, 1] is a compact metric space it follows from Proposition 1.18 that the above functions are in fact the homeomorphisms of [0, 1] with Γ. We are now ready for our formal definition. Recall that given any two metric spaces X and Y, *Hom* (X, Y) denotes the set of all homeomorphisms of X with Y.

* See, for example, G. F. Simmons, *Introduction to Topology and Modern Analysis*, (McGraw-Hill, 1963) Appendix 2.

An *arc* is a subset Γ of $\mathbf{C}$ that is homeomorphic with the closed interval $[0, 1]$. In other words, a subset of Γ of $\mathbf{C}$ is an arc if and only if the set *Hom* $([0, 1], \Gamma)$ is not empty. We say that an arc Γ is *represented* by each element of *Hom* $([0, 1], \Gamma)$. We also say that each element γ of *Hom* $([0, 1], \Gamma)$ is a *representative function* for Γ.

Note that a subset Γ of $\mathbf{C}$ is an arc if and only if it is the range of a continuous function that is one-to-one on $[0, 1]$. In particular, given $\alpha, \beta \in \mathbf{C}$, $\alpha \neq \beta$, $[\alpha, \beta]$ is an arc. To see this we simply take

$$\gamma(t) = (1 - t)\alpha + t\beta \qquad (t \in [0, 1]).$$

As a further example take

$$\gamma(t) = \alpha + r \exp(\pi i t) \qquad (t \in [0, 1])$$

and we obtain a semi-circular arc. The student should have no difficulty in producing many other simple examples.

We begin our study of arcs by analysing the set *Hom* $([0, 1], \Gamma)$. It turns out that there is a very simple description of this set. In what follows, the set *Hom* $([0, 1], [0, 1])$ will be of special importance; we denote it by Φ for short. The lemma below shows that the elements of Φ are (strictly) monotonic functions. In view of Proposition 1.18, Φ thus consists of the continuous strictly monotonic functions on $[0, 1]$ that have range $[0, 1]$. (In fact, Φ consists of the strictly monotonic functions on $[0, 1]$ that have range $[0, 1]$; see Problem 5.1.)

Lemma 5.1. *Each element of Φ is a monotonic function.*

Proof. Let $\varphi \in \Phi$ and suppose $\varphi(0) < \varphi(1)$. Suppose there is $t \in (0, 1)$ such that $\varphi(t) < \varphi(0)$. We then have $\varphi(0) \in [\varphi(t), \varphi(1)]$. Since φ is continuous on $[t, 1]$ it follows from the intermediate value theorem that there is $u \in (t, 1)$ with $\varphi(u) = \varphi(0)$. This contradicts the fact that φ is one-to-one and so we must have $\varphi(t) > \varphi(0)$ $(t \in (0, 1))$. A similar argument shows that $\varphi(t) < \varphi(1)$ $(t \in (0, 1))$.

Suppose now that $0 < s < t < 1$ and $\varphi(s) > \varphi(t)$. We then have $\varphi(s) \in (\varphi(t), \varphi(1))$. Arguing as above we obtain $v \in (t, 1)$ such that $\varphi(v) = \varphi(s)$. This is again a contradiction and so $\varphi(s) < \varphi(t)$. We have thus shown that φ is strictly increasing if $\varphi(0) < \varphi(1)$. A similar argument shows that φ is strictly decreasing if $\varphi(0) > \varphi(1)$. The proof is complete.

Given $\varphi \in \Phi$ it follows from the above lemma that $\varphi(0) = 0$ or $\varphi(0) = 1$. It is then clear that φ is increasing if and only if $\varphi(0) = 0$, and φ is decreasing if and only if $\varphi(0) = 1$. We define

$$\Phi_+ = \{\varphi : \varphi \in \Phi, \varphi(0) = 0\}, \qquad \Phi_- = \{\varphi : \varphi \in \Phi, \varphi(0) = 1\}$$

and we then have

$$\Phi = \Phi_+ \cup \Phi_-, \qquad \Phi_+ \cap \Phi_- = \varnothing.$$

Lemma 5.2. *Given* $\gamma \in Hom([0, 1], \Gamma)$,

$$Hom([0, 1], \Gamma) = \{\gamma \circ \varphi : \varphi \in \Phi\}.$$

Proof. Let $\gamma \in Hom([0, 1], \Gamma)$. Given $\varphi \in \Phi$, $\gamma \circ \varphi \in Hom([0, 1], \Gamma)$ by the Corollary to Proposition 1.12. Given $\eta \in Hom([0, 1], \Gamma)$ let $\varphi = \gamma^{-1} \circ \eta$. Since $\gamma \in Hom([0, 1], \Gamma)$ we have $\gamma^{-1} \in Hom(\Gamma, [0, 1])$ and so $\varphi \in \Phi$ by the Corollary to Proposition 1.12. Then

$$\eta = (\gamma \circ \gamma^{-1}) \circ \eta = \gamma \circ (\gamma^{-1} \circ \eta) = \gamma \circ \varphi$$

and the proof is complete.

Given any representative function for an arc Γ we thus have a simple description of the set of all representative functions for Γ. Suppose now that $a, b \in \mathbf{R}$ with $a < b$. Since $[a, b]$ is an arc it is homeomorphic with any other arc Γ. In other words we could equally describe Γ by functions defined on $[a, b]$, namely the functions of $Hom([a, b], \Gamma)$. We shall also say that any function in $Hom([a, b], \Gamma)$ *represents* Γ. Thus a function in $\mathscr{C}([a, b])$ represents an arc if and only if it is one-to-one.

For most purposes of complex analysis it is sufficient to consider only special classes of arcs. An arc Γ is said to be *rectifiable* if there is some representative function γ for Γ that is of bounded variation (see Problem 3.7.). (The student who wishes some information on functions of bounded variation may turn to the Appendix.) Not every arc is rectifiable (see Problem 5.4), but if an arc is rectifiable then every representative function is of bounded variation (see Problem 5.5). We shall see later that rectifiable arcs are those arcs that have a finite length. An arc Γ is said to be *smooth* if some representative function γ in $Hom([a, b], \Gamma)$ is differentiable on (a, b) with γ' bounded and continuous. Not all functions representing a smooth arc have the above property (see Problem 5.6), but it follows from Problem 3.7(iii) that a smooth arc is rectifiable. An arc Γ is said to be *piecewise smooth* if some representative function γ in $Hom([a, b], \Gamma)$ is differentiable on (a, b) except at a finite number of points with γ' bounded and continuous (on the domain of definition of γ', of course). It is easily seen that a piecewise smooth arc is the union of a finite number of smooth arcs and is also rectifiable.

To illustrate the above definitions, note that any arc of the form $[\alpha, \beta]$ ($\alpha, \beta \in \mathbf{C}$, $\alpha \neq \beta$) is smooth. Now let Γ be the semi-circular arc represented by

$$\gamma(t) = \alpha + r \exp(\pi i t) \qquad (t \in [0, 1]).$$

Then γ is differentiable on (0, 1) and

$$\gamma'(t) = \pi i r \exp(\pi i t) \qquad (t \in (0, 1))$$

so that Γ is smooth. To obtain a simple example of a piecewise smooth arc let $\alpha_1, \alpha_2, \alpha_3$ be distinct points of $\mathbf{C}$ such that $\alpha_3 \notin \{t\alpha_1 + (1 - t)\alpha_2 : t \in \boldsymbol{R}\}$. Then

$$[\alpha_1, \alpha_2] \cup [\alpha_2, \alpha_3]$$

is easily shown to be a piecewise smooth arc.

PROBLEMS 5

1. Let φ be a strictly monotonic real function on [0, 1] with range [0, 1]. Show that $\varphi \in \Phi$.

2. Give examples of functions in Φ that are differentiable except at (i) one point (ii) n points (iii) a countable infinity of points.

3. Use the Weierstrass continuous nowhere differentiable function to produce a representative function γ for an arc such that γ is nowhere differentiable.

4. If $\gamma(0) = i$ and $\gamma(t) = t \sin\left(\frac{1}{t}\right) + i \exp(t)$ ($t \in (0, 1]$), show that γ represents an arc which is not rectifiable.

5. Let Γ be an arc and let γ be a representative function for Γ which is of bounded variation. Use Lemma 5.2 to show that every representative function for Γ is of bounded variation.

6. Let Γ be a smooth arc. Use Problem 5.2 to produce a representative function for Γ that is not differentiable.

7. Let Γ be the arc $[0, 1] \cup [1, 1 + i]$. Use the function

$$\mu(t) = \begin{cases} 1 - (t - 1)^2 & (t \in [0, 1]) \\ 1 + i(t - 1)^2 & (t \in (1, 2]) \end{cases}$$

to show that Γ is smooth. More generally let Γ be any piecewise smooth arc with representative function γ which has one-sided derivatives at the points where γ is not differentiable. Show that Γ is smooth.

8. Let $\gamma(t) = t^2 \sin\left(\frac{1}{t^2}\right)$ $(t \in (0, 1])$, $\gamma(0) = 0$. Show that γ is differentiable on $[0, 1]$ (one-sided derivatives at the end points) with γ' continuous but unbounded on $(0, 1]$.

9. Let $\gamma \in \mathscr{C}([0, 1])$ be such that γ is differentiable on $(0, 1)$ with γ' bounded and continuous. Show by an example that γ need not have one-sided derivatives at 0 and 1. If $\lim_{t \to 0+} \gamma'(t) = a$, $\lim_{t \to 1-} \gamma'(t) = b$, show that γ has one-sided derivatives at 0, 1 whose values are a, b respectively.

10. Given an arc Γ determine *Hom* (Γ, Γ). Show that *Hom* (Γ, Γ) is a group under composition of mappings and that the group is isomorphic with Φ.

5.2 Oriented arcs

In this section we consider the important problem of specifying a direction on an arc. This concept will be essential when we wish to integrate along arcs. It is intuitively clear that there are two natural directions on an arc, the one opposite to the other. Our problem is to make this intuition meaningful in mathematical language.

Consider first the basic example of an arc, namely $[0, 1]$; in essence it is enough to solve the problem in this case. We might specify a direction on $[0, 1]$ by either of the orderings $\leqslant$ or $\geqslant$. Given an arbitrary arc Γ we could then transfer these orderings from $[0, 1]$ to Γ via a representative function for Γ. This method is perfectly feasible (see Problem 5.11), but we shall follow a simpler approach. Observe that we can distinguish between the orderings $\leqslant$ and $\geqslant$ on $[0, 1]$ by stating which point is to be the 'first point'. Thus 0 is the first point for $\leqslant$ and 1 is the first point for $\geqslant$. It now becomes clear how we should specify direction on an arbitrary arc.

Let Γ be an arbitrary arc, let $\gamma \in$ *Hom* $([0, 1], \Gamma)$, and let $z_0 = \gamma(0)$, $z_1 = \gamma(1)$. Given any $\eta \in$ *Hom* $([0, 1], \Gamma)$ it follows from Lemma 5.2 that

$$\{\eta(0), \eta(1)\} = \{\gamma(0), \gamma(1)\}.$$

We call the points z_0, z_1 the *end points* of Γ. An *oriented arc* is a pair (Γ, z_0) where Γ is an arc and z_0 is one of the end points of Γ. We say that z_0 is the *first point* of the oriented arc; and if z_1 is the other end point of Γ we say that z_1 is the *last point* of the oriented arc. We shall also say that the oriented arc (Γ, z_0) has the *direction from* z_0 *to* z_1. It is convenient in diagrams to illustrate the direction of an oriented arc by means of an arrow pointing from the first point to the last point (see Figure 5.1).

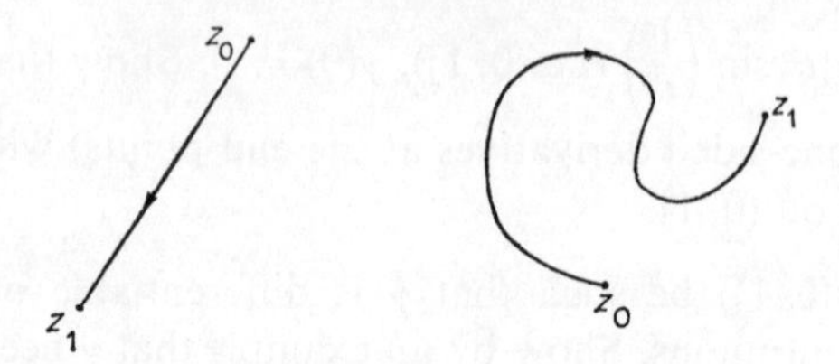

Figure 5.1

Given an oriented arc (Γ, z_0) we say that $\gamma \in Hom\ ([0, 1], \Gamma)$ *represents* (Γ, z_0) if $\gamma(0) = z_0$. It is clear that, for every $\varphi \in \Phi_+$, $\gamma \circ \varphi$ represents (Γ, z_0), since we then have $\gamma \circ \varphi(0) = \gamma(\varphi(0)) = \gamma(0)$. Moreover, if z_1 is the other end point of Γ, then, for every $\varphi \in \Phi_-$, $\gamma \circ \varphi$ represents the oriented arc (Γ, z_1). In view of Lemma 5.2 every element of $Hom\ ([0, 1]. \Gamma)$ represents either (Γ, z_0) or (Γ, z_1). We say that (Γ, z_1) is the *opposite arc* to (Γ, z_0) and we write

$$(\Gamma, z_1) = -(\Gamma, z_0).$$

Given $\gamma \in Hom\ ([0, 1], \Gamma)$ let

$$\gamma_-(t) = \gamma(1 - t) \qquad (t \in [0, 1]).$$

Then $\gamma_- \in Hom\ ([0, 1], \Gamma)$ and clearly γ represents (Γ, z_0) if and only if γ_- represents $-(\Gamma, z_0)$.

It is also convenient to represent oriented arcs by functions defined on intervals other than [0, 1]. More precisely, let $a, b \in \mathbf{R}$ with $a < b$. Given an oriented arc (Γ, z_0) we say that $\lambda \in Hom\ ([a, b], \Gamma)$ *represents* (Γ, z_0) if $\lambda(a) = z_0$. Thus if $\mu \in \mathscr{C}([a, b])$ is one-to-one then $\Gamma = \mu([a, b])$ is an arc and μ represents the oriented arc $(\Gamma, \mu(a))$. (It should be clear to the student that $\mu(a)$, $\mu(b)$ are the end points of Γ.)

More often than not an oriented arc (Γ, z_0) will be denoted simply by Γ for short. It must be clearly understood that when we write Γ for an oriented arc we have in mind one specific direction on Γ. We then write $-\Gamma$ for the opposite arc.

We conclude this section by considering the processes of 'piecing together' and 'chopping up' arcs. Suppose first that Γ_1, Γ_2 are two arcs such that $\Gamma_1 \cap \Gamma_2$ is a single point z_1 that is an end point for both Γ_1 and Γ_2. Let $\Gamma = \Gamma_1 \cup \Gamma_2$. Then Γ is an arc. To see this suppose that z_0 is the other end point of Γ_1. Now choose representative functions γ_1, γ_2 (on [0, 1]) for (Γ_1, z_0), (Γ_2, z_1) respectively. Define γ on [0, 2] by

$$\gamma(t) = \begin{cases} \gamma_1(t) & \text{if } 0 \leqslant t \leqslant 1 \\ \gamma_2(t - 1) & \text{if } 1 < t \leqslant 2. \end{cases}$$

Then γ maps [0, 2] one-to-one onto Γ. It is now sufficient to show that γ is continuous. It is obvious that γ is continuous except possibly at $t = 1$. Since

$$\lim_{t \to 1-} \gamma(t) = \lim_{t \to 1-} \gamma_1(t) = z_1$$
$$\lim_{t \to 1+} \gamma(t) = \lim_{t \to 1+} \gamma_2(t) = z_1$$

it follows that γ is indeed continuous at $t = 1$. We have seen a simple example of this situation in the last section for the case $\Gamma_1 = [\alpha_1, \alpha_2]$, $\Gamma_2 = [\alpha_2, \alpha_3]$ where α_3 does not lie on the infinite line determined by α_1 and α_2.

Consider now the opposite process. Given an oriented arc Γ, we say that oriented arcs Γ_r $(r = 1, 2, \ldots, n)$ form a *direction-preserving decomposition* of Γ if

(i) $\Gamma = \bigcup \{\Gamma_r : r = 1, 2, \ldots, n\}$
(ii) the first point of Γ_1 is the first point of Γ
(iii) the first point of Γ_{r+1} is the last point of Γ_r for $r = 1, \ldots, n - 1$.

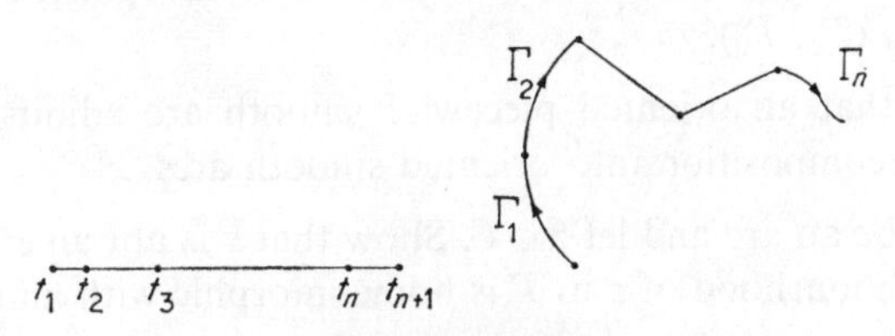

Figure 5.2

In other words the 'pieces' Γ_r are joined 'head to tail' as in Figure 5.2. As an example suppose that Γ is represented by γ and let t_r be such that

$$0 = t_1 < t_2 < \cdots < t_{n+1} = 1.$$

For $r = 1, 2, \ldots, n$ let Γ_r be the oriented arc represented by $\gamma|_{[t_r, t_{r+1}]}$ (see Figure 5.2). Then it is trivial to verify that the oriented arcs Γ_r $(r = 1, 2, \ldots, n)$ from a direction-preserving decomposition of Γ. Suppose conversely that oriented arcs Γ_r $(r = 1, 2, \ldots, n)$ form any direction-preserving decomposition of Γ. Let Γ be represented by $\gamma \in Hom$ ([0, 1], Γ), let z_r be the first point of Γ_r and let $t_r = \gamma^{-1}(z_r)$. It follows easily that

$$0 = t_1 < t_2 < \cdots < t_n < 1$$

and that each Γ_r is represented by $\gamma|_{[t_r, t_{r+1}]}$. We have thus shown that there is a one-to-one correspondence between the direction-preserving decompositions of an oriented arc Γ and the subdivisions of the closed interval [0, 1].

PROBLEMS 5

11. Let Γ be an arc. Given $\gamma \in Hom([0, 1], \Gamma)$ define the relation $\leqslant_\gamma$ on Γ by

$$z_1 \leqslant_\gamma z_2 \quad \text{if} \quad \gamma^{-1}(z_1) \leqslant \gamma^{-1}(z_2).$$

Show that $\leqslant_\gamma$ is a *total* ordering on Γ (i.e. $\leqslant_\gamma$ is a partial ordering such that $z_1, z_2 \in \Gamma$ implies either $z_1 \leqslant_\gamma z_2$ or $z_2 \leqslant_\gamma z_1$) and that the end points of Γ are the first and last points with respect to the ordering. Show also that $\leqslant_{\gamma \circ \varphi} = \leqslant_\gamma$ $(\varphi \in \Phi_+)$ and $\leqslant_{\gamma \circ \varphi}$ is the opposite ordering to $\leqslant_\gamma$ for each $\varphi \in \Phi_-$. Thus $Hom([0, 1], \Gamma)$ induces only two orderings on Γ, the one opposite to the other. If $\leqslant$ is either of these orderings we say that $(\Gamma, \leqslant)$ is an *ordered* arc. (Of course we simply have an oriented arc in different guise.)

12. Let $(\Gamma_1, \leqslant_1)$, $(\Gamma_2, \leqslant_2)$ be ordered arcs. We say that $\eta: \Gamma_1 \to \Gamma_2$ is an *order isomorphism* of Γ_1 with Γ_2 if η maps Γ_1 one-to-one onto Γ_2 and

$$z_1 \leqslant_1 z_2 \quad \Rightarrow \quad \eta(z_1) \leqslant_2 \eta(z_2).$$

Use Problem 5.1 to show that if η is an order isomorphism of Γ_1 with Γ_2 then $\eta \in Hom(\Gamma_1, \Gamma_2)$.

13. Show that an oriented piecewise smooth arc admits a direction-preserving decomposition into oriented smooth arcs.

14. Let Γ be an arc and let $z \in \Gamma$. Show that z is not an end point of Γ iff some neighbourhood of z in Γ is homeomorphic with an open interval of $\mathbf{R}$.

15. Let Γ_1, Γ_2 be arcs such that $\Gamma_1 \cap \Gamma_2 = \{z\}$. Show that $\Gamma_1 \cup \Gamma_2$ is an arc iff z is an end point for both Γ_1 and Γ_2.

16. Let $\mathscr{J}$ denote the set of all arcs and let ρ be defined on $\mathscr{J} \times \mathscr{J}$ by

$$\rho(\Gamma_1, \Gamma_2) = \text{dist}(\Gamma_1, \Gamma_2).$$

Which of the properties of a metric does ρ possess?

17. Let Γ_n $(n \in \mathbf{P})$, Γ be arcs. We say that $\{\Gamma_n\}$ converges to Γ if there are representative functions γ_n for Γ_n, γ for Γ (on $[0, 1]$) such that $\gamma_n \to \gamma$ uniformly on $[0, 1]$. For which representative functions η_n for Γ_n, η for Γ does it follow that $\eta_n \to \eta$ uniformly on $[0, 1]$? Give examples of sequences of distinct arcs that converge to an arc.

5.3 Simple closed curves

A *simple closed curve* is a subset Γ of $\mathbf{C}$ that is homeomorphic with the unit circle $C(0, 1)$. Since any two circles in the complex plane are homeomorphic it is clear that a subset Γ of $\mathbf{C}$ is a simple closed curve if and only if it is homeomorphic with some circle in the complex plane. (The student may object to the use of Γ for both arcs and simple closed curves; but the value of using the same symbol will be evident when we wish to integrate along arcs and closed curves.) Analogy with the case of an arc suggests that we should begin by analysing the set *Hom* $(C(0, 1), \Gamma)$. We shall do this later in the section. We begin by giving two simple characterizations of simple closed curves, the first in terms of representative functions and the second in terms of decompositions into subarcs.

For purposes of integration it is inconvenient to represent simple closed curves by functions defined on the circle $C(0, 1)$. To obtain functions defined on $[0, 1]$ we shall make use of the natural mapping from $[0, 1]$ onto $C(0, 1)$ defined by

$$\omega(t) = \exp(2\pi i t) \qquad (t \in [0, 1]).$$

Observe that ω is continuous on $[0, 1]$, $\omega(0) = \omega(1)$, and ω is one-to-one on $[0, 1)$. Moreover the inverse mapping of ω on $(0, 1)$ is given by $\frac{1}{2\pi i}\log_\pi$ (see §4.5).

Proposition 5.3. *Let Γ be a subset of* $\mathbf{C}$. *Then Γ is a simple closed curve if and only if there is a continuous function h on* $[0, 1]$ *such that Γ is the range of h,* $h(0) = h(1)$ *and h is one-to-one on* $[0, 1)$.

Proof. Suppose that Γ is a simple closed curve so that there exists a homeomorphism η from $C(0, 1)$ onto Γ. Let $h = \eta \circ \omega$. Then h is continuous on $[0, 1]$ and $h(0) = h(1)$. Since ω is one-to-one on $[0, 1)$ and η is one-to-one on $C(0, 1)$ it follows that h is one-to-one on $[0, 1)$. Finally ω has range $C(0, 1)$ and so h has range Γ.

Suppose now that h has the properties stated in the proposition. Let

$$\eta(z) = h\left(\frac{1}{2\pi i}\log_\pi(z)\right) \ (z \in C(0, 1)).$$

Since $\log_\pi$ has range $(0, 2\pi i]$ on $C(0, 1)$ it follows that Γ is the range of η. Since h is one-to-one on $[0, 1)$ and $\log_\pi$ is one-to-one on $C(0, 1)$ it follows that η is one-to-one on $C(0, 1)$. Since $C(0, 1)$ is compact the proof will be complete if we show that η is continuous. Since h is continuous and $\log_\pi$ is continuous on $C(0, 1) \setminus \{1\}$ it is clear that η is continuous except possibly at $z = 1$. Suppose that $\operatorname{Im} z_n \geqslant 0$ and $z_n \to 1$. It follows that

$$\eta(z_n) \to h\left(\frac{1}{2\pi i}\log_\pi(1)\right) = h(0).$$

Suppose now that $\operatorname{Im} z_n < 0$ and $z_n \to 1$. It then follows that

$$\eta(z_n) \to h\left(\frac{1}{2\pi i}\log_{3\pi}(1)\right) = h(1) = h(0).$$

It is now routine to show that if $\{z_n\}$ is any sequence that converges to 1 then $\eta(z_n) \to h(0)$. Thus η is continuous at 1 by Proposition 1.10 and the proof is complete.

Given a simple closed curve Γ we say that *h represents Γ* if $h \in \mathscr{C}([0, 1])$, $h(0) = h(1)$, h is one-to-one on $[0, 1)$ and h has range Γ. Note that we have established in the course of the above proof that there is a one-to-one correspondence between the functions h that represent Γ and the homeomorphisms η of $C(0, 1)$ with Γ, the correspondence being given by $h = \eta \circ \omega$. Given any point $z \in \Gamma$ we may choose a representative function h for Γ such that $h(0) = z$. To see this suppose that $\eta \in Hom\,(C(0, 1), \Gamma)$ and $\eta^{-1}(z) = \omega(\theta)$. Let $h(t) = \eta(\omega(t + \theta))$ $(t \in [0, 1])$. Then $h(0) = z$ and it is clear from the argument in Proposition 5.3 that h represents Γ. It is also clear that the above proposition and remarks remain true if the interval $[0, 1]$ is replaced by any interval $[a, b]$ where $a, b \in \mathbf{R}$, $a < b$. We leave the details to the reader.

Any two distinct points of a circle divide the circle into two arcs, and conversely any circle may be regarded as the union of two such arcs. We now consider the analogues of these statements for an arbitrary simple closed curve. The situation we are about to consider is illustrated in Figure 5.3.

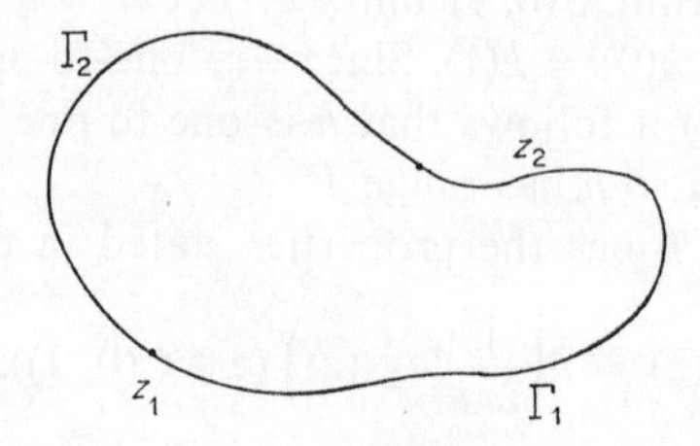

Figure 5.3

Proposition 5.4. (i) *Let Γ be a simple closed curve and let $z_1, z_2 \in \Gamma$, $z_1 \neq z_2$. Then there exist arcs Γ_1, Γ_2 with common end points z_1, z_2, such that $\Gamma = \Gamma_1 \cup \Gamma_2$ and $\Gamma_1 \cap \Gamma_2 = \{z_1, z_2\}$.*

(ii) *Given arcs Γ_1, Γ_2 with common end points z_1, z_2 and with $\Gamma_1 \cap \Gamma_2 = \{z_1, z_2\}$, $\Gamma_1 \cup \Gamma_2$ is a simple closed curve.*

Proof. (i) Choose a representative function h for Γ such that $h(0) = z_1$. Since $z_1 \neq z_2$ there is $u \in (0,1)$ such that $h(u) = z_2$. Let $\gamma_1 = h|_{[0,u]}$, $\gamma_2 =$

$h|_{[u,1]}$. It is clear that γ_1, γ_2 represent arcs Γ_1, Γ_2 respectively and that $\Gamma = \Gamma_1 \cup \Gamma_2$, $\Gamma_1 \cap \Gamma_2 = \{z_1, z_2\}$. Since $h(0) = h(1)$ it is also clear that z_1, z_2 are the end points for both Γ_1 and Γ_2.

(ii) Choose $h_1 \in Hom([0, \frac{1}{2}], \Gamma_1)$ such that $h_1(0) = z_1$, $h_1(\frac{1}{2}) = z_2$; and choose $h_2 \in Hom([\frac{1}{2}, 1], \Gamma_2)$ such that $h_2(\frac{1}{2}) = z_2$, $h_2(1) = z_1$. Define h on $[0, 1]$ by

$$h(t) = \begin{cases} h_1(t) & \text{if } 0 \leqslant t \leqslant \frac{1}{2} \\ h_2(t) & \text{if } \frac{1}{2} < t \leqslant 1. \end{cases}$$

If Γ is the range of h then clearly $\Gamma = \Gamma_1 \cup \Gamma_2$. Evidently h is continuous except possibly at $t = \frac{1}{2}$; and since $h_1(\frac{1}{2}) = h_2(\frac{1}{2})$ it follows readily that h is indeed continuous at $t = \frac{1}{2}$. Since $\Gamma_1 \cap \Gamma_2 = \{z_1, z_2\}$ and Γ_1, Γ_2 are arcs, it follows that h is one-to-one on $[0, 1)$. Since $h(0) = h(1)$ it follows from Proposition 5.3 that Γ is a simple closed curve.

Given a simple closed curve Γ and arcs Γ_1, Γ_2 such that $\Gamma = \Gamma_1 \cup \Gamma_2$ and $\Gamma_1 \cap \Gamma_2$ is the set of end points for Γ_1 and Γ_2, we say that $\{\Gamma_1, \Gamma_2\}$ is a *simple decomposition* of Γ. We also say that Γ_1, Γ_2 are *subarcs* of Γ. The above result implies that there is a one-to-one correspondence between the simple decompositions of Γ and the pairs of distinct points of Γ. The second part of the result also provides a useful technique for constructing simple closed curves. It is clear for example that the union of a semi-circle and its corresponding diameter is a simple closed curve. It is equally clear that we may form simple closed curves by piecing together several appropriate arcs. We illustrate this with an important example.

Example 5.5. *Given distinct points $\alpha_1, \alpha_2, \alpha_3$ not on the same straight line, the set $[\alpha_1, \alpha_2] \cup [\alpha_2, \alpha_3] \cup [\alpha_3, \alpha_1]$ is a simple closed curve.*

Proof. Recall that we have already established that $[\alpha_1, \alpha_2] \cup [\alpha_2, \alpha_3]$ is an arc, whose end points are clearly α_1 and α_3. The result now follows from Proposition 5.4 (ii).

We shall now determine the set of all functions that represent a simple closed curve Γ. This is essentially the same as determining $Hom(C(0, 1), \Gamma)$ in view of the one-to-one correspondence between representative functions and functions in $Hom(C(0, 1), \Gamma)$. Recall that Φ denotes the set of all homeomorphisms of $[0, 1]$ with itself.

Proposition 5.6. *Given a simple closed curve Γ and $\eta \in Hom(C(0, 1), \Gamma)$, the set of all functions that represent Γ is given by*

$$h(t) = \eta(\omega(\theta + \varphi(t))) \qquad (t \in [0, 1])$$

where $\varphi \in \Phi$ and $\theta \in [0, 1)$.

Proof. Given $\varphi \in \Phi$, $\theta \in [0, 1)$ it is clear that the mapping $t \to \theta + \varphi(t)$ is a homeomorphism of $[0, 1]$ with $[\theta, \theta + 1]$. If $h(t) = \eta(\omega(\theta + \varphi(t)))$ $(t \in [0, 1])$ it follows easily that $h \in \mathscr{C}([0, 1])$, $h(0) = h(1)$, h is one-to-one on $[0, 1)$ and h as range Γ, i.e. h represents Γ.

Suppose now that h is any function (on $[0, 1]$) that represents Γ. Choose $\theta \in [0, 1)$ such that $\omega(\theta) = \eta^{-1}(h(0))$. Now define φ on $(0, 1)$ by

$$\varphi(t) = \frac{1}{2\pi i} \log_\pi (\omega(-\theta)\eta^{-1}(h(t))) \qquad (0 < t < 1).$$

It follows from the properties of h, η and $\log_\pi$ that φ is continuous and maps $(0, 1)$ one-to-one onto $(0, 1)$. It follows from the argument used in Lemma 5.1 that φ is monotonic. Suppose that φ is increasing and then define $\varphi(0) = 0$, $\varphi(1) = 1$. Since φ is monotonic and

$$\varphi(0) = \inf \{\varphi(t) : t \in (0, 1)\}$$

it follows that φ is continuous at 0. Similarly φ is continuous at 1, and so $\varphi \in \Phi$. For $t \in [0, 1]$ we now have

$$\omega(\varphi(t)) = \omega(-\theta)\eta^{-1}(h(t))$$

and so

$$h(t) = \eta(\omega(\theta + \varphi(t))).$$

A similar argument holds if φ is decreasing on $(0, 1)$ and so the proof is complete.

The situation in the above proposition is described in Figure 5.4. The crucial distinction between this case and the arc case resides in the pre-

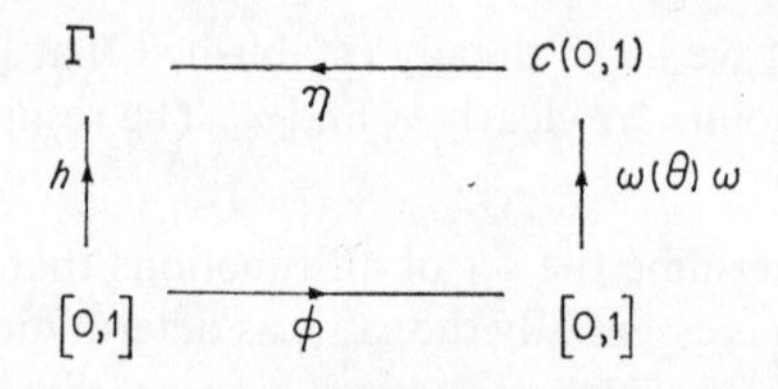

Figure 5.4

sence of the 'rotations' $\omega(\theta)$. As in the arc case, it is sufficient for the purposes of complex analysis to consider only special classes of simple closed curves. A simple closed curve Γ is said to be *rectifiable* if some representative function for Γ is of bounded variation. It is a routine exercise (Problem 5.22) to check that Γ is rectifiable if and only if each representative function is of bounded variation. A simple closed curve Γ is said to be *piecewise smooth* if some representative function γ is differentiable except at a finite number of points with γ' bounded and continuous. If Γ is piecewise smooth it is rectifiable. Moreover, (Problem 5.23) Γ is piecewise smooth if and only if each simple decomposition consists of piecewise smooth arcs. In particular the simple closed curve of Example 5.5 is piecewise smooth.

PROBLEMS 5

18. Let Γ be a simple closed curve, let

$$\Psi = Hom\,(C(0,1), C(0,1)) \quad \text{and let} \quad \Psi_0 = \{\psi : \psi \in \Psi, \psi(1) = 1\}.$$

(i) Given $\eta \in Hom\,(C(0,1), \Gamma)$ show that

$$Hom\,(C(0,1)\,\Gamma) = \{\eta \circ \psi : \psi \in \Psi\}.$$

(ii) Show that $\Psi = \{\omega(\theta)\psi : \psi \in \Psi_0,\ 0 \leqslant \theta < 1\}$.

(iii) Show that there is a one-to-one mapping $\varphi \to \psi_\varphi$ of Φ onto Ψ_0 such that

$$\psi_\varphi \circ \omega = \omega \circ \varphi \qquad (\varphi \in \Phi).$$

(iv) Show that $Hom\,(\Gamma, \Gamma)$ is a group under composition of mappings and that the group is isomorphic with Ψ. Show that the group Φ can be embedded as a subgroup of Ψ. Is it a normal subgroup? The group of rotations of the circle is evidently a subgroup of Ψ. Is it a normal subgroup?

19. Give an example of $h \in \mathscr{C}([0, 1])$ such that $h(0) = h(1)$, h is one-to-one on $(0, 1)$ and the range of h is not a simple closed curve.

20. State and prove the analogue of Proposition 5.3 when $[0, 1]$ is replaced by $[a, b]$, $a, b \in \mathbf{R}$, $a < b$.

21. Specify a representative function for the simple closed curve of Example 5.5 and then show that it is piecewise smooth. Do the same for the simple closed curves sketched in Figure 5.5.

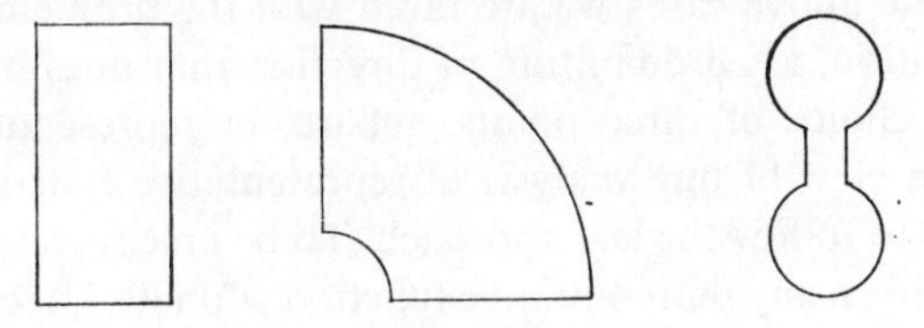

Figure 5.5

22. Show that a simple closed curve Γ is rectifiable iff each representative function is of bounded variation. Show also that Γ is rectifiable iff each simple decomposition consists of rectifiable arcs.

23. Show that a simple closed curve Γ is piecewise smooth iff each simple decomposition consists of piecewise smooth arcs. Show also that Γ is the union of a finite number of smooth subarcs.

24. Let $\gamma \in \mathscr{C}([0, 1])$ be such that $\gamma(0) = \gamma(1)$ and $\gamma^{-1}(\{t\})$ is finite for each $t \in [0, 1]$. What can be said about the range of γ? Can the range be a simple closed curve if $\gamma^{-1}(\{t\})$ contains more than one point for each $t \in [0, 1]$?

5.4 Oriented simple closed curves

In this section, which is the analogue of §5.2, we consider the problem of specifying a direction on a simple closed curve Γ. The problem is harder in this case than in the arc case. The essential difficulty lies in specifying a direction on the circle $C(0, 1)$. In order to see the difficulty we shall consider first this special case.

It is intuitively clear that there are two directions on the circle $C(0, 1)$, namely counterclockwise (or positive) and clockwise (or negative). Suppose we wish to specify the counterclockwise direction. There is no hope of doing this by specifying a 'first point' and 'last point'; indeed, in a natural sense every point of the circle is both a 'first point' and a 'last point' for any direction on the circle. Three possible methods suggest themselves. We might specify the counterclockwise direction on $C(0, 1)$ by writing down the order in which three distinct points should appear. Thus for example the ordered triple $(1, i, -1)$ would determine the counterclockwise direction. Secondly we might specify a direction on some subarc of $C(0, 1)$. Thus for example the semi-circle in the upper half plane with first point 1 and last point -1 would also determine the counterclockwise direction. Thirdly we might specify a direction on $C(0, 1)$ by a choice of representative function. Thus for example the function ω would determine the counterclockwise direction.

In each of the above cases we are faced with the problem of giving an *invariant* definition, i.e. a definition of direction that does not depend on the particular choice of three points, subarc, or representative function respectively. In view of our analysis of representative functions it will be simplest for us to follow the last approach. To be precise, let Γ be a simple closed curve. Given any representative function γ (on $[0, 1]$) for Γ we define

$$[\gamma] = \{\eta \circ \omega(\theta)\omega \circ \varphi \colon \theta \in [0, 1),\ \varphi \in \Phi_+\}$$

where η is the element of $Hom\,(C(0, 1), \Gamma)$ such that $\gamma = \eta \circ \omega$. We need first a simple lemma. Recall that if γ is a function on [0, 1], γ_- is the function on [0, 1] given by $\gamma_- = \gamma \circ \chi$ where $\chi(t) = 1 - t$ $(t \in [0, 1])$. Recall also that $\mathbf{j}(z) = \dfrac{1}{z}$ $(z \in \mathbf{C})$.

Lemma 5.7. *Let* γ *be a representative function on* [0, 1] *for a simple closed curve* Γ. *Then*

(i) $[\gamma] \cap [\gamma_-] = \varnothing$, $[\gamma] \cup [\gamma_-]$ *is the set of all representative functions for* Γ;

(ii) *if* λ *is any representative function on* [0, 1] *for* Γ *then* $\{[\lambda], [\lambda_-]\} = \{[\gamma], [\gamma_-]\}$.

Proof. (i) To prove this part it will be sufficient to show that

$$[\gamma_-] = \{\eta \circ \omega(\theta)\omega \circ \varphi \colon \theta \in [0, 1),\ \varphi \in \Phi_-\},$$

for Proposition 5.6 will then complete the proof. We have

$$\gamma_- = \eta \circ \omega \circ \chi = (\eta \circ \mathbf{j}) \circ \omega$$

and so

$$\begin{aligned} [\gamma_-] &= \{(\eta \circ \mathbf{j}) \circ \omega(\theta)\omega \circ \varphi \colon \theta \in [0, 1),\ \varphi \in \Phi_+\} \\ &= \{\eta \circ \omega(-\theta)\omega \circ (-\varphi) \colon \theta \in [0, 1),\ \varphi \in \Phi_+\} \\ &= \{\eta \circ \omega(\theta)\omega \circ (\chi \circ \varphi) \colon \theta \in [0, 1),\ \varphi \in \Phi_+\}. \end{aligned}$$

It is easy to show that $\Phi_- = \{\chi \circ \varphi \colon \varphi \in \Phi_+\}$ and so (i) is proved.

(ii) Let $\Psi = Hom\,(C(0, 1), C(0, 1))$ and let $\Psi_+(\Psi_-)$ denote the set of $\psi \in \Psi$ of the form $\psi \circ \omega = \omega(\theta)\omega \circ \varphi$ for some $\varphi \in \Phi_+(\varphi \in \Phi_-)$ and $\theta \in [0, 1)$. Clearly $\Psi = \Psi_+ \cup \Psi_-$, $\Psi_+ \cap \Psi_- = \varnothing$. Let $\psi_1, \psi_2 \in \Psi_+$ with $\psi_1 \circ \omega = \omega(\theta_1)\omega \circ \varphi_1$, $\psi_2 \circ \omega = \omega(\theta_2)\omega \circ \varphi_2$. Then there is $\varphi_3 \circ \Phi$, $\theta_3 \in [0, 1)$ such that $\psi_1 \circ \psi_2 \circ \omega = \omega(\theta_3)\omega \circ \varphi_3$. We easily check that

$$\varphi_3(t) = \varphi_1(\theta_2 + \varphi_2(t)) - \varphi_1(\theta_2)\ (t \in [0, \varphi_2^{-1}(1 - \theta_2)])$$

so that $\varphi_3 \in \Phi_+$ and $\psi_1 \circ \psi_2 \in \Psi_+$. Similarly $\Psi_1^{-1} \in \Psi_+$. It follows that $\psi \circ \Psi_+ = \Psi_+(\psi \in \Psi_+)$ and similarly $\psi \circ \Psi_+ = \Psi_-(\psi \in \Psi_-)$.

If $\lambda = \xi \circ \omega$ then $[\lambda] = \xi \circ \Psi_+ \circ \omega$, $[\lambda_-] = \xi \circ \Psi \circ \omega$. Let $\zeta = \eta^{-1} \circ \xi$ so that $\zeta \in \Psi$. Then $[\lambda] = \eta \circ \zeta \circ \Psi_+ \circ \omega$. If $\zeta \in \Psi_+$ then $[\lambda] = [\gamma]$ by the above paragraph and similarly $[\lambda_-] = [\gamma_-]$. If $\zeta \in \Psi_-$ a similar argument gives $[\lambda] = [\gamma_-]$, $[\lambda_-] = [\gamma]$. The proof is complete.

The above lemma partitions the set of all representative functions for Γ into two classes $[\gamma]$, $[\gamma_-]$, and shows that each representative function induces the same partition. We say that each of the classes $[\gamma]$, $[\gamma_-]$ is a *direction* on Γ, and that each is the *opposite* direction to the other. We say that the direction $[\gamma]$ is *determined* by each function in $[\gamma]$. An *oriented simple closed curve* is a pair $(\Gamma, [\gamma])$ where Γ is a simple closed curve and

$[\gamma]$ is one of the two directions on Γ. As in the arc case we shall often write Γ for an oriented simple closed curve, and $-\Gamma$ to indicate the same curve with the opposite direction; it will be understood that, whenever we do this, we have in mind one specific direction on Γ.

A *simple closed contour* is an oriented piecewise smooth simple closed curve. Such curves will figure prominently in the subsequent chapters.

It is well to check that the above definition of direction accords with our intuition. To see this consider the case $\Gamma = C(0, 1)$ with representative function ω. Then

$$[\omega] = \{\omega(\theta)\omega \circ \varphi : \theta \in [0, 1), \varphi \in \Phi_+\}.$$

As t 'traverses' $[0, 1]$ from 0 to 1, $\varphi(t)$ also 'traverses' $[0, 1]$ from 0 to 1, and so $\omega(\theta + \varphi(t))$ 'traverses' $C(0, 1)$ in the counterclockwise direction. Similarly each function in $[\omega_-]$ gives rise to a clockwise 'traversal' of $C(0, 1)$.

Suppose now that $a, b \in \mathbf{R}$, $a < b$ and that $\lambda \in \mathscr{C}([a, b])$ represents a simple closed curve Γ. Let $[\gamma]$ be a direction on Γ. Let

$$\sigma(t) = (1 - t)a + tb \qquad (t \in [0, 1])$$

so that σ is a direction-preserving homeomorphism from $[0, 1]$ onto $[a, b]$. We say that λ represents $(\Gamma, [\gamma])$ if $\lambda \circ \sigma^{-1} \in [\gamma]$. A few pictures will convince the student of the intuitiveness of this definition.

It is again convenient in diagrams to indicate the direction on a curve by placing an arrow on the curve, see Figure 5.6.

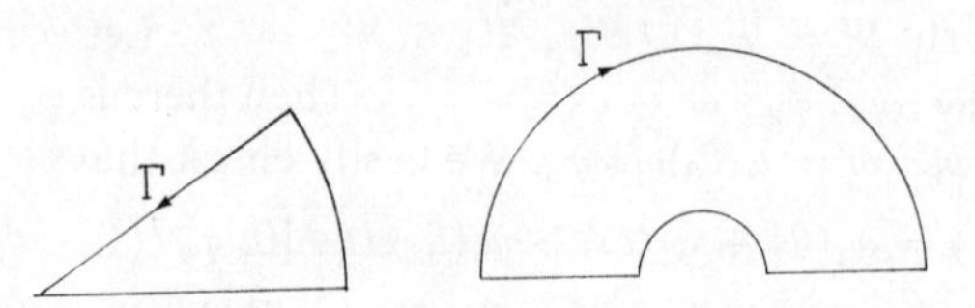

Figure 5.6

It should be understood that, at this point of the discussion, there is no obvious method of defining what we mean by the positive (or counterclockwise) direction on an arbitrary simple closed curve. In the next section we shall mention one possible definition for a special class of curves; in Chapter 8 we shall give a definition appropriate to complex analysis. For very special curves we may use *ad hoc* descriptions of the positive and negative directions (see Problem 5.26). For the very important case of the circle $C(\alpha, r)$ we shall always take the *positive* direction to be determined by the function $\alpha + r\omega$; moreover unless otherwise stated we shall *always* assume that a circle is given the positive direction.

We have already pointed out the possibility of specifying a direction on a simple closed curve by specifying a direction on a subarc. Let us now consider this in the light of the definition we have actually chosen for direction on simple closed curves.

Let Γ be a simple closed curve with a direction given by $[\gamma]$. Let Γ_1 be a subarc of Γ with end points z_1, z_2. Choose $\gamma_1 \in [\gamma]$ such that $\gamma_1(0) = z_1$ and suppose that $z_2 = \gamma_1(u)$ where $u \in (0, 1)$. Either $\Gamma_1 = \gamma_1([0, u])$ or $\Gamma_1 = \gamma_1([u, 1])$. In the first case we say that $(\Gamma, [\gamma])$ *induces* on Γ_1 the direction from z_1 to z_2, i.e. the direction with the first point z_1. In the second case it is clear that the opposite direction is induced on Γ_1. The situation is described in Figure 5.7.

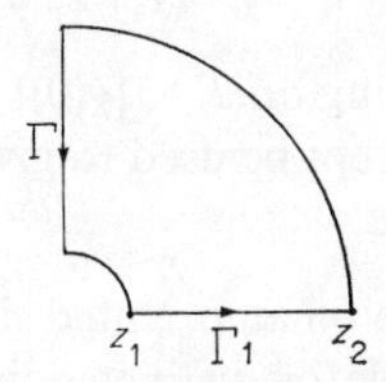

Figure 5.7

Suppose conversely that Γ is a simple closed curve and (Γ_1, z_1) is an oriented subarc of Γ. Let z_2 be the other end point of Γ_1 and let Γ_2 be the subarc of Γ such that $\{\Gamma_1, \Gamma_2\}$ is a simple decomposition of Γ. Choose a representative function γ_1 on $[0, \frac{1}{2}]$ for (Γ_1, z_1) and a representative function γ_2 on $[\frac{1}{2}, 1]$ for (Γ_2, z_2). Now define γ on $[0, 1]$ by

$$\gamma(t) = \begin{cases} \gamma_1(t) & \text{if } 0 \leqslant t \leqslant \frac{1}{2} \\ \gamma_2(t) & \text{if } \frac{1}{2} < t \leqslant 1. \end{cases}$$

It is clear that γ represents Γ. We say that (Γ_1, z_1) *determines* the direction $[\gamma]$ on Γ. Again we have simply described in mathematical language what is perfectly intuitive in Figure 5.7. Consider in particular the simple closed contour in Example 5.5. We may determine a direction on this contour simply by placing an arrow on one of the line segments.

We need one more definition. Given an oriented simple closed curve $(\Gamma, [\gamma])$ we say that oriented subarcs (Γ_r, z_r) $(r = 1, 2, \ldots, n)$ of Γ form a *direction-preserving decomposition* of Γ if

(i) $\Gamma = \bigcup \{\Gamma_r : r = 1, 2, \ldots, n\}$
(ii) Γ induces on Γ_r the direction with first point z_r
(iii) z_{r+1} is the last point of (Γ_r, z_r) $(r = 1, \ldots, n - 1)$ and z_1 is the last point of (Γ_n, z_n).

It is a simple exercise, as in the arc case, to verify that there is a one-to-one correspondence between the subdivisions of the closed interval [0, 1] and those direction-preserving decompositions of $(\Gamma, [\gamma])$ for which $\gamma(0)$ is an end point for some subarc. As a simple example let Γ be as in Example 5.5 with direction determined by $[\alpha_1, \alpha_2]$. Then $[\alpha_1, \alpha_2]$, $[\alpha_2, \alpha_3]$, $[\alpha_3, \alpha_1]$ form a direction-preserving decomposition of Γ.

PROBLEMS 5

25. Let Γ be a simple closed curve with representative function γ. Define the relation $\leqslant_\gamma$ on $\Gamma \setminus \{\gamma(0)\}$ by

$$z_1 \leqslant_\gamma z_2 \quad \text{if} \quad \gamma^{-1}(z_1) \leqslant \gamma^{-1}(z_2).$$

Show that $\leqslant_\gamma$ is a total ordering on $\Gamma \setminus \{\gamma(0)\}$ with neither first point nor last point. How may this concept be used to give an invariant definition of direction on Γ?

26. Show how the direction on an oriented simple closed curve Γ can be described by an ordered triple of distinct points of Γ. In particular the direction on a triangular contour Γ may be described by an ordering of its vertices $\alpha_1, \alpha_2, \alpha_3$. Let the direction be given by $(\alpha_1, \alpha_2, \alpha_3)$, let $\beta = \frac{1}{3}(\alpha_1 + \alpha_2 + \alpha_3)$ and $\theta = \arg(\alpha_1 - \beta)$. We say that Γ is *positively oriented* if

$$\arg_{\theta+\pi}(\alpha_2 - \beta) < \arg_{\theta+\pi}(\alpha_3 - \beta).$$

(i) If Γ is positively oriented show that the same is true for the triangular contours with vertices and directions given by $(\alpha_1, \alpha_2, \alpha_4)$, $(\alpha_4, \alpha_2, \alpha_3)$, $(\alpha_3, \alpha_1, \alpha_4)$, where α_4 is any point inside Γ, i.e. α_4 is of the form

$$\alpha_4 = t_1\alpha_1 + t_2\alpha_2 + t_3\alpha_3$$

where $t_1 > 0$, $t_2 > 0$, $t_3 > 0$, $t_1 + t_2 + t_3 = 1$.

(ii) If $\lambda_1 = \frac{1}{2}(\alpha_2 + \alpha_3)$, $\lambda_2 = \frac{1}{2}(\alpha_3 + \alpha_1)$, $\lambda_3 = \frac{1}{2}(\alpha_1 + \alpha_2)$, show that the triangular contours with vertices and directions given by $(\alpha_1, \lambda_3, \lambda_2)$, $(\alpha_2, \lambda_1, \lambda_3)$, $(\alpha_3, \lambda_2, \lambda_1)$, $(\lambda_1, \lambda_2, \lambda_3)$ are also positively oriented.

27. Prove the assertion in the text about the relation between direction-preserving decompositions and subdivisions of [0, 1].

28. Prove that every simple closed contour admits a direction-preserving decomposition into oriented smooth subarcs.

29. Discuss the analogues of Problems 5.16, 5.17 for simple closed curves.

5.5 The Jordan curve theorem

The Jordan curve theorem (for the complex plane) states that every simple closed curve has an inside and an outside. It is the classical example of a theorem which is intuitively too obvious for words and yet whose proof is extremely difficult. Indeed the original proof given by Jordan was inadequate; only later was a correct proof given by Veblen. Given a picture of some particular simple closed curve it is easy to 'see' what is the inside and what is the outside. The problem is to translate the picture into a mathematical proof.

To be more precise, the Jordan curve theorem states that if Γ is any simple closed curve then the open set $\mathbf{C} \setminus \Gamma$ has two components D_1, D_2 and moreover $b(D_1) = b(D_2) = \Gamma$. One of these domains is bounded and is called the *inside* of Γ or the *Jordan interior* of Γ, denoted by $\text{int}(\Gamma)$; the other domain is unbounded and is called the *outside* of Γ or the *Jordan exterior* of Γ, denoted by $\text{ext}(\Gamma)$. The proof of the theorem in this general form is harder than anything we shall prove in this book; but it so happens that the Jordan curve theorem is not essential for complex analysis. On the other hand it is true that the Jordan curve theorem simplifies considerably the statement of several important theorems. It also helps to provide a more concrete geometrical setting for some theorems and thereby facilitates understanding. We shall therefore prove the theorem for a special class of simple closed curves. This class will be large enough to deal with most situations encountered by the student. Occasionally it is necessary to develop *ad hoc* techniques to deal with other cases.

Let Γ be a simple closed curve and let $\alpha \in \mathbf{C} \setminus \Gamma$. We say that Γ is *starred* with respect to α if for each $z \in \mathbf{C} \setminus \{\alpha\}$ the ray $\{(1-t)\alpha + tz: t \geqslant 0\}$ meets Γ in exactly one point. We then say that Γ is *starlike* and that α is a *star centre* for Γ. It is clear that Γ is starred with respect to α if and only if each point of $\mathbf{C} \setminus \{\alpha\}$ has a unique representation as $\alpha + \lambda(w - \alpha)$ for some $\lambda > 0$, $w \in \Gamma$. Another equivalent formulation is that the mapping $z \to \exp(i \arg(z - \alpha))$ should be one-to-one from Γ onto $C(0, 1)$. It is thus clear that a circle is starred with respect to its centre. Another simple example of a starlike contour is given by the triangular contour of Example 5.5, the centroid being a star centre. Further examples are described in Figure 5.8.

Given that Γ is starred with respect to α, and given $z_0 \in \Gamma$, choose a representative function γ for Γ such that $\gamma(0) = z_0$. If $\theta = \pi + \arg(z_0 - \alpha)$ then the mapping $t \to \arg_\theta(\gamma(t) - \alpha)$ is a continuous one-to-one mapping from $(0, 1)$ onto $(\theta - \pi, \theta + \pi)$ and is thus monotonic. This suggests a method of distinguishing the positive and negative directions on Γ. In fact

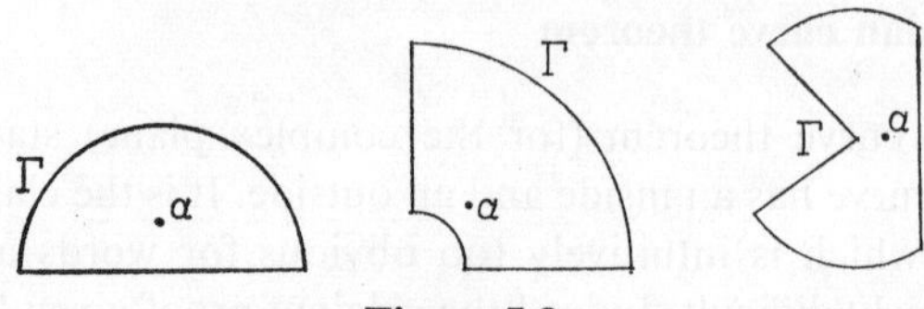

Figure 5.8

we might well define the positive direction as being determined by γ when the mapping $t \to \arg_\theta(\gamma(t) - \alpha)$ is increasing. However, a definition which is more useful for complex analysis will be given in §8.7.

There is ample motivation for this definition of a starlike simple closed curve Γ. We shall see in particular that Γ is the boundary of a bounded starlike domain. The definition also makes it fairly clear what we should take for the inside of Γ. To see this, consider the case of the circle $C(0, 1)$. We know of course that $\Delta(0, 1)$ and $\nabla(0, 1)$ are the components of $\mathbf{C} \setminus C(0, 1)$, $\Delta(0, 1)$ being the inside of $C(0, 1)$. Moreover, $\Delta(0, 1)$ may be considered as the union of all the radii $\{tz: 0 \leqslant t < 1, z \in C(0, 1)\}$.

One further word is in order before we come to the theorem itself. In practice it may be a little tedious to demonstrate that a given simple closed curve is starlike, even though the fact may be perfectly obvious from a pictorial point of view. What we are here doing is to take a geometrical condition on simple closed curves (namely starlikeness) that can be described in analytic terms, and to give an analytic proof of the Jordan curve theorem for such curves. The method of proof is conceptually quite simple. Given a homeomorphism η of Γ with $C(0, 1)$ we show how η may be extended to a homeomorphism of $\mathbf{C}$ with itself (this is a special case of Schoenflies' theorem). We then deduce the required result for Γ from the corresponding known result for $C(0, 1)$. We also deduce some corollaries. In particular we show that the inside of Γ is homeomorphic with $\Delta(0, 1)$.

Theorem 5.8. (Jordan). *Let Γ be a starlike simple closed curve. Then there is a bounded starlike domain D_1 (= int (Γ)) and an unbounded domain D_2 (= ext (Γ)) such that $D_1 \cup D_2 = \mathbf{C} \setminus \Gamma$, $D_1 \cap D_2 = \varnothing$, $b(D_1) = b(D_2) = \Gamma$.*

Proof. Let α be a star centre for Γ and let η be any homeomorphism of Γ with $(C(0, 1)$. Then each point z of $\mathbf{C} \setminus \{\alpha\}$ has a unique representation as $z = \alpha + \lambda(w - \alpha)$ for some $\lambda > 0$, $w \in \Gamma$. Define f on $\mathbf{C}$ by

$$f(\alpha + \lambda(w - \alpha)) = \lambda\eta(w) \qquad (\lambda \geqslant 0, w \in \Gamma).$$

Clearly f maps $\mathbf{C}$ one-to-one onto $\mathbf{C}$ and $f|_\Gamma = \eta$. We show that f is a homeomorphism.

Let $z_n \to z$, where $z_n = \alpha + \lambda_n(w_n - \alpha)$, $z = \alpha + \lambda(w - \alpha)$. If $z = \alpha$, then $\lambda_n(w_n - \alpha) \to 0$. Since Γ is compact and $\alpha \notin \Gamma$ there exist $m, M \in \mathbf{R}$ such that

$$0 < m \leqslant |w - \alpha| \leqslant M \qquad (w \in \Gamma).$$

It follows that $|f(z_n)| = \lambda_n \to 0$ and so $f(z_n) \to f(\alpha)$. Suppose now that $z \neq \alpha$, and let $\theta = \pi + \arg(w - \alpha)$. Choose a representative function γ, for Γ such that $\arg_\theta(\gamma(0) - \alpha) = \theta + \pi$. Then $w = \gamma(t)$ for some $t \in (0, 1)$ and for n sufficiently large $w_n = \gamma(t_n)$ where $t_n \in (0, 1)$. Let $h(t) = \arg_\theta(\gamma(t) - \alpha)$. Since α is a star centre for Γ, h is a homeomorphism of $(0, 1)$ with $(\theta - \pi, \theta + \pi)$. Since $z_n \to z$, we have $\lambda_n(w_n - \alpha) \to \lambda(w - \alpha)$, $\arg_\theta(\lambda_n(w_n - \alpha)) \to \arg_\theta(\lambda(w - \alpha))$, and so $h(t_n) \to h(t)$. Therefore $t_n \to t$ and so $w_n = \gamma(t_n) \to \gamma(t) = w$. Hence

$$\lambda_n = \frac{z_n - \alpha}{w_n - \alpha} \to \frac{z - \alpha}{w - \alpha} = \lambda$$

and so $f(z_n) = \lambda_n \eta(w_n) \to \lambda\eta(w) = f(z)$. This shows that f is continuous on $\mathbf{C}$. The inverse mapping f^{-1} is given by

$$f^{-1}(\lambda z) = \alpha + \lambda(\eta^{-1}(z) - \alpha) \qquad (\lambda \geqslant 0, z \in C(0, 1))$$

and a similar argument shows that f^{-1} is continuous on $\mathbf{C}$.

Let $D_1 = f^{-1}(\Delta(0, 1))$, $D_2 = f^{-1}(\nabla(0, 1))$. Since $\Delta(0, 1)$, $\nabla(0, 1)$ are domains and f is a homeomorphism it follows that D_1, D_2 are domains. We have

$$D_1 = \{\alpha + \lambda(w - \alpha) : 0 \leqslant \lambda < 1, w \in \Gamma\}$$

from which it is clear that D_1 is starred with respect to α and $D_1 \subset \Delta(\alpha, M)$. Finally, Γ is the common boundary of D_1 and D_2 since $C(0, 1)$ is the common boundary of $\Delta(0, 1)$ and $\nabla(0, 1)$.

Corollary. (i) int (Γ) *is homeomorphic with* $\Delta(0, 1)$.

(ii) *Given* $\epsilon > 0$ *there is a starlike domain* $D_\epsilon \supset D_1^-$ *such that* dist $(D_1, \mathbf{C} \setminus D_\epsilon) \leqslant \epsilon$.

Proof. (i) This was proved in the course of the theorem.

(ii) Let $D_\epsilon = f^{-1}\left(\Delta\left(0, 1 + \frac{\epsilon}{M}\right)\right)$. It is clear from the properties of f that D_ϵ is a starlike domain with $D_1^- \subset D_\epsilon$. Finally, if $w \in \Gamma$ then

$$\text{dist}\,(D_1^-, \mathbf{C} \setminus D_\epsilon) \leqslant \left|w - \left\{\alpha + \left(1 + \frac{\epsilon}{M}\right)(w - \alpha)\right\}\right|$$

$$= \frac{\epsilon}{M}|w - \alpha| \leqslant \epsilon.$$

It is convenient to have a short notation for the closure of int (Γ); we write $\Gamma^* = \text{int}(\Gamma)^-$. Care should be taken to distinguish Jordan interior, int (Γ), from topological interior, Γ^0 (cf. Problem 5.31).

It was asserted (without proof) in §2.2 that every bounded starlike domain is homeomorphic with $\Delta(0, 1)$. The corollary above proves this result for domains that are the interior of a starlike simple closed curve; such domains are called *starlike Jordan domains.* On the other hand (Problem 5.32) there are bounded starlike domains whose boundary is not a simple closed curve.

PROBLEMS 5

30. Give the details of the following outline of an alternative proof of Theorem 5.8.

(i) Let

$$D_1 = \{(1 - t)\alpha + tw: 0 \leqslant t < 1, w \in \Gamma\},$$
$$D_2 = \{(1 - t)\alpha + tw: t > 1, w \in \Gamma\}.$$

Show that D_1, D_2 are polygonally connected.

(ii) Show that $\Delta(\alpha, \epsilon) \subset D_1$ where $\epsilon = \text{dist}(\alpha, \Gamma)$. Given $\beta \in D_1 \setminus \{\alpha\}$, show that β lies in some open sector (of a disc) that is contained in D_1. Thus D_1 is open. Show similarly that D_2 is open.

(iii) Show that D_1 is a bounded starlike domain and that $b(D_1) = b(D_2) = \Gamma$.

31. Let Γ be a starlike simple closed curve.

(i) Show that α is a star centre for Γ iff α is a star centre for int (Γ).

(ii) If A is a polygonal line joining some point of int (Γ) to some point of ext (Γ), show that A intersects Γ.

(iii) Show that (as a subset of **C**) $\Gamma^0 = \varphi$. Prove the result without the restriction that Γ be starlike.

(iv) If D is a starlike domain and $\Gamma \subset D$ show that int $(\Gamma) \subset D$.

32. Let $D = \Delta(0, 1) \setminus [0, 1]$. Show that D is a bounded starlike domain whose boundary is not a simple closed curve. Show that D is homeomorphic with $\Delta(0, 1) \cap \{z: \text{Re}\, z < 0\}$ and so with $\Delta(0, 1)$.

33. Let D be a bounded starlike domain such that $b(D)$ is a simple closed curve. Show that $b(D)$ is starlike.

5.6 Contour integration

The concept of the integral of a complex function along an oriented arc or closed curve is central to the development of complex analysis that we follow in this book.* The definition that we give here could apply to arbitrary rectifiable arcs and closed curves, but for purposes of application it is sufficient to consider the piecewise smooth case. We shall require some results from the theory of (complex) Riemann-Stieltjes integration. We do not wish to interrupt our progress at this point by proving these results; but since this topic may not be familiar to all our readers we give in the Appendix a definition of the Riemann-Stieltjes integral and a proof of the theorem stated below. In fact, we shall quickly convert the Riemann-Stieltjes integral to a Riemann integral; but we emphasize that the Riemann-Stieltjes integral is necessary in order to give an invariant definition of contour integration, i.e. to show that the value of the integral is independent of the choice of representative function. Recall that $\mathscr{BV}([a, b])$ denotes the set of all complex functions of bounded variation on $[a, b]$ (Problem 3.7).

Theorem 5.9. *Given* $f \in \mathscr{C}([a, b])$ *and* $g \in \mathscr{BV}([a, b])$, f *is Riemann-Stieltjes integrable with respect to* g, *and the following statements hold.*

(i) *If* $a = t_0 < t_1 < \cdots < t_n = b$, *then*

$$\int_a^b f\,\mathrm{d}g = \sum_{r=1}^{n} \int_{t_{r-1}}^{t_r} f\,\mathrm{d}g.$$

(ii) *If* φ *is a continuous strictly increasing mapping from* $[c, d]$ *onto* $[a, b]$ *then*

$$\int_a^b f\,\mathrm{d}g = \int_c^d f \circ \varphi\,\mathrm{d}(g \circ \varphi).$$

(iii) *If* $g \in \mathscr{C}([a, b])$ *is differentiable on* (a, b) *except at a finite number of points with* g' *bounded and continuous then*

$$\int_a^b f\,\mathrm{d}g = \int_a^b fg',$$

where the second integral is the (complex) Riemann integral.

Remark. The above function g' (and so also fg') is not defined at a finite number of points of $[a, b]$. We may extend the function g' by taking it to be zero at these points if we wish, but it is well known in the theory of Riemann integration that the value of the integral would remain unchanged whatever values we assigned to g' at the above finite number of points.

* There are treatments of complex analysis in which integration scarcely appears at all, e.g. Whyburn, *Topological Analysis*.

For the rest of the chapter Γ will always denote either an oriented piecewise smooth arc or a simple closed contour, and γ will denote a representative function for Γ which is differentiable on (0, 1) except at a finite number of points with γ' bounded and continuous. Given $f \in \mathscr{C}(\Gamma)$ we thus have $f \circ \gamma \in \mathscr{C}([0, 1])$ and $\gamma \in \mathscr{BV}([0, 1])$ and so $f \circ \gamma$ is Riemann-Stieltjes integrable with respect to γ.

Suppose first that Γ is an arc. If γ_1 is any function representing Γ then we have $\gamma_1 = \gamma \circ \varphi$ for some $\varphi \in \Phi_+$. By Theorem 5.9 (ii) we now have

$$\int_0^1 f \circ \gamma_1 \, d\gamma_1 = \int_0^1 f \circ \gamma \circ \varphi \, d(\gamma \circ \varphi) = \int_0^1 f \circ \gamma \, d\gamma.$$

Suppose now that Γ is a contour and that $\gamma = \eta \circ \omega$. If γ_1 is any function representing Γ we have $\gamma_1 = \eta \circ \omega(\theta)\omega \circ \varphi$ for some $\varphi \in \Phi_+$ and θ such that $0 \leqslant \theta < 1$. Let $h = \eta \circ \omega(\theta)\omega$. It is readily seen that $h \in \mathscr{BV}([0, 1])$ and so using (repeatedly) parts (i) and (ii) of Theorem 5.9 we obtain

$$\begin{aligned}
\int_0^1 f \circ \gamma_1 \, d\gamma_1 &= \int_0^1 f \circ h \circ \varphi \, d(h \circ \varphi) = \int_0^1 f \circ h \, dh \\
&= \int_0^{1-\theta} f \circ \eta \circ \omega(t + \theta) \, d(\eta \circ \omega(t + \theta)) \\
&\quad + \int_{1-\theta}^1 f \circ \eta \circ \omega(t + o) \, d(\eta \circ \omega(t + o)) \\
&= \int_\theta^1 f \circ \eta \circ \omega(u) \, d(\eta \circ \omega(u)) + \int_0^\theta f \circ \eta \circ \omega(v) \, d(\eta \circ \omega(v)) \\
&= \int_0^1 f \circ \eta \circ \omega \, d(\eta \circ \omega) \\
&= \int_0^1 f \circ \gamma \, d\gamma.
\end{aligned}$$

We may now define the *integral* of *f along* Γ by

$$\int_\Gamma f = \int_0^1 f \circ \gamma_1 \, d\gamma_1$$

where γ_1 is any function representing Γ. By the above reasoning the integral is *well-defined*, i.e. the value of the integral is independent of the choice of representative function. In particular, using the given function γ and Theorem 5.9 (iii) we obtain

$$\int_\Gamma f = \int_0^1 (f \circ \gamma)\gamma'.$$

It is sometimes convenient to display a variable in the integral by writing it as

$$\int_\Gamma f(z)\,dz.$$

We illustrate the definition by two simple cases. Suppose first that Γ is the arc $[0, 1]$ represented by $\gamma(t) = t\,(t \in [0, 1])$. Given $f \in \mathscr{C}(\Gamma)$ we then have

$$\int_\Gamma f = \int_0^1 (f \circ \gamma)\gamma' = \int_0^1 f(t)\,dt.$$

In other words, in this case the integral is nothing other than the usual Riemann integral—as expected. Suppose now that Γ is the circle $C(0, 1)$ represented by the function ω. Given $f \in \mathscr{C}(\Gamma)$ we then have

$$\int_\Gamma f = \int_0^1 (f \circ \gamma)\gamma' = 2\pi i \int_0^1 f(\mathrm{e}^{2\pi it})\,\mathrm{e}^{2\pi it}\,dt.$$

In many situations it is more natural to represent Γ by a function on some closed interval $[a, b]$ rather than on $[0, 1]$. Suppose that $\lambda \in \mathscr{C}[a, b]$ represents Γ and is such that λ is differentiable on (a, b) except at a finite number of points with λ' bounded and continuous. If

$$\sigma(t) = (1 - t)a + tb \qquad (t \in [0, 1])$$

we may easily verify that $\lambda \circ \sigma$ represents Γ. Using Theorem 5.9 we now obtain for $f \in \mathscr{C}(\Gamma)$

$$\int_\Gamma f = \int_0^1 f \circ \lambda \circ \sigma\,\mathrm{d}(\lambda \circ \sigma) = \int_a^b f \circ \lambda\,\mathrm{d}\lambda = \int_a^b (f \circ \lambda)\lambda'.$$

As an illustration let

$$\Gamma = [0, R] \cup [R, R + iS]$$

and let

$$\lambda(t) = \begin{cases} tR & \text{if } 0 \leqslant t \leqslant R \\ R + i(t - R) & \text{if } R < t \leqslant R + S. \end{cases}$$

Given $f \in \mathscr{C}(\Gamma)$ we then have

$$\begin{aligned} \int_\Gamma f &= \int_0^{R+S} (f \circ \lambda)\lambda' \\ &= \int_0^R f(t)\,\mathrm{d}t + \int_0^S f(R + it)i\,\mathrm{d}t. \end{aligned}$$

We shall now obtain some of the basic properties of the integral. Most of these properties have obvious counterparts in the theory of Riemann integration.

CI 1. $\displaystyle\int_\Gamma \lambda f = \lambda \int_\Gamma f \qquad (f \in \mathscr{C}(\Gamma), \lambda \in \mathbf{C}).$

Proof. $\displaystyle\int_\Gamma \lambda f = \int_0^1 \lambda(f \circ \gamma)\gamma' = \lambda \int_0^1 (f \circ \gamma)\gamma' = \lambda \int_\Gamma f.$

CI 2. $\displaystyle\int_\Gamma (f + g) = \int_\Gamma f + \int_\Gamma g \qquad (f, g \in \mathscr{C}(\Gamma)).$

Proof.
$$\int_\Gamma (f + g) = \int_0^1 ((f + g) \circ \gamma)\gamma' = \int_0^1 (f \circ \gamma)\gamma' + \int_0^1 (g \circ \gamma)\gamma'$$
$$= \int_\Gamma f + \int_\Gamma g.$$

CI 3. $\displaystyle\int_{-\Gamma} f = -\int_\Gamma f \qquad (f \in \mathscr{C}(\Gamma)).$

Proof.
$$\int_{-\Gamma} f = \int_0^1 (f \circ \gamma_-)\gamma'_- = \int_0^1 f(\gamma(1 - t))(-\gamma'(1 - t))\, \mathrm{d}t$$
$$= -\int_0^1 f(\gamma(u))\gamma'(u)\, \mathrm{d}u = -\int_\Gamma f.$$

CI 4. *If Γ_r $(r = 1, 2, \ldots, n)$ form a direction-preserving decomposition of Γ,*

$$\int_\Gamma f = \sum_{r=1}^n \int_{\Gamma_r} f \qquad (f \in \mathscr{C}(\Gamma)).$$

Proof. We may choose a representative function γ for Γ such that $\gamma(0)$ is the first point of Γ_1. Let t_r $(r = 0, 1, \ldots, n)$ be such that $t_0 = 0$ and $\gamma(t_r)$ is the last point of Γ_r $(r = 1, 2, \ldots, n)$. Then Γ_r may be represented by $\gamma|_{[t_{r-1}, t_r]}$ $(r = 1, 2, \ldots, n)$. Using Theorem 5.9 (i) we now obtain

$$\int_\Gamma f = \int_0^1 f \circ \gamma\, \mathrm{d}\gamma = \sum_{r=1}^n \int_{t_{r-1}}^{t_r} f \circ \gamma\, \mathrm{d}\gamma$$
$$= \sum_{r=1}^n \int_{\Gamma_r} f.$$

For the next property it is necessary to introduce the concept of the length of Γ. A *subdivision* of $[0, 1]$ is a set of the form $P = \{t_r : r = 0, 1, \ldots, n\}$ where $0 = t_0 < t_1 < \cdots < t_n = 1$. We denote by $\mathscr{P}$ by the set of all subdivisions of $[0, 1]$. Given such a subdivision P of $[0, 1]$ we define

$$T(\gamma, P) = \sum_{r=1}^{n} |\gamma(t_r) - \gamma(t_{r-1})|$$
$$V(\gamma) = \sup \{T(\gamma, P) : P \in \mathscr{P}\}.$$

Since $\gamma \in \mathscr{BV}([0, 1])$ we certainly have $0 \leqslant V(\gamma) < +\infty$.

Suppose that Γ is an arc and that γ_1 is any representative function for Γ. Then $\gamma_1 = \gamma \circ \varphi$ for some $\varphi \in \Phi_+$. Given $P \in \mathscr{P}$ with $P = \{t_r : r = 0, 1, \ldots, n\}$ we easily see that $\{\varphi(t_r) : r = 0, 1, \ldots, n\} \in \mathscr{P}$. It follows that

$$V(\gamma_1) = \sup \{T(\gamma \circ \varphi, P) : P \in \mathscr{P}\} \leqslant \sup \{T(\gamma, P) : P \in \mathscr{P}\} = V(\gamma).$$

Since $\varphi^{-1} \in \Phi_+$ we also have $\{\varphi^{-1}(t_r) : r = 0, 1, \ldots, n\} \in \mathscr{P}$ and since $t = \varphi \circ \varphi^{-1}(t)$ it now follows that

$$\begin{aligned} V(\gamma) &= \sup \{T(\gamma \circ \varphi \circ \varphi^{-1}, P) : P \in \mathscr{P}\} \\ &\leqslant \sup \{T(\gamma \circ \varphi, P) : P \in \mathscr{P}\} \\ &= V(\gamma_1). \end{aligned}$$

We have thus shown that $V(\gamma_1) = V(\gamma)$ for any function γ_1 that represents Γ. (Observe that we have in particular given a proof of Problem 5.5.) We leave the reader to verify that this is also true in the contour case.

We may now define the *length* of Γ by

$$l_\Gamma = V(\gamma_1)$$

where γ_1 is any representative function for Γ. By the above reasoning the length of Γ is thus well defined. Choosing the given representative function we then obtain

$$l_\Gamma = \int_0^1 |\gamma'|.$$

The proof of this last statement is given in the Appendix.

Given a non-empty subset S of $\mathbf{C}$ and given $f : S \to \mathbf{C}$ we shall use the following abbreviated notation:

$$\sup_S |f| = \sup \{|f(z)| : z \in S\}.$$

CI 5. $\left| \int_\Gamma f \right| \leqslant l_\Gamma \sup_\Gamma |f| \qquad (f \in \mathscr{C}(\Gamma)).$

Proof.
$$\begin{aligned} \left| \int_\Gamma f \right| &= \left| \int_0^1 (f \circ \gamma)\gamma' \right| \leqslant \int_0^1 |f \circ \gamma| \, |\gamma'| \\ &\leqslant \sup \{|f(\gamma(t))| : t \in [0, 1]\} \int_0^1 |\gamma'| \\ &= l_\Gamma \sup_\Gamma |f|. \end{aligned}$$

Given $f \in \mathscr{C}(D)$ recall that $g \in \mathscr{A}(D)$ is a primitive of f if $g' = f$.

CI 6. *If $f \in \mathscr{C}(D)$ with primitive g and if $\Gamma \subset D$,*

$$\int_\Gamma f = g(\gamma(1)) - g(\gamma(0)).$$

In particular, if Γ is a simple closed contour,

$$\int_\Gamma f = 0.$$

Proof.
$$\int_\Gamma f = \int_0^1 (f \circ \gamma)\gamma' = \int_0^1 (g' \circ \gamma)\gamma' = \int_0^1 (g \circ \gamma)'$$
$$= g(\gamma(1)) - g(\gamma(0)).$$

To illustrate **CI 6** suppose first that Γ is an arc with first point z_0 and last point z_1. Then for $n = 0, 1, 2, \ldots$, we have

$$\int_\Gamma z^n \,\mathrm{d}z = \frac{1}{n+1}\{z_1^{n+1} - z_0{}^{n+1}\}.$$

If Γ is a contour and p is any polynomial function then we have

$$\int_\Gamma p = 0.$$

Moreover, if the contour Γ is such that $0 \notin \Gamma$ we have

$$\int_\Gamma \frac{1}{z^{n+1}}\,\mathrm{d}z = 0 \qquad (n \in \mathbf{P}).$$

On the other hand if we take Γ to be $C(0, 1)$ then we obtain

$$\int_{C(0,1)} \frac{1}{z}\,\mathrm{d}z = \int_0^1 (\mathbf{j} \circ \omega)\omega' = 2\pi i \int_0^1 \mathbf{1} = 2\pi i.$$

CI 7. *Let $f_n \in \mathscr{C}(\Gamma)$ $(n \in \mathbf{P})$ and let*

$$\lim_{n\to\infty} f_n = f \quad (\textit{uniformly on } \Gamma).$$

Then $f \in \mathscr{C}(\Gamma)$ and

$$\int_\Gamma f = \lim_{n\to\infty} \int_\Gamma f_n.$$

Proof. It follows from Proposition 1.14 that $f \in \mathscr{C}(\Gamma)$. Since $\{f_n\}$ converges to f uniformly on Γ, given $\epsilon > 0$ there is $N \in \mathbf{P}$ such that

$$\sup_\Gamma |f_n - f| \leqslant \epsilon \qquad (n > N).$$

Using **CI 1, 2** and **5** we now obtain

$$\left|\int_\Gamma f_n - \int_\Gamma f\right| = \left|\int_\Gamma (f_n - f)\right|$$
$$\leqslant l_\Gamma \sup_\Gamma |f_n - f|$$
$$\leqslant l_\Gamma \epsilon \qquad (n > N).$$

It is now clear that

$$\int_\Gamma f = \lim_{n\to\infty} \int_\Gamma f_n.$$

Remark. **CI 7** asserts that $f \to \int_\Gamma f$ is a continuous mapping from the metric space $\mathscr{C}(\Gamma)$ to $\mathbf{C}$.

PROBLEMS 5

34. Evaluate the following integrals.

$$\int_{[0,\alpha]} \operatorname{Re} z\, dz, \qquad \int_{C(0,1)} \operatorname{Im} z\, dz, \qquad \int_{C(0,2)} \frac{dz}{z^2+1}, \qquad \int_{C(0,1)} \bar{z}^n\, dz.$$

35. Show that the function $z \to z \log(z)$ has a primitive on $\mathbf{C} \setminus N$ and evaluate

$$\int_{[0,i]} z \log(z)\, dz.$$

36. Let $f \in \mathscr{A}(D)$ be such that $f' \in \mathscr{C}(D)$ and $f(D) \subset \mathbf{C} \setminus N$. Show that

$$\int_\Gamma f'/f = 0$$

for any simple closed contour $\Gamma \subset D$.

37. Let f be a power series function on $\Delta(\alpha, \rho)$. Show that

$$\int_\Gamma f = 0$$

for every simple closed contour $\Gamma \subset \Delta(\alpha, \rho)$.

38. Let Γ be a simple closed contour starred with respect to α, and let $\{r_n\} \subset (0, 1)$ be such that $r_{n+1} \geqslant r_n$ $(n \in \mathbf{P})$ and $\lim_{n\to\infty} r_n = 1$. For each $n \in \mathbf{P}$ let

$$\gamma_n(t) = r_n \gamma(t) + (1 - r_n)\alpha \qquad (t \in [0, 1]).$$

Show that each γ_n represents a starlike simple closed contour, Γ_n say. If $K = \Gamma^* \setminus \text{int}(\Gamma_1)$ show that

$$\int_\Gamma f = \lim_{n\to\infty} \int_{\Gamma_n} f \qquad (f \in \mathscr{C}(K)).$$

39. Show that

$$\lim_{R\to+\infty} \int_{C(0,R)} \frac{z\,dz}{z^3+1} = 0, \qquad \lim_{R\to+\infty} \int_{[-R,-R+i]} \frac{z^2 \exp(z)}{z+1}\,dz = 0.$$

40. Develop an elementary integration theory for arcs and contours that consist of unions of line segments and arcs of circles. Such a theory is in fact adequate for almost all complex function theory; but the ideas of this chapter provide an easy introduction to certain topics on the topology of the complex plane and also a useful application of the Riemann-Stieltjes integral.

6

CAUCHY'S THEOREM FOR STARLIKE DOMAINS

Cauchy's theorem (in some form) is the most important single theorem in complex analysis. Almost every deep result that we shall prove depends in one way or another on Cauchy's theorem. Roughly speaking the theorem states that if f is analytic on a domain containing a simple closed contour Γ and its Jordan interior then

$$\int_{\Gamma} f = 0.$$

No attempt is made in the present book to prove the most general form of the theorem. We content ourselves by proving a version of the theorem sufficient for most applications. The version that we consider has the merit of admitting a straightforward proof, and also illustrating the spirit of more general proofs. Geometrical language is occasionally used in this chapter to facilitate understanding, but the student who so wishes may express everything in this chapter in purely analytical language.

6.1 Cauchy's theorem for triangular contours

In this section we shall prove Cauchy's theorem for the case in which the simple closed contour is triangular, i.e. as in Example 5.5. This preliminary case admits an elegant proof that is due to Pringsheim. The proof gives remarkably deep penetration into what is involved in any version of the theorem, namely (a) the completeness of **C** (b) **CI 1–6** (c) the topological nature of the given simple closed contour in relation to the domain in which the given function is analytic.

By a triangle in the complex plane we mean the convex hull (see Problem 2.14) of three distinct points that do not lie on a straight line. Thus given $\alpha_1, \alpha_2, \alpha_3 \in \mathbf{C}$, not all on the same straight line, the associated triangle is given by

$$T = \{\lambda_1\alpha_1 + \lambda_2\alpha_2 + \lambda_3\alpha_3 : \lambda_1, \lambda_2, \lambda_3 \geqslant 0, \lambda_1 + \lambda_2 + \lambda_3 = 1\}.$$

It is straightforward to verify the obvious fact that

$$b(T) = [\alpha_1, \alpha_2] \cup [\alpha_2, \alpha_3] \cup [\alpha_3, \alpha_1].$$

We have already seen (Example 5.5) that $b(T)$ is a simple closed contour; we shall denote it by $\Gamma(T)$. Given $\alpha_1, \alpha_2, \alpha_3$ (in that order) we agree to give $\Gamma(T)$ the direction determined by the subarc $[\alpha_1, \alpha_2]$. Thus if $f \in \mathscr{C}(\Gamma(T))$ then

$$\int_{\Gamma(T)} f = \int_{[\alpha_1,\alpha_2]} f + \int_{[\alpha_2,\alpha_3]} f + \int_{[\alpha_3,\alpha_1]} f.$$

For the purposes of this section it is irrelevant whether $\Gamma(T)$ has the positive or negative direction since of course

$$\int_{\Gamma(T)} f = 0 \Leftrightarrow \int_{-\Gamma(T)} f = 0.$$

Theorem 6.1. *Given a triangle T in a domain D,*

$$\int_{\Gamma(T)} f = 0 \qquad (f \in \mathscr{A}(D)).$$

Proof. Let T have vertices $\alpha_1, \alpha_2, \alpha_3$ (in that order) and sides of length a, b, c. Given $f \in \mathscr{A}(D)$ let

$$A = \left|\int_{\Gamma(T)} f\right|.$$

We split T into four smaller triangles as follows. Let

$$\beta_1 = \tfrac{1}{2}(\alpha_2 + \alpha_3), \qquad \beta_2 = \tfrac{1}{2}(\alpha_3 + \alpha_1), \qquad \beta_3 = \tfrac{1}{2}(\alpha_1 + \alpha_2)$$

and let T_1^r $(r = 1, 2, 3, 4)$ be the triangles with vertices $(\alpha_1, \beta_3, \beta_2)$, $(\alpha_2, \beta_1, \beta_3)$, $(\alpha_3, \beta_2, \beta_1)$, $(\beta_1, \beta_2, \beta_3)$ respectively (see Figure 6.1).

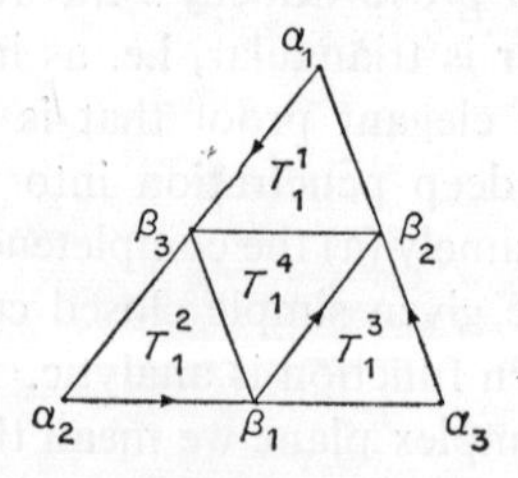

Figure 6.1

By **CI 4** we have

$$\int_{\Gamma(T)} f = \int_{[\alpha_1,\beta_3]} f + \int_{[\beta_3,\alpha_2]} f + \int_{[\alpha_2,\beta_1]} f + \int_{[\beta_1,\alpha_3]} f + \int_{[\alpha_3,\beta_2]} f + \int_{[\beta_2,\alpha_1]} f.$$

By **CI 3** we have

$$\int_{[\beta_j,\beta_k]} f + \int_{[\beta_k,\beta_j]} f = 0 \qquad (j, k = 1, 2, 3; j \neq k).$$

It follows that

$$\int_{\Gamma(T)} f = \sum_{r=1}^{4} \int_{\Gamma(T_1^r)} f.$$

There is some r in $\{1, 2, 3, 4\}$ such that

$$A \leqslant 4\left|\int_{\Gamma(T_1^r)} f\right|$$

for otherwise we should have the contradiction

$$\left|\int_{\Gamma(T)} f\right| < A.$$

We denote the triangle T_1^r by T_1 and observe that T_1 has sides of length $\frac{1}{2}a$, $\frac{1}{2}b$, $\frac{1}{2}c$. For notational convenience we write T_0 for T.

Suppose now that we have obtained triangles $T_1, T_2, \ldots, T_k$ such that

(i) $T_{m+1} \subset T_m$ $(m = 0, 1, \ldots, k-1)$,
(ii) T_m has sides length $2^{-m}a$, $2^{-m}b$, $2^{-m}c$ $(m = 1, 2, \ldots, k)$,
(iii) $A \leqslant 4^m\left|\int_{\Gamma(T_m)} f\right|$ $(m = 1, 2, \ldots, k)$.

Split T_k into four triangles by the method used for T_0. Arguing as above we then see that one of the triangles, T_{k+1} say, has the property that

$$A \leqslant 4^k\left|\int_{\Gamma(T_k)} f\right| \leqslant 4^{k+1}\left|\int_{\Gamma(T_{k+1})} f\right|.$$

Since $T_{k+1} \subset T_k$ and T_{k+1} has sides length $2^{-(k+1)}a$, $2^{-(k+1)}b$, $2^{-(k+1)}c$, it follows from the principle of induction that there is a sequence of triangles $\{T_n\}$ such that

(i) $T_{n+1} \subset T_n$ $(n \in \mathbf{P})$,
(*ii*) T_n has sides length $2^{-n}a$, $2^{-n}b$, $2^{-n}c$ $(n \in \mathbf{P})$,
(iii) $A \leqslant 4^n\left|\int_{\Gamma(T_n)} f\right|$ $(n \in \mathbf{P})$.

It is readily verified that

$$\operatorname{diam}(T_n) \leqslant 2^{-n}(a + b + c) \qquad (n \in \mathbf{P})$$

so that $\lim_{n\to\infty} \operatorname{diam}(T_n) = 0$. By Proposition 2.3, $\bigcap\{T_n : n \in \mathbf{P}\}$ consists of a single point, α say, and then $\alpha \in T \subset D$.

Since D is open there is $\rho < 0$ with $\Delta(\alpha, \rho) \subset D$. Since f is differentiable at α there is $\eta : \Delta(\alpha, \rho) \to \mathbf{C}$ such that

$$f(z) = f(\alpha) + (z - \alpha)\{f'(\alpha) + \eta(z)\} \qquad (z \in \Delta(\alpha, \rho))$$

where η is continuous at α with $\eta(\alpha) = 0$. Since f is continuous on $\Delta'(\alpha, \rho)$ and

$$\eta(z) = \frac{f(z) - f(\alpha)}{z - \alpha} - f'(\alpha) \qquad (z \in \Delta'(\alpha, \rho))$$

it follows that η is continuous on $\Delta(\alpha, \rho)$. Given $\epsilon > 0$ there is δ such that $0 < \delta \leqslant \rho$ and

$$|\eta(z)| < \epsilon \qquad (z \in \Delta(\alpha, \delta)).$$

Appealing to Proposition 2.3 again we may choose $m \in \mathbf{P}$ such that $T_m \subset \Delta(\alpha, \delta)$. Using **CI 2** and **6** we obtain

$$\begin{aligned}\int_{\Gamma(T_m)} f(z)\,\mathrm{d}z &= \int_{\Gamma(T_m)} \{f(\alpha) + (z - \alpha)f'(\alpha)\}\,\mathrm{d}z + \int_{\Gamma(T_m)} (z - \alpha)\eta(z)\,\mathrm{d}z \\ &= \int_{\Gamma(T_m)} (z - \alpha)\eta(z)\,\mathrm{d}z.\end{aligned}$$

Finally by **CI 5** we have

$$\begin{aligned}0 \leqslant A \leqslant 4^m\left|\int_{\Gamma(T_m)} f\right| &\leqslant 4^m\left|\int_{\Gamma(T_m)} (z - \alpha)\eta(z)\,\mathrm{d}z\right| \\ &\leqslant 4^m l_{\Gamma(T_m)} \sup\{|z - \alpha|\,|\eta(z)| : z \in \Gamma(T_m)\} \\ &\leqslant 4^m 2^{-m}(a + b + c)2^{-m}(a + b + c)\epsilon \\ &= (a + b + c)^2\epsilon.\end{aligned}$$

Since ϵ was an arbitrary positive number we conclude that $A = 0$ and so the proof is complete.

It is possible to use a combinatorial argument to show that Cauchy's theorem holds for polygonal contours Γ such that $\Gamma^* \subset D$ (although the technical details are by no means simple). Given an arbitrary simple closed contour Γ with $\Gamma^* \subset D$ (note the tacit assumption of the Jordan curve theorem) and given $\epsilon > 0$, it is possible to show that there is a polygonal contour Γ_1 such that $\Gamma_1^* \subset D$ and

$$\left| \int_{\Gamma} f - \int_{\Gamma_1} f \right| < \epsilon.$$

This leads to a general form of Cauchy's theorem; but we shall follow a much simpler approach in the next section.

6.2 Cauchy's theorem for starlike domains

In this section we shall prove Cauchy's theorem for the case in which the domain D is starlike and Γ is any simple closed contour contained in D. In view of **CI 6** it is sufficient to prove that any function f in $\mathscr{A}(D)$ has a primitive on D. Analogy with real analysis suggests that we might obtain a primitive of f by integrating f in some sense. Given $\alpha \in D$ we wish to integrate f along some arc joining α to an arbitrary point of D. To understand something of the difficulty for the case of an arbitrary domain we consider an example in which we know it is impossible to find a primitive.

Let D be the domain $\mathbf{C} \setminus \{0\}$ and let $f = \mathbf{j}$. We know from Example 4.17 that $\mathbf{j}$ has no primitive on $\mathbf{C} \setminus \{0\}$. Take $\alpha = 1$ as the reference point for integrating $\mathbf{j}$. Along which arc should we integrate from 1 to -1? There are at least two natural candidates. Let Γ_1 be represented by $\gamma_1(t) = \exp(it)$ $(t \in [0, \pi])$. Then

$$\begin{aligned} \int_{\Gamma_1} \mathbf{j} &= \int_0^{\pi} \exp(-it) i \exp(it)\, dt \\ &= \pi i. \end{aligned}$$

Let Γ_2 be represented by $\gamma_2(t) = \exp(-it)$ $(t \in [0, \pi])$. Then

$$\begin{aligned} \int_{\Gamma_2} \mathbf{j} &= \int_0^{\pi} \exp(it)(-i) \exp(-it)\, dt \\ &= -\pi i. \end{aligned}$$

This indicates something of the difficulty in general and suggests that trouble always occurs around 'holes' in the domain.

Recall that a domain D is starlike if there is some point α in D such that $[\alpha, z] \subset D$ wherever $z \in D$. Such domains have of course no 'holes'. Moreover it is clear in this case that we should integrate along the arc $[\alpha, z]$. As we shall see in the theorem below, this reduces the problem of finding a primitive of $f \in \mathscr{A}(D)$ to an application of Cauchy's theorem for a triangle.

Theorem 6.2. *If D is a starlike domain then each f in $\mathscr{A}(D)$ has a primitive on D.*

Proof. Let α be a star centre for D and define g on D by

$$g(z) = \int_{[\alpha,z]} f \qquad (z \in D).$$

(We take $\int_{[\alpha,\alpha]} f = 0$, of course.) Let $\beta \in D \setminus \{\alpha\}$. Since D is open there is $r > 0$ such that $\Delta(\beta, r) \subset D$. Given $h \in \Delta'(0, r)$ we have $\beta + h \in D$ and so

$$g(\beta + h) = \int_{[\alpha,\beta+h]} f.$$

Clearly $[\beta, \beta + h] \subset \Delta(\beta, r) \subset D$, and so $[\alpha, w] \subset D$ for each w in $[\beta, \beta + h]$ since D is starred with respect to α. If $\alpha, \beta, \beta + h$ do not lie on the same straight line then the triangle with vertices $\alpha, \beta, \beta + h$ is contained in D (see Figure 6.2.). It follows from Theorem 6.1 that

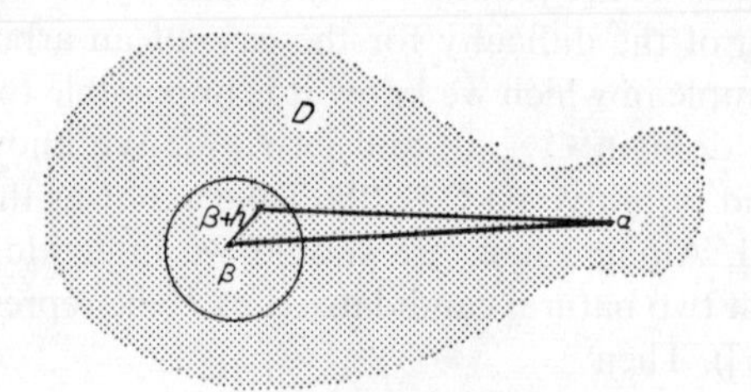

Figure 6.2

$$\int_{[\alpha,\beta]} f + \int_{[\beta,\beta+h]} f + \int_{[\beta+h,\alpha]} f = 0.$$

This statement is also clear by **CI 3** in the case when $\alpha, \beta, \beta + h$ lie on the same line. Therefore

$$g(\beta + h) - g(\beta) = \int_{[\beta,\beta+h]} f.$$

This latter statement is obvious in the case $\beta = \alpha$.

Given $\beta \in D, \Delta(\beta, r) \subset D, h \in \Delta'(0, r)$ we now have (using **CI 6, 2** and **5**)

$$\begin{aligned}\left|\frac{g(\beta + h) - g(\beta)}{h} - f(\beta)\right| &= \left|\frac{1}{h}\int_{[\beta,\beta+h]} f - \frac{f(\beta)}{h}\int_{[\beta,\beta+h]} \mathbf{1}\right| \\ &= \left|\frac{1}{h}\int_{[\beta,\beta+h]} \{f(z) - f(\beta)\}\, dz\right| \\ &\leqslant \sup\{|f(z) - f(\beta)| : z \in [\beta, \beta + h]\}.\end{aligned}$$

Since f is differentiable at β, it is also continuous at β. Given $\epsilon > 0$ there is then $\delta > 0$ such that $\delta \leqslant r$ and

$$|f(z) - f(\beta)| < \tfrac{1}{2}\epsilon \qquad (z \in \Delta(\beta, \delta)).$$

Given $h \in \Delta'(0, \delta)$ we now have

$$\sup\{|f(z) - f(\beta)| : z \in [\beta, \beta + h]\} \leqslant \sup\{|f(z) - f(\beta)| : z \in \Delta(\beta, \delta)\}$$
$$\leqslant \tfrac{1}{2}\epsilon < \epsilon.$$

This shows that g is differentiable at β with $g'(\beta) = f(\beta)$. Since β was any point of D this shows that g is a primitive of f as required.

Corollary. *Let D be any domain and let $f \in \mathscr{A}(D)$. Then f has a primitive locally on D, i.e. given $\alpha \in D$ there is $\delta > 0$ with $\Delta(\alpha, \delta) \subset D$ and $g \in \mathscr{A}(\Delta(\alpha, \delta))$ such that g is a primitive of $f|_{\Delta(\alpha,\delta)}$.*

Proof. Since D is open, given $\alpha \in D$ there is $\delta > 0$ such that $\Delta(\alpha, \delta) \subset D$. Since $\Delta(\alpha, \delta)$ is starlike and $f|_{\Delta(\alpha,\delta)} \in \mathscr{A}(\Delta(\alpha, \delta))$ the result follows from the theorem.

In the light of the above corollary it is instructive to consider again the function $\mathbf{j}$ on $\mathbf{C} \setminus \{0\}$. Certainly $\mathbf{j}$ has a primitive locally on $\mathbf{C} \setminus \{0\}$; in fact given $\Delta(\alpha, r) \subset \mathbf{C} \setminus \{0\}$, $\log_\theta|_{\Delta(\alpha,r)}$ is a primitive of $\mathbf{j}|_{\Delta(\alpha,r)}$ where $\theta = \pi + \arg(\alpha)$. But there is no way of 'fitting together' these local primitives to form a primitive on $\mathbf{C} \setminus \{0\}$.

We are now ready to state and prove Cauchy's theorem for the case of starlike domains. The proof is clear now that we have done the work in Theorems 6.1 and 6.2. This is not to detract from the depth of this remarkable theorem; the subsequent chapters will pay ample tribute to the fruitfulness of the theorem.

Theorem 6.3 (Cauchy). *If D is a starlike domain and Γ is any simple closed contour contained in D then*

$$\int_\Gamma f = 0 \qquad (f \in \mathscr{A}(D)).$$

Proof. By Theorem 6.2 f has a primitive g on D. Then by **CI 6**

$$\int_\Gamma f = \int_\Gamma g' = 0 \qquad (f \in \mathscr{A}(D))$$

for any simple closed contour Γ contained in D.

There is a sharper form of Cauchy's theorem in which the simple closed contour Γ can lie in the boundary of D (provided f is also continuous on D^-). This will be discussed in the next section, but one further comment is in order about the value of generalizations of Cauchy's theorem. If one is interested in Cauchy's theorem for itself, then one ought to investigate the most general possible form. If one wishes to use Cauchy's theorem only as a tool to obtain the results of complex function theory, one ought to know that simple forms of the theorem are then sufficient. Indeed, virtually every function theoretic result may be obtained by applying Cauchy's theorem for circles and for contours that are made up of line segments parallel to the axes. The version of Cauchy's theorem that we have chosen to prove is equally sufficient.

6.3 Applications

This section is concerned with some simple direct applications of Theorems 6.2 and 6.3. The first three applications are of interest in their own right; the last two are technical tools which will be used in the next chapter.

1. We can now extend some of the results in §4.5 on branches-of-log. Let D be a starlike domain such that $0 \notin D$. The student may establish that there is some $\theta \in \mathbf{R}$ such that $D \subset \mathbf{C} \setminus N_\theta$. It then follows that $\log_\theta|_D$ is a branch-of-log on D. A simple alternative proof is as follows. Since $\mathbf{j} \in \mathscr{A}(D)$, Theorem 6.2 implies that $\mathbf{j}$ has a primitive g on D. By Proposition 4.15 g differs by a constant from a branch-of-log on D.

There is a generalization of the concept of branch-of-log that is useful as a tool for some theorems (e.g. for one proof of the famous Riemann mapping theorem, see page 243). Let D be any domain and let $f \in \mathscr{A}(D)$ be such that $0 \notin f(D)$. We say that $g \in \mathscr{A}(D)$ is an *analytic-logarithm* of f if

$$\exp(g(z)) = f(z) \qquad (z \in D).$$

If f is the identity function on D then the concept of analytic-logarithm coincides with that of branch-of-log. If $f(D)$ is such that there is a branch-of-log on a domain containing $f(D)$, say h, then $h \circ f$ is clearly an analytic-logarithm of f. On the other hand there may well be an analytic-logarithm of f when there is no branch-of-log on any domain containing $f(D)$. To see this take $D = \mathbf{C} \setminus N$ and $f(z) = z^2$. It is an elementary exercise to show that $f(D) = \mathbf{C} \setminus \{0\}$. We proved in Example 4.17 that there is no branch-of-log on $\mathbf{C} \setminus \{0\}$. Now define

$$g(z) = 2 \log(z) \qquad (z \in \mathbf{C} \setminus N).$$

Then $g \in \mathscr{A}(\mathbf{C} \setminus N)$ and

$$\begin{aligned} \exp(g(z)) &= \exp(\log(z) + \log(z)) \\ &= \exp(\log(z)) \cdot \exp(\log(z)) \\ &= z^2 \qquad (z \in \mathbf{C} \setminus N) \end{aligned}$$

so that g is indeed an analytic-logarithm of f on $\mathbf{C} \setminus N$.

We now give a simple result on the existence of analytic-logarithms.

Proposition 6.4. *Let D be a starlike domain and let $f \in \mathscr{A}(D)$ be such that $0 \notin f(D)$ and $f' \in \mathscr{A}(D)$. Then f has an analytic-logarithm on D.*

Proof. Since $0 \notin f(D)$ we have that $f'/f \in \mathscr{A}(D)$ and so by Theorem 6.2, f'/f has a primitive on D, say h. Let $k = \dfrac{1}{f}(\exp \circ h)$ and we obtain

$$k' = \frac{1}{f}(\exp \circ h)\cdot\frac{f'}{f} + \left(\frac{-f'}{f^2}\right)\exp \circ h = 0.$$

Therefore k is constant on D, say $k = \beta\mathbf{1}$. Since exp has no zeros, $\beta \neq 0$, and since exp has range $\mathbf{C}\setminus\{0\}$, there is $\alpha \in \mathbf{C}$ such that $\exp(\alpha) = \beta$. It follows that $h - \alpha\mathbf{1} \in \mathscr{A}(D)$ and for each $z \in D$

$$\exp(h(z) - \alpha) = \beta f(z)\exp(-\alpha) = f(z).$$

Thus $h - \alpha\mathbf{1}$ is an analytic-logarithm of f on D.

Remark. We shall see in Theorem 8.1 that $f' \in \mathscr{A}(D)$ whenever $f \in \mathscr{A}(D)$.

2. It is occasionally useful to have a slightly stronger version of Cauchy's theorem than the one considered in Theorem 6.3. Let Γ be a starlike simple closed contour and let $f \in \mathscr{C}(\Gamma^*)$. If f is analytic on int (Γ) then $\int_\Gamma f = 0$. The proof of this sharpened version of Cauchy's theorem is outlined in Problem 6.2 and consists essentially in approximating Γ suitably from the inside. We illustrate the ideas involved by considering a very simple case.

Example 6.5. *Let $D = \{z\colon \operatorname{Re} z > 0\}$ and let $f \in \mathscr{C}(D^-)$ be analytic on D. Let Γ be the boundary of the rectangle K with vertices $0, i, 1 + i, 1$, with direction determined by the subarc $[0, i]$. Then $\int_\Gamma f = 0$.*

Proof. Given $0 < \delta < 1$ let Γ_δ be the boundary of the rectangle with vertices δ, $\delta + i$, $1 + i$, 1, with direction determined by the subarc $[\delta, \delta + i]$ (see Figure 6.3). Since D is a starlike domain and $\Gamma_\delta \subset D$,

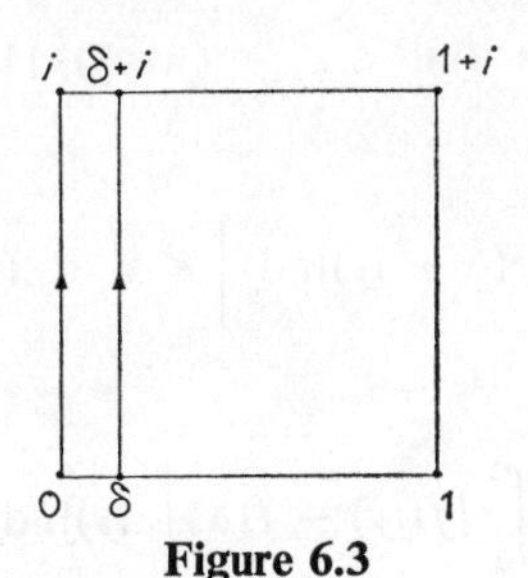

Figure 6.3

Theorem 6.3 gives

$$\int_{\Gamma_\delta} f = 0 \qquad (0 < \delta < 1).$$

We complete the proof by showing that

$$\int_\Gamma f = \lim_{\delta \to 0+} \int_{\Gamma_\delta} f.$$

Since f is continuous on the compact set K it is bounded and uniformly continuous on K. Suppose $|f(z)| \leqslant M$ $(z \in K)$. From **CI 3, 4** we obtain

$$\int_\Gamma f - \int_{\Gamma_\delta} f = \int_{[0,i]} f + \int_{[i,\delta+i]} f - \int_{[\delta,\delta+i]} f - \int_{[0,\delta]} f.$$

By **CI 5** we have

$$\left| \int_{[i,\delta+i]} f \right| \leqslant \delta M$$

so that

$$\lim_{\delta \to 0+} \int_{[i,\delta+i]} f = 0.$$

A similar argument gives

$$\lim_{\delta \to 0+} \int_{[0,\delta]} f = 0.$$

Further,

$$\int_{[0,i]} f - \int_{[\delta,\delta+i]} f = \int_0^1 \{f(iy) - f(\delta + iy)\} i \, \mathrm{d}y.$$

Given $\epsilon > 0$ it follows from the uniform continuity of f on K that there is $\delta_1 > 0$ such that

$$|f(z) - f(w)| < \epsilon, \qquad (z, w \in K, |z - w| < \delta_1).$$

In particular we have

$$|f(iy) - f(\delta + iy)| < \epsilon \qquad (y \in [0, 1], 0 < \delta < \delta_1)$$

and so

$$\left| \int_0^1 \{f(iy) - f(\delta + iy)\} i \, \mathrm{d}y \right| \leqslant \epsilon \qquad (0 < \delta < \delta_1).$$

This establishes that

$$\lim_{\delta \to 0+} \int_0^1 \{f(iy) - f(\delta + iy)\} i \, \mathrm{d}y = 0$$

and hence

$$\int_{\Gamma} f = \lim_{\delta \to 0+} \int_{\Gamma_\delta} f$$

as required.

3. We show next how Cauchy's theorem may be used to evaluate certain definite integrals in terms of other (known) integrals. As above we illustrate the general principle by a simple example; further examples are given in the problems.

Example 6.6. *Given $b > 0$ we have*

$$\int_0^\infty e^{-x^2} \cos(2bx)\, dx = e^{-b^2} \int_0^\infty e^{-x^2}\, dx,$$

$$\int_0^\infty e^{-x^2} \sin(2bx)\, dx = e^{-b^2} \int_0^b e^{x^2}\, dx.$$

Proof. It is well known from real analysis that each of the above integrals converges absolutely. Given $n \in \mathbf{P}$ let Γ_n be the boundary of the rectangle with vertices $0, n, n + ib, ib$ with direction determined by the subarc $[0, n]$ (see Figure 6.4.).

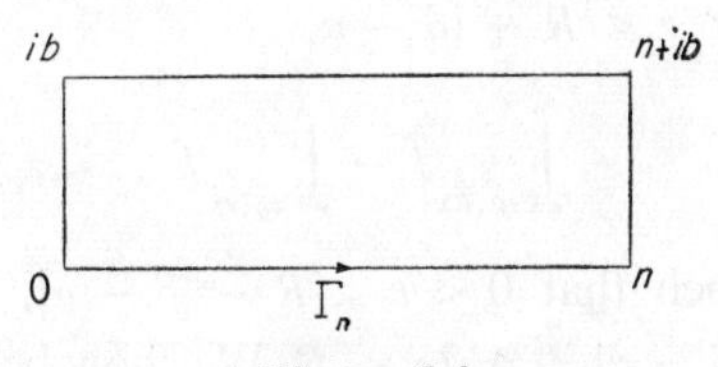

Figure 6.4

Let $f(z) = \exp(-z^2)$ $(z \in \mathbf{C})$. Since f is an entire function Theorem 6.3 gives

$$\int_{\Gamma_n} f = 0 \qquad (n \in \mathbf{P}).$$

By **CI 4** we have

$$\int_0^n e^{-x^2}\, dx + \int_{[n, n+ib]} f + \int_n^0 e^{-(x+ib)^2}\, dx + \int_b^0 e^{-(iy)^2}\, i\, dy = 0$$

and so

$$\int_0^n e^{-x^2}\, dx - e^{b^2} \int_0^n e^{-x^2} \cos(2bx)\, dx - i\, e^{b^2} \int_0^n e^{-x^2} \sin(2bx)\, dx$$
$$+ i \int_0^b e^{x^2}\, dx = - \int_{[n, n+ib]} f. \qquad (*)$$

By **CI 5** we have

$$\left|\int_{[n,n+ib]} f\right| \leqslant b \sup\{|\exp(-z^2)| : z \in [n, n+ib]\}$$
$$= b \sup\{e^{-n^2+y^2} : y \in [0, b]\}$$
$$= b\, e^{b^2} e^{-n^2}$$

so that

$$\lim_{n\to\infty} \int_{[n,n+ib]} f = 0.$$

Let $n \to \infty$ in (*) and then take real and imaginary parts to obtain the required results.

4. There is a formulation of Cauchy's theorem which states (roughly) that if $f \in \mathscr{A}(D)$ and $\Gamma \subset D$ then $\int_\Gamma f$ remains unchanged if Γ is 'continuously deformed' in D to Γ_1. The result below is a very special case of this and will be a key tool in the next chapter when we discuss functions that are analytic on a disc. Recall that circles are always given the positive direction.

Proposition 6.7. *Let* $f \in \mathscr{A}(D)$ *and let* $\bar{\Delta}(\alpha, R) \setminus \{\beta\} \subset D$ *where* $\beta \in \Delta(\alpha, R)$. *Then for* $0 < r < R - |\beta - \alpha|$

$$\int_{C(\alpha,R)} f = \int_{C(\beta,r)} f.$$

Proof. Given r such that $0 < r < R - |\beta - \alpha|$, choose ρ such that $0 < \sqrt{2}\rho < r$ and $|\beta - \alpha| + \sqrt{2}\rho < R$. We show below that

$$\int_{C(\alpha,R)} f = \int_{C(\beta,\rho)} f.$$

On taking $\alpha = \beta$, we obtain similarly

$$\int_{C(\beta,r)} f = \int_{C(\beta,\rho)} f$$

and the proof is then complete.

Let $\kappa, \lambda, \mu, \nu$ be as in Figure 6.5. Let Γ be the union of $[\kappa, \lambda]$, the minor arc from λ to μ, $[\mu, \nu]$, and the minor arc from ν to κ. It is readily verified that Γ is a simple closed contour; let its direction be determined by the minor arc from λ to μ. Let

$$w = \beta + \frac{\beta - \alpha}{|\beta - \alpha|}(\rho + i\rho).$$

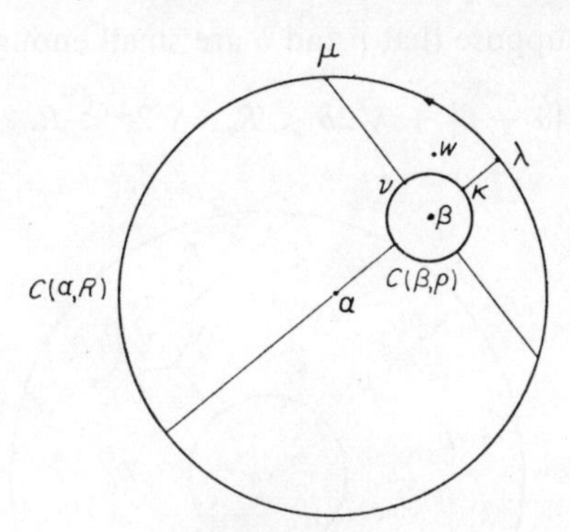

Figure 6.5

Then $|w - \alpha| \leqslant |\beta - \alpha| + \sqrt{2}\rho < R$, and $|w - \beta| = \sqrt{2}\rho > \rho$, so that $w \notin \Gamma$. The tangents from w to the circle $C(\beta, \rho)$ meet the circle at κ, ν. It follows that Γ is starred with respect to w (the doubting student should supply the analytical details). If $\epsilon = \text{dist}\,(\Gamma, (\mathbf{C} \setminus D) \cup \{\beta\})$ then Proposition 1.21 (iii) gives $\epsilon > 0$. Let D_ϵ be the starlike domain given by the Corollary to Theorem 5.8. Then $\Gamma \subset D_\epsilon \subset D$ and Theorem 6.3 gives $\int_\Gamma f = 0$. A similar argument applies to the three other simple closed contours of Figure 6.5 that are shaped as Γ. Addition of the four associated integrals now leads, via **CI 3, 4** to

$$\int_{C(\alpha,R)} f = \int_{C(\beta,\rho)} f$$

as required.

5. We consider finally a slight generalization of Proposition 6.7 which will be a key tool in the next chapter when we discuss functions that are analytic on a punctured disc, i.e. a domain of the form $\Delta'(\alpha, \rho)$.

Proposition 6.8. *Let* $f \in \mathscr{A}(D)$ *and let* $\bar{\Delta}(\alpha, R) \setminus \{\alpha, \beta\} \subset D$ *where* $\beta \in \Delta'(\alpha, R)$. *Then for* $0 < \delta < R - |\beta - \alpha|$, $0 < r < |\beta - \alpha| - \delta$

$$\int_{C(\beta,\delta)} f = \int_{C(\alpha,R)} f - \int_{C(\alpha,r)} f.$$

Proof. Let δ, r be such that $0 < \delta < R - |\beta - \alpha|$, $0 < r < |\beta - \alpha| - \delta$ (see Figure 6.6). Proposition 6.7 gives

$$\int_{C(\beta,\delta)} f = \int_{C(\beta,s)} f \qquad (0 < s \leqslant \delta)$$

$$\int_{C(\alpha,r)} f = \int_{C(\alpha,t)} f \qquad (0 < t \leqslant r).$$

We may therefore suppose that r and δ are small enough so that

$$|\alpha - \beta| + \sqrt{2}\delta < R, \quad \sqrt{2}r < R.$$

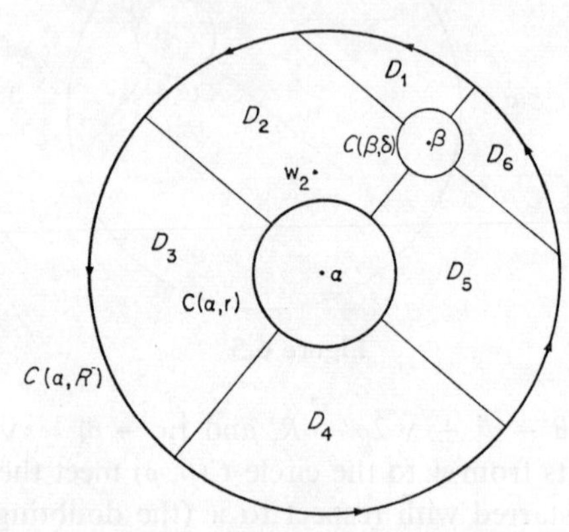

Figure 6.6

Let the domains D_k ($k = 1, \ldots, 6$) be as in Figure 6.6 and let $\Gamma_k = b(D_k)$. It is readily verified that each Γ_k is a simple closed contour; let Γ_k have the direction determined by its intersection with $C(\alpha, R)$. By arguing as in Proposition 6.7 we obtain

$$\int_{\Gamma_k} f = 0 \qquad (k = 1, 3, 4, 6).$$

Let

$$w_2 = \alpha + \frac{\beta - \alpha}{|\beta - \alpha|}(r + ir).$$

It is routine (though tedious) to verify that Γ_2 is starred with respect to w_2. Again arguing as in Proposition 6.7 we obtain

$$\int_{\Gamma_2} f = 0$$

and similarly

$$\int_{\Gamma_4} f = 0.$$

Combining the six integrals and using **CI 3, 4** we conclude that

$$\int_{C(\alpha,R)} f = \int_{C(\alpha,r)} f + \int_{C(\beta,\delta)} f$$

as required.

PROBLEMS 6

1. Let D be a domain and let $f \in \mathscr{A}(D)$ be such that $0 \notin f(D)$. If $g \in \mathscr{C}(D)$ and $\exp \circ g = f$, show that g is an analytic-logarithm of f. Show also that $\{g + 2k\pi i\mathbf{1}: k \in \mathbf{Z}\}$ is the set of all analytic-logarithms of f.

2. Let Γ be a starlike simple closed contour and let $f \in \mathscr{C}(\Gamma^*)$ be analytic on $\operatorname{int}(\Gamma)$. Use Problem 5.38 and Theorem 6.3 to show that $\int_\Gamma f = 0$.

3. Discuss the following 'proof' of Cauchy's theorem.
Let $h(\lambda) = \lambda \int_\Gamma f(\lambda z)\, \mathrm{d}z$ ($\lambda \in [0, 1]$). Then $h \in \mathscr{C}([0, 1])$ and

$$\begin{aligned} h'(\lambda) &= \int_\Gamma \{f(\lambda z) + \lambda z f'(\lambda z)\}\, \mathrm{d}z \\ &= \int_\Gamma \frac{\mathrm{d}}{\mathrm{d}z}\{z f(\lambda z)\}\, \mathrm{d}z \\ &= 0 \qquad (\lambda \in (0, 1)). \end{aligned}$$

Therefore $\int_\Gamma f = h(1) = h(0) = 0$.

4. Let $D = \{z: \operatorname{Re} z > 0\}$ and let $f \in \mathscr{C}(D^-)$ be analytic on D. If there is $M > 0$ such that

$$|f(z)| \leqslant \frac{M}{|z|^2} \qquad (z \in D^- \cap \nabla(0, 1))$$

show that $\int_{-\infty}^{\infty} f(iy)\, \mathrm{d}y = 0$.

5. Let θ_1, θ_2 be such that $-\pi < \theta_1 < 0 < \theta_2 < \pi$, and let the domain $D = \{z: \theta_1 < \arg(z) < \theta_2\}$. Let $f \in \mathscr{C}(D^-)$ be analytic on D. If there is $M > 0$ such that

$$|f(z)| \leqslant \frac{M}{|z|^2} \qquad (z \in D \cap \nabla(0, 1))$$

show that

$$\int_0^\infty f(e^{i\lambda}x)\, \mathrm{d}x = e^{-i\lambda} \int_0^\infty f(x)\, \mathrm{d}x \qquad (\theta_1 \leqslant \lambda \leqslant \theta_2).$$

(i) Given that $\int_0^\infty e^{-x^2}\, \mathrm{d}x = \frac{1}{2}\sqrt{\pi}$, evaluate

$$\int_0^\infty e^{-x^2} \cos(x^2)\, \mathrm{d}x, \qquad \int_0^\infty e^{-x^2} \sin(x^2)\, \mathrm{d}x \qquad \left(\textit{Hint}. \text{ Take } \lambda = \frac{\pi}{8}\right).$$

(ii) Take $f(z) = (\exp(i\alpha z))(1+z)^{-2}$, $\lambda = \dfrac{\pi}{2}$ to show that

$$\int_0^\infty \frac{\cos(\alpha x)}{(1+x)^2}\,dx + \alpha\int_0^\infty \frac{e^{-\alpha y}}{1+y^2}\,dy = 1 \qquad (\alpha \geqslant 0).$$

6. If $f(z) = \exp(-z^2)$ show that

$$\lim_{R\to+\infty}\int_0^{\pi/2} f(Re^{i\theta})i\,Re^{i\theta}\,d\theta = 0$$

and deduce that

$$\int_0^\infty \cos(x^2)\,dx = \int_0^\infty \sin(x^2)\,dx = \frac{1}{2}\sqrt{\frac{\pi}{2}}.$$

7. Establish the following results by integrating the indicated function around a suitable rectangular contour.

(i) $$\int_0^\infty \frac{1-b^2+x^2}{(1-b^2+x^2)^2+4b^2x^2}\,dx = \frac{\pi}{2} \qquad (-1 < b < 1)$$

$$\int_0^\infty \frac{x}{(1-b^2+x^2)^2+4b^2x^2}\,dx = \frac{1}{4b}\log\frac{1+b}{1-b}$$

$$(-1 < b < 1,\, b \neq 0). \qquad \left(f(z) = \frac{1}{1+z^2}.\right)$$

(ii) $$\int_{-\infty}^\infty e^{-x^2}\,\mathrm{Im}\,\{e^{-2ix}p(x+i)\}\,dx = 0$$

for any polynomial function p with real coefficients. ($f(z) = \exp(-z^2)p(z)$.)

8. Suppose $\bar{\Delta}(\beta, r) \subset \Delta(\alpha, R)$. If f is continuous on $\bar{\Delta}(\alpha, R) \setminus \Delta(\beta, r)$ and analytic on $\Delta(\alpha, R) \setminus \bar{\Delta}(\beta, r)$ show that

$$\int_{C(\alpha,R)} f = \int_{C(\beta,r)} f.$$

9. Suppose $\bar{\Delta}(\beta, \delta) \subset A(\alpha; R, r)$. If f is continuous on $\bar{A}(\alpha; R, r) \setminus \Delta(\beta, \delta)$ and analytic on $A(\alpha; R, r) \setminus \bar{\Delta}(\beta, \delta)$, show that

$$\int_{C(\beta,\delta)} f = \int_{C(\alpha,R)} f - \int_{C(\alpha,r)} f.$$

10. Let p be a polynomial function with $\deg(p) \geqslant 2$. If the zeros of p all lie in $\Delta(0, M)$, use Problem 8 above and **CI 5** to show that

$$\int_{C(0,M)} \frac{1}{p} = 0.$$

7

LOCAL ANALYSIS

In this chapter we obtain some of the properties of functions that are analytic on an open disc or an open disc punctured at its centre. The results may thus be used to describe the local behaviour of a function that is analytic on some domain; for given any point of the domain we may choose an open disc about the point that is contained in the domain. It is for this reason that we have entitled the chapter 'local analysis'. Most of the results of this chapter are preparatory for the study in Chapter 8 of the global behaviour of a function that is analytic on some domain. We shall see there that topological difficulties may arise in the global analysis if the domain has a complicated shape. These topological difficulties do not arise in this chapter because of the simple nature of open discs.

We show that a function which is analytic on an open disc is a power series function, and that a function which is analytic on an open disc punctured at its centre is a quasi power series function. The latter result leads to an analysis of the isolated singularities of a complex function. The key tool throughout is the use of certain integral formulae due to Cauchy.

7.1 Cauchy's integral formulae

We begin this section by establishing the famous Cauchy integral formula for a function analytic on a disc. The formula enables us to recover the values of the function inside a circle by an appropriate integration around the circle. With the aid of the formula we can then establish that such functions are infinitely differentiable and that their derivatives are also given by integral formulae.

Theorem 7.1. (Cauchy). *Let* $f \in \mathscr{A}(\Delta(\alpha, R))$ *and let* $0 < r < R$. *Then*

$$f(z) = \frac{1}{2\pi i} \int_{C(\alpha,r)} \frac{f(w)}{w - z} \, \mathrm{d}w \qquad (z \in \Delta(\alpha, r)).$$

Proof. Let $z \in \Delta(\alpha, r)$, let $D = \Delta(\alpha, R) \setminus \{z\}$ and let

$$F(w) = \frac{f(w)}{w - z} \qquad (w \in D).$$

It is clear that $F \in \mathscr{A}(D)$. Since $z \in \Delta(\alpha, r)$ we may choose $\rho_0 > 0$ such that $\bar{\Delta}(z, \rho_0) \subset \Delta(\alpha, r)$. For $0 < \rho < \rho_0$, Proposition 6.7 gives

$$\begin{aligned}\int_{C(\alpha,r)} F &= \int_{C(z,\rho)} F \\ &= \int_{C(z,\rho)} \frac{f(w) - f(z)}{w - z}\, \mathrm{d}w + f(z) \int_{C(z,\rho)} \frac{\mathrm{d}w}{w - z}\end{aligned}$$

and thus

$$\begin{aligned}\left| \int_{C(\alpha,r)} F - 2\pi i f(z) \right| &= \left| \int_{C(z,\rho)} \frac{f(w) - f(z)}{w - z}\, \mathrm{d}w \right| \\ &\leqslant 2\pi\rho \sup \left\{ \frac{|f(w) - f(z)|}{\rho} : w \in C(z, \rho) \right\} \\ &= 2\pi \sup\{|f(w) - f(z)| : w \in C(z, \rho)\}.\end{aligned}$$

Since f is continuous at z, given $\epsilon > 0$ there is $\delta > 0$ such that $\delta < \rho_0$ and $|f(w) - f(z)| < \epsilon$ $(w \in \Delta(z, \delta))$. For $0 < \rho < \delta$ we now have

$$\sup\{|f(w) - f(z)| : w \in C(z, \rho)\} \leqslant \epsilon$$

and so

$$\left| \int_{C(\alpha,r)} F - 2\pi i f(z) \right| \leqslant 2\pi\epsilon \qquad (\epsilon > 0).$$

It follows that

$$2\pi i f(z) = \int_{C(\alpha,r)} \frac{f(w)}{w - z}\, \mathrm{d}w.$$

Since z was any point of $\Delta(\alpha, r)$, the proof is complete.

Corollary. $f(\alpha) = \dfrac{1}{2\pi} \displaystyle\int_0^{2\pi} f(\alpha + r\mathrm{e}^{i\theta})\, \mathrm{d}\theta.$

Proof. To evaluate the integral stated in the theorem take $w(\theta) = \alpha + r\,\mathrm{e}^{i\theta}$ $(\theta \in [0, 2\pi])$. Then

$$\begin{aligned}f(\alpha) &= \frac{1}{2\pi i} \int_{C(\alpha,r)} \frac{f(w)}{w - \alpha}\, \mathrm{d}w \\ &= \frac{1}{2\pi i} \int_0^{2\pi} \frac{f(\alpha + r\mathrm{e}^{i\theta})}{r\,\mathrm{e}^{i\theta}}\, ir\,\mathrm{e}^{i\theta}\, \mathrm{d}\theta \\ &= \frac{1}{2\pi} \int_0^{2\pi} f(\alpha + r\mathrm{e}^{i\theta})\, \mathrm{d}\theta.\end{aligned}$$

The theorem asserts that the values of f on $\Delta(\alpha, r)$ are given by an 'averaging process' on its values on $C(\alpha, r)$. The corollary shows in particular that the value at the centre of the circle is simply the mean of the values on the circle.

Another simple consequence of Cauchy's formula is as follows. Suppose that $f, g \in \mathscr{A}(\Delta(\alpha, R))$ and suppose that f and g agree on $C(\alpha, r)$. Then f, g also agree on $\Delta(\alpha, r)$. An extremely useful generalization of this observation will be given in Theorem 8.4.

As a less trivial application of Cauchy's formula we show that if $f \in \mathscr{A}(\Delta(\alpha, R))$ then f is infinitely differentiable. Clearly we want to differentiate 'under the integral sign' in Cauchy's formula. We begin with a lemma.

Lemma 7.2. *Let g be a continuous complex function on $C(\alpha, r)$, let $n \in \mathbf{P}$ and let*

$$h(z) = \int_{C(\alpha,r)} \frac{g(w)}{(w - z)^n}\, \mathrm{d}w \qquad (z \in \Delta(\alpha, r)).$$

Then $h \in \mathscr{A}(\Delta(\alpha, r))$ and

$$h'(z) = \int_{C(\alpha,r)} \frac{ng(w)}{(w - z)^{n+1}}\, \mathrm{d}w \qquad (z \in \Delta(\alpha, r)).$$

Proof. Let $z_0 \in \Delta(\alpha, r)$. Choose $\rho < r$ such that $z_0 \in \Delta(\alpha, \rho)$ and let $z_1 \in \Delta(\alpha, \rho)$. Given $w \in C(\alpha, r)$ we have

$$\left|\frac{1}{(w - z_1)^n} - \frac{1}{(w - z_0)^n}\right| = \left|\frac{(w - z_0)^n - (w - z_1)^n}{(w - z_0)^n(w - z_1)^n}\right|$$

$$= \left|\frac{(z_1 - z_0)}{(w - z_0)^n(w - z_1)^n}\{(w - z_0)^{n-1} + (w - z_0)^{n-2}(w - z_1) + \cdots + (w - z_1)^{n-1}\}\right|$$

$$\leqslant \frac{|z_1 - z_0|}{(r - \rho)^{2n}}\, n(2r)^{n-1}.$$

Since g is continuous on the compact set $C(\alpha, r)$, it is bounded, say $|g| \leqslant M$. It follows from **CI 5** that

$$|h(z_1) - h(z_0)| = \left|\int_{C(\alpha,r)} \left\{\frac{g(w)}{(w - z_1)^n} - \frac{g(w)}{(w - z_0)^n}\right\} \mathrm{d}w\right|$$

$$\leqslant 2\pi r \frac{Mn(2r)^{n-1}}{(r - \rho)^{2n}}\, |z_1 - z_0|.$$

This holds in particular for $z_1 \in \Delta(z_0, \rho - |z_0|)$ and so it is clear that h is continuous at z_0.

We now have

$$\frac{h(z_1) - h(z_0)}{z_1 - z_0} = \sum_{j=1}^{n} \int_{C(\alpha,r)} \frac{g(w)}{(w - z_0)^{n-j+1}(w - z_1)^j}\, dw.$$

Applying the above continuity result to each integral we see that h is differentiable at z_0 and

$$h'(z_0) = \sum_{j=1}^{n} \int_{C(\alpha,r)} \frac{g(w)}{(w - z_0)^{n+1}}\, dw$$
$$= \int_{C(\alpha,r)} \frac{ng(w)}{(w - z_0)^{n+1}}\, dw.$$

Since z_0 was any point of $\Delta(\alpha, r)$, the proof is now complete.

Theorem 7.3 (Cauchy). *Let* $f \in \mathscr{A}(\Delta(\alpha, R))$. *Then* f *is infinitely differentiable on* $\Delta(\alpha, R)$, *and for* $n \in \mathbf{P}$, $0 < r < R$,

$$f^{(n)}(z) = \frac{n!}{2\pi i} \int_{C(\alpha,r)} \frac{f(w)}{(w - z)^{n+1}}\, dw \qquad (z \in \Delta(\alpha, r)).$$

Proof. Let $z \in \Delta(\alpha, r)$, where $0 < r < R$. By Theorem 7.1

$$f(z) = \frac{1}{2\pi i} \int_{C(\alpha,r)} \frac{f(w)}{w - z}\, dw.$$

Since f is continuous on $C(\alpha, r)$ it follows from Lemma 7.2 that

$$f'(z) = \frac{1}{2\pi i} \int_{C(\alpha,r)} \frac{f(w)}{(w - z)^2}\, dw.$$

We now proceed by induction. Suppose that f has k derivatives on $\Delta(\alpha, r)$ and that

$$f^{(k)}(z) = \frac{k!}{2\pi i} \int_{C(\alpha,r)} \frac{f(w)}{(w - z)^{k+1}}\, dw \qquad (z \in \Delta(\alpha, r)).$$

It follows from Lemma 7.2 that $f^{(k)}$ is analytic on $\Delta(\alpha, r)$ and

$$f^{(k+1)}(z) = \frac{(k + 1)!}{2\pi i} \int_{C(\alpha,r)} \frac{f(w)}{(w - z)^{k+2}}\, dw \qquad (z \in \Delta(\alpha, r)).$$

The result of the theorem now follows by induction.

The above theorem marks the beginning of an increasing divergence between the results of real and complex analysis. Recall that a real differentiable function on $(-1, 1)$ need not be twice differentiable at any point of $(-1, 1)$. In fact let g be a continuous real function on $(-1, 1)$ that is nowhere differentiable,* and let

$$f(x) = \int_0^x g(t)\,\mathrm{d}t \qquad (x \in (-1, 1)).$$

By the fundamental theorem of calculus f is differentiable on $(-1, 1)$ and $f' = g$, so that f is nowhere twice differentiable. This situation cannot occur for functions that are analytic on a disc. In other words, the condition that a function be analytic on a disc is much more restrictive than might be indicated by experience with real analysis.

PROBLEMS 7

1. Let $f \in \mathscr{C}(\bar{\Delta}(\alpha, R))$ be analytic on $\Delta(\alpha, R)$. Show that

$$f(z) = \frac{1}{2\pi i}\int_{C(\alpha,R)} \frac{f(w)}{w - z}\,\mathrm{d}w \qquad (z \in \Delta(\alpha, R)).$$

Does the analogous formula hold for the derivatives of f?

2. Let $K(w, z) = \dfrac{1}{2\pi i(w - z)}$ $(w - z \neq 0)$, and let $\Gamma = C(\alpha, r)$. Given $f \in \mathscr{C}(\Gamma)$ define Tf on $\mathbf{C} \setminus \Gamma$ by

$$(Tf)(z) = \int_\Gamma K(w, z) f(w)\,\mathrm{d}w \qquad (z \in \mathbf{C} \setminus \Gamma).$$

Show that Tf is analytic on $\mathbf{C} \setminus \Gamma$, and vanishes on ext (Γ). The mapping $f \to Tf$ is an example of an *integral transformation*. The function K is called the *kernel* of the transformation. Other important examples include the Laplace and Fourier transforms, cf. Examples 7.5, 8.21.

3. Let $g = T\mathbf{j}$, where T is as above. Show that

$$\lim_{\rho \to r-} g(\alpha + \rho e^{i\theta}) \neq \frac{1}{\alpha + re^{i\theta}}.$$

(cf. Theorem 9.18)

4. Show that

$$\int_0^{2\pi} \log(1 + re^{i\theta})\,\mathrm{d}\theta = 0 \qquad (0 < r < 1)$$

and deduce that

$$\int_0^{\pi/2} \log \sin x\,\mathrm{d}x = -\frac{\pi}{2}\log 2.$$

* See, for example, E. C. Titchmarsh, *The Theory of Functions* (Oxford, second edition), §11.22.

7.2 Taylor expansions

In this section we use the Cauchy integral formulae to derive the Taylor expansions of functions that are analytic on a disc. We consider separately the finite and infinite expansions. The finite expansions are more useful for applications.

The following well-known identities will figure significantly in the proofs.

K.1. $$\frac{1}{w-z} = \sum_{j=0}^{n-1} \frac{(z-\alpha)^j}{(w-\alpha)^{j+1}} + \frac{(z-\alpha)^n}{(w-\alpha)^n(w-z)}.$$

K.2. $$\frac{1}{w-z} = \sum_{j=0}^{\infty} \frac{(z-\alpha)^j}{(w-\alpha)^{j+1}} \qquad (|w-\alpha| > |z-\alpha|)$$

the convergence being uniform for $|w - \alpha| \geqslant \rho$ provided that $\rho > |z - \alpha|$.

Theorem 7.4 (Taylor). *Let $f \in \mathscr{A}(\Delta(\alpha, R))$ and let $n \in \mathbf{P}$. Then there is $f_n \in \mathscr{A}(\Delta(\alpha, R))$ such that $f_n(\alpha) = \frac{1}{n!}f^{(n)}(\alpha)$ and*

$$f(z) = f(\alpha) + \sum_{j=1}^{n-1} \frac{f^{(j)}(\alpha)}{j!}(z-\alpha)^j + (z-\alpha)^n f_n(z) \qquad (z \in \Delta(\alpha, R)).$$

Proof. Define f_n on $\Delta(\alpha, R)$ by $f_n(\alpha) = \frac{1}{n!}f^{(n)}(\alpha)$ and

$$f_n(z) = \frac{1}{(z-\alpha)^n}\left\{f(z) - f(\alpha) - \sum_{j=1}^{n-1} \frac{f^{(j)}(\alpha)}{j!}(z-\alpha)^j\right\} \qquad (z \in \Delta'(\alpha, R)).$$

It is clear that f_n is analytic on $\Delta'(\alpha, R)$ and it is now sufficient to show that f_n is differentiable at α. Using Theorems 7.1, 7.3 and **K.1**, we have for $z \in \Delta(\alpha, r)$ $(0 < r < R)$

$$\begin{aligned} f(z) &= \frac{1}{2\pi i}\int_{C(\alpha,r)} \frac{f(w)}{w-z}\,\mathrm{d}w \\ &= \sum_{j=0}^{n-1} \frac{(z-\alpha)^j}{2\pi i}\int_{C(\alpha,r)} \frac{f(w)}{(w-\alpha)^{j+1}}\,\mathrm{d}w \\ &\qquad + \frac{(z-\alpha)^n}{2\pi i}\int_{C(\alpha,r)} \frac{f(w)}{(w-\alpha)^n(w-z)}\,\mathrm{d}w \\ &= f(\alpha) + \sum_{j=1}^{n-1} \frac{f^{(j)}(a)}{j!}(z-\alpha)^j + (z-\alpha)^n g_n(z) \end{aligned}$$

where

$$g_n(z) = \frac{1}{2\pi i}\int_{C(\alpha,r)} \frac{f(w)}{(w-\alpha)^n(w-z)}\,\mathrm{d}w \qquad (z \in \Delta(\alpha, r)).$$

It is clear that $g_n(z) = f_n(z)$ $(z \in \Delta'(\alpha, r))$, and Theorem 7.3 gives $g_n(\alpha) = f_n(\alpha)$. It is also clear from Lemma 7.2 that g_n is differentiable at α. Therefore f_n is differentiable at α, as required.

Corollary. *If f is an entire function the above result holds with $\Delta(\alpha, R)$ replaced by $\mathbf{C}$.*

As an application of the usefulness of the finite Taylor expansion we shall show that the finite Laplace transform of an integrable function is an entire function.

Example 7.5. *Let $a, b \in \mathbf{R}$, $a < b$, let f be integrable on $[a, b]$ and let*

$$F(z) = \int_a^b e^{-zt} f(t)\, dt \qquad (z \in \mathbf{C}).$$

Then F is an entire function.

Proof. Since exp is an entire function, it follows from the above corollary that there is an entire function g such that

$$\exp(w) = 1 + w + w^2 g(w) \qquad (w \in \mathbf{C}).$$

Given $z, h \in \mathbf{C}$, $h \neq 0$, we have

$$\begin{aligned} \frac{F(z+h) - F(z)}{h} &= \int_a^b \frac{e^{-zt}e^{-ht} - e^{-zt}}{h} f(t)\, dt \\ &= \int_a^b \frac{e^{-ht} - 1}{-ht}(-t)\, e^{-zt} f(t)\, dt \\ &= \int_a^b [1 - htg(-ht)](-t)\, e^{-zt} f(t)\, dt \end{aligned}$$

and therefore

$$\left|\frac{F(z+h) - F(z)}{h} + \int_a^b t\, e^{-zt} f(t)\, dt\right| = |h| \left|\int_a^b t^2 e^{-zt} g(-ht) f(t)\, dt\right|.$$

Since g is entire, it is bounded on any bounded subset of $\mathbf{C}$. In particular if we restrict h to lie in $\Delta'(0, 1)$ we may obtain $M_z > 0$ such that

$$\left|\int_a^b t^2 e^{-zt} g(-ht) f(t)\, dt\right| \leqslant M_z \int_a^b |f(t)|\, dt.$$

It follows that F is differentiable at z. Since z was any point of $\mathbf{C}$, F is an entire function and

$$F'(z) = \int_a^b -t\, e^{-zt} f(t)\, dt \qquad (z \in \mathbf{C}).$$

Observe that we have again justified the differentiation 'under the integral sign' (cf. Problem 7.9).

Theorem 7.6 (Taylor). *Let* $f \in \mathscr{A}(\Delta(\alpha, R))$. Then

(i) $f(z) = f(\alpha) + \sum_{j=1}^{\infty} \frac{f^{(j)}(\alpha)}{j!}(z-\alpha)^j \qquad (z \in \Delta(a, R))$

the series converging uniformly on $\bar{\Delta}(\alpha, r)$ $(0 < r < R)$;

(ii) if $f(z) = \sum_{j=0}^{\infty} \alpha_j (z-\alpha)^j$ $(z \in \Delta(\alpha, R))$, *then*

$$\alpha_0 = f(\alpha), \qquad \alpha_j = \frac{f^{(j)}(\alpha)}{j!} \qquad (j \in \mathbf{P}).$$

Proof. (i) Let $z \in \Delta(\alpha, R)$ so that there is r such that $|z - \alpha| < r < R$. It follows from **K.2** that

$$\frac{1}{w - z} = \sum_{j=0}^{\infty} \frac{(z-\alpha)^j}{(w-\alpha)^{j+1}} \qquad (w \in C(\alpha, r))$$

the convergence being uniform on $C(\alpha, r)$. Using Theorems 7.1, 7.3 and **CI 7** we now have

$$\begin{aligned} f(z) &= \frac{1}{2\pi i}\int_{C(\alpha,r)} \frac{f(w)}{w-z}\,dw \\ &= \frac{1}{2\pi i}\int_{C(\alpha,r)} f(w)\sum_{j=0}^{\infty}\frac{(z-\alpha)^j}{(w-\alpha)^{j+1}}\,dw \\ &= \sum_{j=0}^{\infty}(z-\alpha)^j \frac{1}{2\pi i}\int_{C(\alpha,r)}\frac{f(w)}{(w-\alpha)^{j+1}}\,dw \\ &= f(\alpha) + \sum_{j=1}^{\infty}\frac{f^{(j)}(\alpha)}{j!}(z-\alpha)^j. \end{aligned}$$

The required uniform convergence follows from Proposition 4.8.

(ii) This follows from Theorem 4.10.

Corollary. *If f is an entire function the above result holds with $\Delta(\alpha, R)$ replaced by* **C**.

The expansion of f in part (i) above is called the *Taylor expansion* of f at α, and the numbers α_j $\left(= \frac{f^{(j)}(\alpha)}{j!}\right)$ are called the *Taylor coefficients* of f at α.

We have now shown that any function which is analytic on a disc is a power series function. Moreover the power series expansion is unique. This last remark is not simply of academic interest, as the following example demonstrates.

Example 7.7.

$$\frac{1}{1-z-z^2} = \sum_{n=0}^{\infty} \alpha_n z^n \qquad \left(z \in \Delta\left(0, \frac{\sqrt{5}-1}{2}\right)\right)$$

where the α_n are the Fibonacci numbers *defined by*

$$\alpha_0 = \alpha_1 = 1, \qquad \alpha_{n+1} = \alpha_n + \alpha_{n-1} \qquad (n \in \mathbf{P}).$$

Moreover,

$$\alpha_n = \frac{1}{\sqrt{5}}\left[\left(\frac{1+\sqrt{5}}{2}\right)^{n+1} - \left(\frac{1-\sqrt{5}}{2}\right)^{n+1}\right] \qquad (n \in \mathbf{P}).$$

Proof. We have $1 - z - z^2 = 0$ iff $z = \lambda$ or μ where $\lambda = \frac{1}{2}(\sqrt{5}-1)$, $\mu = -\frac{1}{2}(\sqrt{5}+1)$. Thus if

$$f(z) = \frac{1}{1-z-z^2} \qquad (z \in \Delta(0, \lambda))$$

then $f \in \mathscr{A}(\Delta(0, \lambda))$ and so by Theorem 7.6

$$\frac{1}{1-z-z^2} = \sum_{n=0}^{\infty} \alpha_n z^n \qquad (z \in \Delta(0, \lambda))$$

where the α_n are the Taylor coefficients of f at 0. In particular $\alpha_0 = f(0) = 1$. To calculate the remaining α_n by direct differentiation is a formidable task. We proceed instead by the method of 'undetermined coefficients'.

We have

$$1 = (1 - z - z^2)(\alpha_0 + \alpha_1 z + \alpha_2 z^2 + \cdots) \qquad (z \in \Delta(0, \lambda)).$$

By the uniqueness of the Taylor expansion for the constant function **1** we may equate the coefficients of z^n on both sides. This gives

$$\alpha_0 = \alpha_1 = 1, \qquad \alpha_{n+1} = \alpha_n + \alpha_{n-1} \qquad (n \in \mathbf{P})$$

as required. For $z \in \Delta(0, \lambda)$ we also have

$$\begin{aligned} f(z) &= \frac{-1}{(z-\lambda)(z-\mu)} \\ &= \frac{1}{\mu - \lambda}\left[\frac{1}{z-\lambda} - \frac{1}{z-\mu}\right] \\ &= \frac{1}{\sqrt{5}}\left[\sum_{n=0}^{\infty} \frac{z^n}{\lambda^{n+1}} - \sum_{n=0}^{\infty} \frac{z^n}{\mu^{n+1}}\right] \end{aligned}$$

The uniqueness of the Taylor expansion of f now gives for each $n \in \mathbf{P}$

$$\begin{aligned} a_n &= \frac{1}{\sqrt{5}}\left[\frac{1}{\lambda^{n+1}} - \frac{1}{\mu^{n+1}}\right] \\ &= \frac{1}{\sqrt{5}}\left[\left(\frac{2(\sqrt{5}+1)}{5-1}\right)^{n+1} - \left(\frac{-2(\sqrt{5}-1)}{5-1}\right)^{n+1}\right] \\ &= \frac{1}{\sqrt{5}}\left[\left(\frac{1+\sqrt{5}}{2}\right)^{n+1} - \left(\frac{1-\sqrt{5}}{2}\right)^{n+1}\right] \end{aligned}$$

as required.

PROBLEMS 7

5. Let $f: \mathbf{R} \to \mathbf{R}$ be defined by $f(0) = 0$ and $f(x) = e^{-1/x^2}$ $(x \neq 0)$. Show that f is infinitely differentiable on $\mathbf{R}$ and that $f^{(n)}(0) = 0$ $(n \in \mathbf{P})$. Thus the Taylor expansion of f at 0 agrees with f only at the point 0.

6. (i) Given that $f \in \mathscr{A}(\Delta(0, R))$ is real valued on $(-R, R)$ show that the Taylor coefficients of f at 0 are real and that

$$f(\bar{z}) = \overline{f(z)} \qquad (z \in \Delta(0, R)).$$

If further f takes pure imaginary values on $(-iR, iR)$ show that $f(-z) = -f(z)$ $(z \in \Delta(0, R))$.

(ii) Given $f \in \mathscr{A}(\Delta(0, R))$ show that there exist $f_1, f_2 \in \mathscr{A}(\Delta(0, R))$, real valued on $(-R, R)$ and such that

$$f = f_1 + if_2.$$

7. Let

$$f(z) = \sum_{n=0}^{\infty} \alpha_n (z - \alpha)^n \qquad (z \in \Delta(\alpha, R))$$

and let

$$M(r) = \max\{|f(z)| : z \in C(\alpha, r)\} \qquad (0 < r < R)$$

(i) (Cauchy's coefficient inequalities.) Show that

$$|\alpha_n| \leqslant \frac{M(r)}{r^n} \qquad (0 < r < R, n = 0, 1, 2, \ldots).$$

(ii) (Parseval's identity.) Show that

$$\frac{1}{2\pi}\int_0^{2\pi} |f(\alpha + re^{i\theta})|^2 \, d\theta = \sum_{n=0}^{\infty} |\alpha_n|^2 r^{2n} \qquad (0 < r < R)$$

and deduce that

$$\sum_{n=0}^{\infty} |\alpha_n|^2 r^{2n} \leqslant M(r)^2 \qquad (0 < r < R).$$

(iii) Suppose there is $t \in \mathbf{R}$ such that

$$|f(z)| \leqslant (R - |z - \alpha|)^t \qquad (z \in \Delta(\alpha, R)).$$

If $t > 0$ show that $f = 0$. If $t \leqslant 0$ show that

$$|\alpha_n| < \frac{\mathrm{e}}{R^{n-t}(n+1)^t} \qquad (n = 0, 1, 2, \ldots)$$

(*Hint*. For the final part use (i) with $r = \dfrac{n}{n+1} R$.)

8. Prove Example 7.5 by using the infinite expansion for e^{-zt}.

9. Let Γ be an oriented piecewise smooth arc or a simple closed contour. Let H be continuous on $\Gamma \times \Delta(\alpha, R)$ and such that the function $z \to H(w, z)$ is analytic on $\Delta(\alpha, R)$ for each $w \in \Gamma$. If

$$f(z) = \int_\Gamma H(z, w)\,\mathrm{d}w \qquad (z \in \Delta(\alpha, R))$$

show that $f \in \mathscr{A}(\Delta(\alpha, R))$ and

$$f'(z) = \int_\Gamma \frac{\partial H}{\partial z}(w, z)\,\mathrm{d}w \qquad (z \in \Delta(\alpha, R)).$$

10. Given $t \in (-1, 1)$ show that

$$p^{-\frac{1}{2}}(1 - 2tz + z^2) = 1 + \sum_{n=1}^{\infty} P_n(t) z^n \qquad (z \in \Delta(0, 1))$$

where P_n is a polynomial function of degree n. (The polynomials P_n are called the *Legendre polynomials*.)

11. Let $f(z) = \exp(z^2) \int_{[0,z]} \exp(-w^2)\,\mathrm{d}w$ $(z \in \mathbf{C})$. Show that f is an entire function and that

$$f'(z) = 1 + 2zf(z) \qquad (z \in \mathbf{C}).$$

Deduce that

$$f(z) = \sum_{n=1}^{\infty} \frac{n!2^{2n-1}}{(2n)!} z^{2n-1} \qquad (z \in \mathbf{C}).$$

7.3 The Laurent expansion

In this section we shall derive the Laurent expansion of a function that is analytic on a disc punctured at its centre, say $\Delta'(\alpha, R)$. We shall see in particular that such functions are quasi power series functions and that they may be represented as the sum of two functions, one analytic on $\Delta(\alpha, R)$ and the other analytic on $\mathbf{C} \setminus \{\alpha\}$.

The following analogue of **K.2** will figure significantly in the proof.

K.3 $$-\frac{1}{w-z} = \sum_{j=0}^{\infty} \frac{(w-\alpha)^j}{(z-\alpha)^{j+1}} \qquad (|w-\alpha| < |z-\alpha|)$$

the convergence being uniform for $|w - \alpha| \leqslant \rho$ provided that

$$\rho < |z - \alpha|.$$

Theorem 7.8 (Laurent). *Given $f \in \mathscr{A}(\Delta'(\alpha, R))$ the following statements hold.*

(i) $$f(z) = \sum_{-\infty}^{\infty} \alpha_n (z-\alpha)^n \qquad (z \in \Delta'(\alpha, R))$$

where

$$\alpha_n = \frac{1}{2\pi i} \int_{C(\alpha, r)} \frac{f(w)}{(w-\alpha)^{n+1}} \, \mathrm{d}w \qquad (0 < r < R, n \in \mathbf{Z})$$

the series converging uniformly on any closed subannulus of $\Delta'(\alpha, R)$.

(ii) *There exist $f_1 \in \mathscr{A}(\Delta(\alpha, R))$, $f_2 \in \mathscr{A}(\mathbf{C} \setminus \{\alpha\})$ such that*

$$f(z) = f_1(z) + f_2(z) \qquad (z \in \Delta'(\alpha, R)).$$

(iii) *If $f(z) = \sum_{-\infty}^{\infty} \beta_n (z-\alpha)^n$ $(z \in \Delta'(\alpha, R))$ then $\beta_n = \alpha_n$ $(n \in \mathbf{Z})$.*

Proof. (i) Given $z \in \Delta'(\alpha, R)$ choose R_1, R_2 such that

$$0 < R_1 < |z - \alpha| < R_2 < R$$

and then choose $\delta > 0$ such that $\bar{\Delta}(z, \delta) \subset A(\alpha; R_2, R_1)$. By Theorem 7.1 and Proposition 6.8 we have

$$\begin{aligned} f(z) &= \frac{1}{2\pi i} \int_{C(z,\delta)} \frac{f(w)}{w-z} \, \mathrm{d}w \\ &= \frac{1}{2\pi i} \int_{C(\alpha,R_2)} \frac{f(w)}{w-z} \, \mathrm{d}w - \frac{1}{2\pi i} \int_{C(\alpha,R_1)} \frac{f(w)}{w-z} \, \mathrm{d}w. \end{aligned}$$

By the argument of Theorem 7.3 we have

$$\frac{1}{2\pi i}\int_{C(\alpha,R_2)}\frac{f(w)}{w-z}\,\mathrm{d}w = \sum_{n=0}^{\infty} p_n\,(z-\alpha)^n$$

where

$$p_n = \frac{1}{2\pi i}\int_{C(\alpha,R_2)}\frac{f(w)}{(w-\alpha)^{n+1}}\,\mathrm{d}w \qquad (n = 0, 1, 2, \ldots).$$

By a similar argument, using **K. 3** instead of **K. 2** we obtain

$$-\frac{1}{2\pi i}\int_{C(\alpha,R_1)}\frac{f(w)}{w-z}\,\mathrm{d}w = \sum_{m=1}^{\infty} q_m(z-\alpha)^{-m}$$

where

$$q_m = \frac{1}{2\pi i}\int_{C(\alpha,R_1)} f(w)(w-\alpha)^{m-1}\,\mathrm{d}w \qquad (m = 1, 2, \ldots).$$

Choose any r in $(0, R)$ and define

$$\alpha_n = \frac{1}{2\pi i}\int_{C(\alpha,r)}\frac{f(w)}{(w-\alpha)^{n+1}}\,\mathrm{d}w \qquad (n \in \mathbf{Z}).$$

It follows from Proposition 6.7 that α_n is independent of the choice of r in $(0, R)$. In particular

$$\alpha_n = p_n \qquad (n = 0, 1, 2, \ldots)$$
$$\alpha_n = q_{-n} \qquad (n = 1, 2, \ldots)$$

and therefore

$$f(z) = \sum_{-\infty}^{\infty} \alpha_n(z-\alpha)^n.$$

Since z was any point of $\Delta'(\alpha, R)$ this establishes the pointwise convergence. The uniform convergence follows from the remark on page 89.

(ii) It is clear from the above that the series $\sum_0^{\infty} \alpha_n(z-\alpha)^n$ converges for each point of $\Delta'(\alpha, R)$, so that we may define f_1 on $\Delta(\alpha, R)$ by

$$f_1(z) = \sum_{n=0}^{\infty} \alpha_n(z-\alpha)^n \qquad (z \in \Delta(\alpha, R)).$$

Then $f_1 \in \mathscr{A}(\Delta(\alpha, R))$ by Theorem 4.10. Similarly we may define f_2 on $\mathbf{C} \setminus \{\alpha\}$ by

$$f_2(z) = \sum_{n=1}^{\infty} \alpha_{-n}(z-\alpha)^{-n} \qquad (z \in \mathbf{C} \setminus \{\alpha\})$$

and then $f_2 \in \mathscr{A}(\mathbf{C} \setminus \{\alpha\})$. Part (i) above now gives

$$f(z) = f_1(z) + f_2(z) \qquad (z \in \Delta'(\alpha, R)).$$

(iii) Choose any r such that $0 < r < R$. Since the quasi power series $\sum_{-\infty}^{\infty} \beta_n(w - \alpha)^n$ converges pointwise on $\Delta'(\alpha, R)$ it converges uniformly on $C(\alpha, r)$. Using **CI 7** we now obtain

$$\begin{aligned} \alpha_n &= \frac{1}{2\pi i} \int_{C(\alpha,r)} \frac{f(w)}{(w - \alpha)^{n+1}} \, dw \\ &= \frac{1}{2\pi i} \int_{C(\alpha,r)} (w - \alpha)^{-n-1} \sum_{-\infty}^{\infty} \beta_m(w - \alpha)^m \, dw \\ &= \sum_{-\infty}^{\infty} \frac{\beta_m}{2\pi i} \int_{C(\alpha,r)} (w - \alpha)^{m-n-1} \, dw \\ &= \beta_n \end{aligned}$$

for each $n \in \mathbf{Z}$.

The expansion of f in $\Delta'(\alpha, R)$ given in part (i) above is called the *Laurent expansion* of f at α and the numbers α_n are the corresponding *Laurent coefficients* at α. A slightly more general result, applicable to any annulus, is given in Problem 7.12. The function f_2 given in part (ii) above is called the *principal part* of f at α, and is denoted by $S(f; \alpha)$. It is clear that the function f may be extended to be analytic on $\Delta(\alpha, R)$ provided that $S(f; \alpha) = 0$. A sufficient condition for this is that f be bounded on $\Delta'(\alpha, R)$.

Proposition 7.9. *If $f \in \mathscr{A}(\Delta'(\alpha, R))$ is bounded then $S(f; \alpha) = 0$.*

Proof. Suppose $|f| \leqslant M$ on $\Delta'(\alpha, R)$. Let α_n be the Laurent coefficients of f at α. For any r such that $0 < r < R$ we have

$$\alpha_{-n} = \frac{1}{2\pi i} \int_{C(\alpha,r)} f(w)(w - \alpha)^{n-1} \, dw \qquad (n \in \mathbf{P})$$

and so by **CI 5**

$$|\alpha_{-n}| \leqslant \frac{1}{2\pi} 2\pi r M r^{n-1} = M r^n \qquad (n \in \mathbf{P}).$$

Since this is true for arbitrary small r we conclude that $\alpha_{-n} = 0$ $(n \in \mathbf{P})$ and so $S(f; \alpha) = 0$.

As a simple application of the above result suppose that $f \in \mathscr{C}(\bar{\Delta}(\alpha, R))$ is analytic on $\Delta'(\alpha, R)$. Then f is analytic on $\Delta(\alpha, R)$. Indeed, since $\bar{\Delta}(\alpha, R)$ is compact, f is certainly bounded on $\Delta'(\alpha, R)$ and so $S(f; \alpha) = 0$. This means that f has an extension from $\Delta'(\alpha, R)$ to $\Delta(\alpha, R)$ that is analytic on $\Delta(\alpha, R)$. Since the extension must be continuous at α, it must agree with f at α. This situation again contrasts with real analysis. For example if $g(x) = |x|$ $(x \in [-1, 1])$, then g is continuous on $[-1, 1]$ and differentiable on $(-1, 1) \setminus \{0\}$, but is not differentiable at 0.

The principal part $S(f; \alpha)$ will be studied in more detail in the next section. We conclude this section with an example which illustrates the usefulness of the fact that the Laurent expansion is unique.

Example 7.10.

$$\exp\left(z + \frac{1}{z}\right) = \alpha_0 + \sum_{n=1}^{\infty} \alpha_n(z^n + z^{-n}) \qquad (z \in \mathbf{C} \setminus \{0\})$$

where α_n $(n = 0, 1, 2, \ldots)$ *is given by*

$$\alpha_n = \frac{1}{\pi}\int_0^{\pi} e^{2\cos\theta} \cos n\theta \, d\theta = \sum_{j=0}^{\infty} \frac{1}{(n+j)!\,j!}$$

Proof. If $f(z) = \exp\left(z + \dfrac{1}{z}\right)$ $(z \in \mathbf{C} \setminus \{0\})$, then $f \in \mathscr{A}(\mathbf{C} \setminus \{0\})$ and so by Laurent's theorem

$$f(z) = \sum_{-\infty}^{\infty} \beta_n z^n \qquad (z \in \mathbf{C} \setminus \{0\})$$

where

$$\beta_n = \frac{1}{2\pi i}\int_{C(0,1)} \frac{f(w)}{w^{n+1}} \, dw = \frac{1}{2\pi i}\int_0^{2\pi} \exp(e^{i\theta} + e^{-i\theta})\, e^{-in\theta}\, i \, d\theta = \frac{1}{2\pi}\int_0^{2\pi} e^{2\cos\theta}\, e^{-in\theta}\, d\theta.$$

Since

$$\bar{\beta}_n = \frac{1}{2\pi}\int_0^{2\pi} e^{2\cos\theta}\, e^{in\theta}\, d\theta = \frac{1}{2\pi}\int_0^{2\pi} e^{2\cos(2\pi - \varphi)}\, e^{in(2\pi - \varphi)}\, d\varphi = \frac{1}{2\pi}\int_0^{2\pi} e^{2\cos\varphi}\, e^{-in\varphi}\, d\varphi = \beta_n$$

it follows that β_n is real and $\beta_{-n} = \bar{\beta}_n = \beta_n$.

Moreover for $n = 0, 1, 2, \ldots,$

$$\begin{aligned}\beta_n &= \frac{1}{2\pi}\int_0^{2\pi} e^{2\cos\theta} \cos n\theta \, d\theta \\ &= \frac{1}{\pi}\int_0^{\pi} e^{2\cos\theta} \cos n\theta \, d\theta \\ &= \alpha_n.\end{aligned}$$

It remains to obtain the series expression for α_n. For each $z \in \mathbf{C} \setminus \{0\}$ we have

$$f(z) = \exp(z) \exp\left(\frac{1}{z}\right) = \left(1 + \sum_{n=1}^{\infty} \frac{z^n}{n!}\right)\left(1 + \sum_{n=1}^{\infty} \frac{1}{n! z^n}\right).$$

By Proposition 4.6 we may take the convolution of the two power series above. Since the Laurent expression is unique we may compare coefficients of z^n to obtain

$$\alpha_n = \sum_{j=0}^{\infty} \frac{1}{(n+j)!\,j!} \qquad (n = 0, 1, 2, \ldots).$$

PROBLEMS 7

12. Given $f \in \mathscr{A}(A(\alpha; R, r))$ show that

$$f(z) = \sum_{-\infty}^{\infty} \alpha_n (z - \alpha)^n \qquad (z \in A(\alpha; R, r))$$

where

$$\alpha_n = \frac{1}{2\pi i}\int_{C(\alpha,\rho)} \frac{f(w)}{(w - \alpha)^{n+1}}\, dw \qquad (r < \rho < R, n \in \mathbf{Z}).$$

The above expansion is called the *Laurent expansion* of f in $A(\alpha; R, r)$.

State and prove the analogues of parts (ii) and (iii) of Theorem 7.8. Establish a similar result for functions in $\mathscr{A}(\nabla(\alpha, R))$.

13. Let f be as in Example 7.7. Obtain the Laurent expansions of f in $A\left(0; \frac{\sqrt{5}+1}{2}, \frac{\sqrt{5}-1}{2}\right)$ and in $\nabla\left(0, \frac{\sqrt{5}+1}{2}\right)$. Comment on the fact that the Laurent coefficients are different in the two cases.

14. If $f \in \mathscr{A}(\nabla(0, R))$ is bounded show that f is of the form

$$f(z) = \sum_{n=0}^{\infty} \beta_n z^{-n} \qquad (z \in \nabla(0, R)).$$

15. Let $f \in \mathscr{A}(\mathbf{C} \setminus \{0\})$ be such that

$$\overline{f(z)} = f\left(\frac{1}{z}\right) \qquad (z \in \mathbf{C} \setminus \{0\}).$$

Show that f is real valued on $C(0, 1)$ and that the Laurent coefficients of f at 0 satisfy $\bar{\alpha}_n = \alpha_{-n}$ $(n \in \mathbf{Z})$.

16. Given $u \in \mathbf{C}$ let f, g be defined on $\mathbf{C} \setminus \{0\}$ by $f(z) = \exp\left(uz + \dfrac{u}{z}\right)$, $g(z) = \exp\left(uz - \dfrac{u}{z}\right)$.

(i) Show that

$$f(z) = \alpha_0 + \sum_{n=1}^{\infty} \alpha_n(z^n + z^{-n}) \quad (z \in \mathbf{C} \setminus \{0\})$$

where for $n = 0, 1, 2, \ldots$

$$\alpha_n = \frac{1}{\pi} \int_0^{\pi} \exp(2u \cos\theta) \cos n\theta \, \mathrm{d}\theta = u^n \sum_{j=0}^{\infty} \frac{u^{2j}}{(n+j)!\,j!}.$$

(ii) Show that

$$g(z) = \beta_0 + \sum_{n=1}^{\infty} \beta_n(z^n + (-z)^{-n}) \qquad (z \in \mathbf{C} \setminus \{0\})$$

where for $n = 0, 1, 2, \ldots$

$$\beta_n = \frac{1}{\pi} \int_0^{\pi} \cos(n\theta - 2u \sin\theta) \, \mathrm{d}\theta = u^n \sum_{j=0}^{\infty} \frac{(-1)^j u^{2j}}{(n+j)!\,j!}.$$

17. Let $f \in \mathscr{A}(A(0; R, r))$ where $r < 1 < R$ and let $g(\theta) = f(\mathrm{e}^{i\theta})$ $(\theta \in \mathbf{R})$ so that g is periodic with period 2π. The (complex) *Fourier coefficients* of g are defined by

$$c_n = \frac{1}{2\pi} \int_{-\pi}^{\pi} g(\theta)\, \mathrm{e}^{-in\theta} \, \mathrm{d}\theta \qquad (n \in \mathbf{Z}).$$

Show that the Fourier coefficients c_n are simply the Laurent coefficients of f at 0. Thus if $f \in \mathscr{A}(\Delta(0, R))$ then the negative Fourier coefficients of g all vanish, i.e. $c_{-n} = 0$ $(n \in \mathbf{P})$. A converse of this remark is discussed in Theorem 9.18.

7.4 Isolated singularities

In this section we shall classify isolated singularities of complex functions and discuss some of their elementary properties. Recall that α is an isolated singularity of $f\colon D \to \mathbf{C}$ if f is analytic on some punctured disc

$\Delta'(\alpha, R)$ but not differentiable at α. There are many standard functions which have singularities that are not isolated. For example the set of singularities of log consists of $\{x: x \in \mathbf{R}, x \leqslant 0\}$ and consequently no singularity of log is isolated. We make no attempt to analyse the behaviour of functions near non-isolated singularities, for the simple reason that we have no elementary technique adequate to the task. On the other hand the Laurent expansion provides a basic tool for investigating isolated singularities.

It is convenient at this point to make some simple remarks about zeros of functions. Given $f \in \mathscr{A}(D)$ we say that $\alpha \in D$ is a *zero* of f if $f(\alpha) = 0$. We say that α is a zero of *order n* if

$$f(\alpha) = 0, \qquad f^{(j)}(\alpha) = 0 \qquad (j = 1, \ldots, n-1), \qquad f^{(n)}(\alpha) \neq 0.$$

A zero of order 1 is sometimes called a *simple zero*. If $f \in \mathscr{A}(\Delta(\alpha, R))$ and α is a zero of order n it follows from Theorem 7.4 that

$$f(z) = (z - \alpha)^n g(z) \qquad (z \in \Delta(\alpha, R))$$

where $g \in \mathscr{A}(\Delta(\alpha, R))$ and $g(\alpha) \neq 0$.

Suppose now that $f \in \mathscr{A}(\Delta'(\alpha, R))$ and that f is not differentiable at α, so that α is an isolated singularity of f. Let

$$S(f; \alpha)(z) = \sum_{n=1}^{\infty} \alpha_{-n}(z - \alpha)^{-n} \qquad (z \in \Delta'(\alpha, R))$$

be the principal part of f at α, given by Laurent's theorem. The coefficient α_{-1} is of particular significance; it is called the *residue* of f at α and is denoted by Res $(f; \alpha)$. Observe that

$$\operatorname{Res}(f; \alpha) = \frac{1}{2\pi i} \int_{C(\alpha, r)} f(w)\, dw \qquad (0 < r < R).$$

Isolated singularities are classified according to the nature of the principal part $S(f; \alpha)$.

(i) α is a *removable singularity* of f if $\alpha_{-n} = 0$ $(n \in \mathbf{P})$.

(ii) α is a *pole* of f, of *order* n_0, if $\alpha_{-n} = 0$ $(n > n_0)$, $\alpha_{-n_0} \neq 0$.

(iii) α is an *isolated essential singularity* of f if there is an infinite subset A of $\mathbf{P}$ such that $\alpha_{-n} \neq 0$ $(n \in A)$.

The case of a removable singularity is of least interest. In fact the singularity can be removed by changing the value of f at α, and then f becomes a power series function on $\Delta(\alpha, R)$. In view of Proposition 7.9 an isolated singularity α is removable if and only if f is bounded on some punctured neighbourhood of α.

We shall see below that there is a close relationship between zeros and poles. A pole of order 1 is sometimes called a *simple pole.* For example, 0 is a simple pole of the function $z \to z + \frac{1}{z}$.

As a simple example of an isolated essential singularity let $f(z) = \exp\left(\frac{1}{z}\right)$ $(z \in \mathbf{C} \setminus \{0\})$. Then 0 is an isolated essential singularity of f. There is little significance in the actual value that is assigned to a function at an isolated essential singularity. Indeed we shall not trouble to assign a value on most occasions, as in the example above. The situation is slightly different for the case of a pole. We shall see below that it is convenient to assign the value ∞ at any pole of f.

The above definitions can all be applied to the point at infinity by considering the behaviour of the function $f \circ \mathbf{j}$ at 0. Thus f has a *pole* at ∞ if and only if $f \circ \mathbf{j}$ has a pole at 0. For example, a polynomial function of degree n has a pole of order n at ∞, and any entire function that is not a polynomial has an isolated essential singularity at ∞.

The theorem below gives two useful characterizations of poles. The first characterization is awkward to state but it operates very simply in practice.

Theorem 7.11. *Let* $f: \Delta(\alpha, R) \to \mathbf{C}$ *be analytic on* $\Delta'(\alpha, R)$.

(i) *If f has a pole of order m at α there is $0 < \delta < R$ such that $\frac{1}{f}$ has an extension from $\Delta'(\alpha, \delta)$ to a function in $\mathscr{A}(\Delta(\alpha, \delta))$ with a zero of order m at α.*

(ii) *If f is regular at α with a zero of order m at α then $\frac{1}{f}$ has a pole of order m at α.*

(iii) *α is a pole of f if and only if $\lim_{z \to \alpha} f(z) = \infty$.*

Proof. (i) If f has a pole of order m at α then f is of the form

$$f(z) = \sum_{n=-m}^{\infty} \alpha_n (z - \alpha)^n \qquad (z \in \Delta'(\alpha, R))$$

where $\alpha_{-m} \neq 0$. Thus f is of the form

$$f(z) = (z - \alpha)^{-m} g(z) \qquad (z \in \Delta'(\alpha, R))$$

where $g \in \mathscr{A}(\Delta(\alpha, R))$ and $g(\alpha) = \alpha_{-m} \neq 0$. Since g is continuous at α there is $\delta > 0$ such that $|g(z)| \geqslant \frac{1}{2}|g(\alpha)|$ $(z \in \Delta(\alpha, \delta))$. We may thus define

$$\left(\frac{1}{f}\right)(z) = \frac{(z - \alpha)^m}{g(z)} \qquad (z \in \Delta(\alpha, \delta))$$

and it is clear that $\frac{1}{f} \in \mathscr{A}(\Delta(\alpha, \delta))$. Moreover since $g(\alpha) \neq 0$ it is easily verified that α is a zero of order m of $\frac{1}{f}$.

(ii) If f has a zero of order m at α then f is of the form

$$f(z) = (z - \alpha)^m g(z) \qquad (z \in \Delta(\alpha, R))$$

where $g \in \mathscr{A}(\Delta(\alpha, R))$ and $g(\alpha) \neq 0$. Since g is continuous at α there is $\delta > 0$ such that $g(z) \neq 0$ $(z \in \Delta(\alpha, \delta))$. Therefore $\frac{1}{g} \in \mathscr{A}(\Delta(\alpha, \delta))$ and so by Theorem 7.4

$$\left(\frac{1}{g}\right)(z) = \sum_{n=0}^{\infty} \beta_n (z - \alpha)^n \qquad (z \in \Delta(\alpha, \delta))$$

where $\beta_0 \neq 0$. This gives

$$\left(\frac{1}{f}\right)(z) = \sum_{n=-m}^{\infty} \beta_{n+m} (z - \alpha)^n \qquad (z \in \Delta'(\alpha, \delta)).$$

By the uniqueness of the Laurent expansion we conclude that $\frac{1}{f}$ has a pole of order m at α.

(iii) Let α be a pole of f, say of order m. In the notation of part (i) we have

$$\begin{aligned} |f(z)| &= |(z - \alpha)^{-m} g(z)| \\ &\geqslant \tfrac{1}{2}|g(\alpha)|\,|z - \alpha|^{-m} \qquad (z \in \Delta'(\alpha, \delta)) \end{aligned}$$

and so $\lim_{z \to \alpha} f(z) = \infty$.

Suppose now that $\lim_{z \to \alpha} f(z) = \infty$. Then in particular there is $\delta > 0$ such that $|f(z)| > 1$ $(z \in \Delta'(\alpha, \delta))$. Thus $\frac{1}{f}$ is analytic and bounded (by 1) on $\Delta'(\alpha, \delta)$ and so has an extension to $\Delta(\alpha, \delta)$, h say, such that h is analytic on $\Delta(\alpha, \delta)$. Since

$$h(\alpha) = \lim_{z \to \alpha} \frac{1}{f(z)} = 0$$

α is a zero of h. If α is not of finite order then $g^{(n)}(\alpha) = 0$ $(n \in \mathbf{P})$ and so by Theorem 7.6

$$g(z) = g(\alpha) + \sum_{n=1}^{\infty} \frac{g^{(n)}(\alpha)}{n!} (z - \alpha)^n = 0 \qquad (z \in \Delta(\alpha, \delta)).$$

This is a contradiction and hence α must be a zero of finite order. It follows from part (ii) above that α is a pole of f.

In view of part (iii) above it is convenient to define $f(\alpha) = \infty$ if α is a pole of f. It then follows that f is continuous at α as a mapping into $\mathbf{C}^\infty$. We shall see in the next chapter that it is important to be able to calculate the residue of a function at a pole. In many situations it is awkward to obtain the actual Laurent expansion at the pole. The result below gives a simple technique for evaluating residues, provided that we know the order of the pole.

Proposition 7.12. (i) *Let $f: \Delta(\alpha, R) \to \mathbf{C}$ have a simple pole at α. Then*

$$\operatorname{Res}(f; \alpha) = \lim_{z \to \alpha} (z - \alpha) f(z).$$

If, in particular, f is of the form $\frac{h}{k}$, $h, k \in \mathscr{A}(\Delta(\alpha, R))$, $h(\alpha) \neq 0$, $k(\alpha) = 0$, $k'(\alpha) \neq 0$, then

$$\operatorname{Res}(f; \alpha) = \frac{h(\alpha)}{k'(\alpha)}.$$

(ii) *Let $f: \Delta(\alpha, R) \to \mathbf{C}$ have a pole of order m (>1) at α and let $g(z) = (z - \alpha)^m f(z)$ $(z \in \Delta(\alpha, R))$. Then*

$$\operatorname{Res}(f; \alpha) = \frac{1}{(m-1)!} g^{(m-1)}(\alpha).$$

Proof. (i) We have

$$(z - \alpha) f(z) = (z - \alpha) \sum_{n=-1}^{\infty} \alpha_n (z - \alpha)^n$$

and therefore

$$\operatorname{Res}(f; \alpha) = \alpha_{-1} = \lim_{z \to \alpha} (z - \alpha) f(z).$$

Now let f be of the form $\frac{h}{k}$ as indicated. Then

$$\operatorname{Res}(f: \alpha) = \lim_{z \to \alpha} \frac{h(z)}{(k(z) - k(\alpha))/(z - \alpha)}$$
$$= \frac{h(\alpha)}{k'(\alpha)}.$$

(ii) If α is a pole of order m then f is of the form

$$f(z) = \sum_{n=-m}^{\infty} \alpha_n (z - \alpha)^n$$

and so

$$g(z) = \sum_{n=0}^{\infty} \alpha_{n-m} (z - \alpha)^n.$$

It follows from the uniqueness of the Taylor expansion that

$$\operatorname{Res}(f; \alpha) = \alpha_{-1} = \frac{1}{(m-1)!} g^{(m-1)}(\alpha).$$

We illustrate the above ideas and some further techniques in the following interesting example.

Example 7.13. *For $n = 0, 1, 2, \ldots$, let*

$$f_n(z) = \frac{1}{z^n(\exp(z) - 1)} \qquad (z \in \Delta'(0, 2\pi)).$$

Then f_n has a pole of order $n + 1$ at 0.

Proof. It is clear that $f_n \in \mathscr{A}(\Delta'(0, 2\pi))$. If $g(z) = \exp(z) - 1$, then $g \in \mathscr{A}(\Delta(0, 2\pi))$, $g(0) = 0$, $g'(0) = 1$. It follows from Theorem 7.11 (ii) that f_n has a pole of order $n + 1$ at 0. Thus far is straightforward, but the computation of the residues is by no means as easy. The case $n = 0$ presents no difficulty. In fact by Proposition 7.12 (i)

$$\operatorname{Res}(f_0; 0) = \frac{1}{\exp'(0)} = 1.$$

It follows that

$$\frac{z}{\exp(z) - 1} = \sum_{n=0}^{\infty} \alpha_n z^n \qquad (z \in \Delta'(0, 2\pi))$$

where $\alpha_0 = 1$. Thus

$$f_m(z) = \sum_{n=0}^{\infty} \alpha_n z^{n-m-1} \qquad (z \in \Delta'(0, 2\pi))$$

and so by the uniqueness of the Laurent expansion

$$\operatorname{Res}(f_m; 0) = \alpha_m \qquad (m \in \mathbf{P}).$$

We now determine the α_m by the method of undetermined coefficients. Since

$$z = \sum_{n=1}^{\infty} \frac{z^n}{n!} \sum_{n=0}^{\infty} \alpha_n z^n \qquad (z \in \Delta(0, 2\pi))$$

we may take the convolution of the above power series and compare coefficients of z^n. This gives

$$\alpha_{n-1} + \frac{1}{2!}\alpha_{n-2} + \cdots + \frac{1}{n!}\alpha_0 = 0 \qquad (n = 2, 3, 4, \ldots) \qquad (*)$$

from which the coefficients α_m may be determined inductively. In particular $\alpha_1 = -\frac{1}{2}$, $\alpha_2 = \frac{1}{12}$. It is not at all obvious from (*) that $\alpha_{2n+1} = 0$ $(n \in \mathbf{P})$. To see this we use a simple trick. Since

$$\frac{z}{\exp(z) - 1} = \sum_{n=0}^{\infty} \alpha_n z^n \qquad (z \in \Delta'(0, 2\pi))$$

we have

$$\frac{-z}{\exp(-z) - 1} = \sum_{n=0}^{\infty} \alpha_n(-z)^n \qquad (z \in \Delta'(0, 2\pi))$$

and therefore

$$-z = \sum_{n=0}^{\infty} 2\alpha_{2n+1} z^{2n+1} \qquad (z \in \Delta'(0, 2\pi)).$$

The uniqueness of the Taylor expansion now gives $\alpha_1 = -\frac{1}{2}$ and $\alpha_{2n+1} = 0$ $(n \in \mathbf{P})$.

Let $B_n = n!\alpha_n$ $(n = 0, 1, 2, \ldots)$. It is clear from (*) that each B_n is a rational number. The numbers B_n are called the *Bernouilli numbers** and they make several interesting appearances in analysis (see, for example, Problems 7.25, 7.26, 8.47). It is easily verified that the Bernouilli numbers satisfy the following equations.

$$\frac{z}{\exp(z) - 1} = \sum_{n=0}^{\infty} \frac{B_n}{n!} z^n \qquad (z \in \Delta'(0, 2\pi))$$

$$\sum_{j=0}^{n-1} \binom{n}{j} B_j = 0 \qquad (n = 2, 3, \ldots).$$

To complete this section it remains to study the case of an isolated essential singularity. We already know that a function cannot be bounded on any neighbourhood of an isolated essential singularity, and neither can the function converge to ∞ as z approaches the singularity. In fact the behaviour of the function near the singularity is rather surprising, as we shall see below.

Theorem 7.14 (Casorati-Weierstrass). *Let $f \in \mathscr{A}(\Delta'(\alpha, R))$ have an isolated essential singularity at α. Then $f(\Delta'(\alpha, \delta))$ is dense in $\mathbf{C}$ for $0 < \delta \leqslant R$, i.e. given $z \in \mathbf{C}$, $\epsilon > 0$, there is $w \in \Delta'(\alpha, \delta)$ such that $|f(w) - z| < \epsilon$.*

Proof. Suppose the result is false. Then there is $0 < \delta_0 \leqslant R$, $z_0 \in \mathbf{C}$, $\epsilon_0 > 0$ such that

$$|f(w) - z_0| \geqslant \epsilon_0 \qquad (w \in \Delta'(\alpha, \delta_0)).$$

* The student should note that some authors define the numbers $(-1)^{n+1}(2n)!\alpha_{2n}$ as the 'Bernouilli numbers'.

We may therefore define g on $\Delta'(\alpha, \delta_0)$ by

$$g(w) = \frac{1}{f(w) - z_0} \qquad (w \in \Delta'(\alpha, \delta_0)).$$

Evidently g is analytic and bounded by $1/\epsilon_0$ on $\Delta'(\alpha, \delta_0)$. It follows from Proposition 7.9 that g has a unique extension to a function $\hat{g}$ that is analytic on $\Delta(\alpha, \delta_0)$. If $\hat{g}(\alpha) \neq 0$, then $f - z_0\mathbf{1}$ has a removable singularity at α and so f has a removable singularity at α. If $\hat{g}(\alpha) = 0$, it follows from Theorem 7.11 (ii) that $f - z_0\mathbf{1}$ has a pole at α and so f has a pole at α. Both these possibilities lead to a contradiction and therefore the theorem is established.

Corollary. *If $f \in \mathscr{A}(\nabla(0, R))$ has an isolated essential singularity at ∞, then $f(\nabla(0, \delta))$ is dense in $\mathbf{C}$ for $\delta \geqslant R$.*

Proof. Let $\delta \geqslant R$. Since $f \circ \mathbf{j}$ has an isolated essential singularity at 0 it follows from the theorem above that

$$\{f(\nabla(0, \delta))\}^- = \left\{(f \circ \mathbf{j})\left(\Delta\left(0, \frac{1}{\delta}\right)\right)\right\}^- = \mathbf{C}.$$

Picard established a much stronger result than the above theorem, namely that each $f(\Delta'(\alpha, \delta))$ is either the whole complex plane or the whole complex plane except for one point. We shall not attempt to prove Picard's theorem in this book, but we give an illustrative example below. Picard's theorem is a most striking result and its original publication stimulated much research in complex analysis. Observe that the Casorati-Weierstrass theorem may be expressed as follows. Given $\beta \in \mathbf{C}$ there is a sequence $\{z_n\}$ in $\Delta'(\alpha, \delta)$ such that

$$\lim_{n \to \infty} z_n = \alpha, \qquad \lim_{n \to \infty} f(z_n) = \beta.$$

This gives some indication of the nature of the discontinuity at α.

Example 7.15. *If $f(z) = \exp\left(\frac{1}{z}\right)$ $(z \in \mathbf{C} \setminus \{0\})$ then $f(\Delta'(0, \delta)) = \mathbf{C} \setminus \{0\}$ for each $\delta > 0$.*

Proof. Clearly f has an isolated essential singularity at 0, and f never takes the value 0. Let $\delta > 0$ and let $\beta \in \mathbf{C} \setminus \{0\}$. Given $r > 0$ we have

$$f(re^{i\theta}) = \exp\left(\frac{1}{r}\,e^{-i\theta}\right) = e^{\frac{1}{r}\cos\theta}\, e^{-\frac{i}{r}\sin\theta}.$$

We wish to choose r, θ such that

$$e^{\frac{1}{r}\cos\theta} = |\beta|, \qquad -\frac{1}{r}\sin\theta \equiv \arg(\beta) \qquad (\text{mod } 2\pi).$$

Let

$$r_n = \{(\arg(\beta) + 2n\pi)^2 + (\log|\beta|)^2\}^{-1/2}$$

and then choose θ_n so that r_n, θ_n satisfy the equations above. We then have $f(r_n e^{i\theta_n}) = \beta$ $(n \in \mathbf{P})$ and if n is sufficiently large $r_n e^{i\theta_n} \in \Delta'(0, \delta)$. This completes the proof.

The student may care to visualize the surface in $\mathbf{R}^3$ given by

$$\left\{\left(z, \left|\exp\left(\frac{1}{z}\right)\right|\right) : z \in \mathbf{C}\right\}.$$

PROBLEMS 7

18. If f has an isolated singularity at α with Laurent coefficients α_n, show that $\lim_{n\to\infty} |\alpha_{-n}|^{1/n} = 0$.

19. If f has a pole of order n at ∞ show that $S(f; \infty)$ is a polynomial function of degree n.

20. Let f have a pole of order m at α and let p be a polynomial function of degree n. Show that $p \circ f$ has a pole of order mn at α.

21. Show that an isolated singularity of f cannot be a pole of $\exp \circ f$. Hence or otherwise show that an isolated singularity is removable if and only if $\operatorname{Re} f$ or $\operatorname{Im} f$ is bounded above or below on some neighbourhood of the singularity.

22. Let the functions below be defined on $\Delta'(0, 1)$. Show that each function has a pole at 0, and determine the order and residue in each case.

(i) $f(z) = \dfrac{z^2 + 1}{z^2(z^4 + 1)}$

(ii) $f(z) = z^{-(n+1)} \sin(\pi z) \quad (n \in \mathbf{P})$

(iii) $f(z) = z \operatorname{cosec}^2(\pi z)$

(iv) $f(z) = \dfrac{\log(1 + z)}{z(\exp(z) - 1)}$

(v) $f(z) = \dfrac{\exp(z)}{(\exp(z) - 1)^2}$.

23. Determine the poles and residues of tan, cot, cosec, sec. Deduce the corresponding results for the hyperbolic functions.

24. Let $f \in \mathscr{A}(\Delta'(0, 1))$ have a simple pole at 0. Given $k \in \mathbf{C}$ show that there is $\delta > 0$ such that

$$\frac{f'(z)}{f(z) - k} = -\frac{1}{z} + \sum_{n=1}^{\infty} u_n(k) z^{n-1} \qquad (z \in \Delta'(0, \delta))$$

where u_n is a polynomial function of degree n (a *Faber polynomial*).

25. Use Example 7.13 to obtain the following expansions in $\Delta\left(0, \frac{\pi}{2}\right)$.

(i) $$z \cot(z) = \sum_{n=0}^{\infty} (-1)^n \frac{2^{2n} B_{2n}}{(2n)!} z^{2n}.$$

(ii) $$\tan(z) = \sum_{n=1}^{\infty} (-1)^{n-1} \frac{2^{2n}(2^{2n} - 1) B_{2n}}{(2n)!} z^{2n-1}$$

(iii) $$z \operatorname{cosec}(z) = \sum_{n=0}^{\infty} (-1)^{n-1} \frac{(2^{2n} - 2) B_{2n}}{(2n!)} z^{2n}.$$

26. Given $k \in \mathbf{C}$ show that

$$\frac{z \exp(kz)}{\exp(z) - 1} = \sum_{n=0}^{\infty} \frac{B_n(k)}{n!} z^n \qquad (z \in \Delta'(0, 2\pi))$$

where $B_n(k) = \sum_{j=0}^{n} \binom{n}{j} B_{n-j} k^j$. Prove the following:

(i) $B_n'(k) = n B_{n-1}(k)$

(ii) $B_n(k + 1) - B_n(k) = n k^{n-1}$

(iii) $\sum_{j=1}^{m} j^n = \frac{1}{n+1} \{B_{n+1}(m + 1) - B_{n+1}\} \qquad (n = 2, 3, \ldots).$

27. Show that

$$\sec(z) = \sum_{n=0}^{\infty} (-1)^n \frac{E_{2n}}{(2n)!} z^{2n} \qquad \left(z \in \Delta'\left(0, \frac{\pi}{2}\right) \tfrac{1}{2}\pi\right)$$

where $E_0 = 1$ and

$$\sum_{j=0}^{n} \binom{2n}{2j} E_{2j} = 0 \qquad (n \in \mathbf{P}).$$

The numbers E_{2n} are called the *Euler numbers*; show that they are all integers.

28. If $f(z) = \exp\left(\frac{z + 1}{z - 1}\right)$ $(z \in \mathbf{C} \setminus \{1\})$, show that f is bounded on $\Delta(0, 1)$ and has an isolated essential singularity at 1. If $g(z) = (z - 1) f(z)$ $(z \in \mathbf{C})$, show that g is continuous on $\bar{\Delta}(0, 1)$, analytic on $\Delta(0, 1)$, and has an isolated essential singularity at 1.

29. Suppose that $f \in \mathscr{A}(\Delta'(\alpha, R))$ has an isolated essential singularity at α, and that $f(z) \neq 0$ $(z \in \Delta'(\alpha, R))$. Show that $1/f$ has an isolated essential singularity at α. If f is of the form

$$f(z) = \sum_{n=1}^{\infty} \alpha_{-n}(z - \alpha)^{-n}$$

show that

$$\operatorname{Res}\left(\frac{1}{f}; \alpha\right) = \alpha_{-2}^2 - \frac{\alpha_{-3}}{\alpha_{-1}}.$$

30. Assume the existence of a function $\lambda \in \mathscr{A}(\mathbf{C} \setminus \{0, 1\})$ such that $\operatorname{Im} \lambda(z) \geqslant 0$ $(z \in \mathbf{C} \setminus \{0, 1\})$. Suppose f is an entire function such that $\alpha, \beta \notin f(\mathbf{C})$, $\alpha \neq \beta$. Let $g(z) = \dfrac{1}{\beta - \alpha}\{f(z) - \alpha\}$ $(z \in \mathbf{C})$ and show that g is entire and $0, 1 \notin g(\mathbf{C})$. Show that $\exp \circ (i\lambda) \circ g$ is a bounded entire function. Use Theorem 7.3 to show that $\exp \circ (i\lambda) \circ g$ is constant and hence g, f are constant. This establishes the *little* Picard theorem, namely that if f is an entire function then $f(\mathbf{C})$ is either $\mathbf{C}$ or $\mathbf{C}$ less one point.

8

GLOBAL ANALYSIS

In this chapter we study the global behaviour of functions that are analytic on some domain. The present chapter is essentially the culmination of the previous chapters and contains most of the elementary results of complex function theory. The results are too numerous to be described in a general introduction and will be discussed at the beginning of each section. The results in the first five sections follow from the local analysis results together with arguments involving connectedness and compactness. A further tool is developed in §6, namely the topological index (or winding number) of a simple closed contour with respect to points not on the contour. This leads to the famous Cauchy residue theorem and numerous applications.

8.1 Taylor expansions revisited

In this section we discuss the global analogues of the local Taylor theorems 7.4, 7.6. The finite expansion carries over completely whereas the infinite expansion fails to have a complete analogue. This is no surprise of course since power series functions are defined only on discs. On the other hand the results on infinite expansions shed some interesting light on real analysis.

Theorem 8.1. *Let D be any domain and let $f \in \mathscr{A}(D)$. Then f is infinitely differentiable on D, and given $\alpha \in D$, $n \in \mathbf{P}$, there is $f_n \in \mathscr{A}(D)$ such that*

$f_n(\alpha) = \frac{1}{n!} f^{(n)}(\alpha)$ *and*

$$f(z) = f(\alpha) + \sum_{j=1}^{n-1} \frac{f^{(j)}(\alpha)}{j!}(z-\alpha)^j + (z-\alpha)^n f_n(z) \qquad (z \in D).$$

Proof. Given $\alpha \in D$ there is $R > 0$ such that $\Delta(\alpha, R) \subset D$. Then $f|_{\Delta(\alpha, R)}$ is analytic on $\Delta(\alpha, R)$ and so infinitely differentiable on $\Delta(\alpha, R)$ by Theorem 7.3. This shows that f is infinitely differentiable on D. Given $n \in \mathbf{P}$ define $f_n: D \to \mathbf{C}$ by $f_n(\alpha) = \dfrac{1}{n!} f^{(n)}(\alpha)$ and

$$f_n(z) = (z - \alpha)^{-n}\left\{f(z) - f(\alpha) - \sum_{j=1}^{n-1} \frac{f^{(j)}(\alpha)}{j!}(z - \alpha)^j\right\} \qquad (z \in D \setminus \{\alpha\}).$$

It is clear that f_n is analytic on $D \setminus \{\alpha\}$, and it follows from Theorem 7.4 that f_n is analytic on $\Delta(\alpha, R)$. This shows that $f_n \in \mathscr{A}(D)$, and so the proof is complete.

Corollary. *If α is a zero of f of order n then*

$$f(z) = (z - \alpha)^n f_n(z) \qquad (z \in D)$$

where $f_n \in \mathscr{A}(D)$, $f_n(\alpha) \neq 0$.

We now consider the infinite Taylor expansion. Given $f \in \mathscr{A}(D)$ and $\alpha \in D$ the statement

$$f(z) = f(\alpha) + \sum_{j=1}^{\infty} \frac{f^{(j)}(\alpha)}{j!}(z - \alpha)^j \qquad (z \in D)$$

is evidently false in general. To see this, take $D = \mathbf{C} \setminus N$, $f = \log$, $\alpha = 1$, so that $\log \in \mathscr{A}(D)$. We know from Example 4.18 that the Taylor expansion of log at 1 fails to converge outside $\Delta(1, 1)$. Thus $\Delta(1, 1)$ is the largest open disc about the point 1 on which log is represented by its Taylor expansion at 1. Given any point α of $\mathbf{C} \setminus N$ we may readily compute the Taylor expansion of log at α (see Problem 4.28 (i)) and then verify that the largest disc about α on which log is represented by its Taylor expansion at α is given by $\Delta(\alpha, r)$, where $r = \text{dist}(\alpha, N)$. In general, the examination of power series can be a tedious affair. The result below gives a very simple means of determining the largest open disc on which a function is represented by its Taylor expansion. The result becomes even more significant when we allow singularities in the domain under consideration.

It is convenient at this point to introduce two notational conventions. If $S = \varnothing$ we define dist (α, S) to be $+\infty$, and if $r = +\infty$ we define $\Delta(\alpha, r)$ to be $\mathbf{C}$.

Proposition 8.2. *Let $f: D \to \mathbf{C}$ be regular at $\alpha \in D$ and let S be the set of singularities of f. Let*

$$\delta_1 = \text{dist}(\alpha, \mathbf{C} \setminus D), \qquad \delta_2 = \text{dist}(\alpha, S), \qquad \delta = \min(\delta_1, \delta_2).$$

Then $\Delta(\alpha, \delta)$ is the largest open disc about α on which f is represented by its Taylor expansion at α.

Proof. It follows from the definition of δ that $\Delta(\alpha, \delta) \subset D$ and that f is analytic on $\Delta(\alpha, \delta)$. By Theorem 7.6 (or its Corollary, if $\delta = +\infty$) we now have

$$f(z) = f(\alpha) + \sum_{n=1}^{\infty} \frac{f^{(n)}(\alpha)}{n!}(z - \alpha)^n \qquad (z \in \Delta(\alpha, \delta)).$$

Let $\Delta(\alpha, \rho)$ be any disc on which f is represented by its Taylor expansion at α. Since f is defined only on D we must have $\Delta(\alpha, \rho) \subset D$ and hence $\rho \leqslant \delta_1$. If $\rho > \delta_2$ it follows that $S \cap \Delta(\alpha, \delta) \neq \varnothing$, and so f has a singularity in $\Delta(\alpha, \rho)$. This contradicts Theorem 4.10, and so $\rho \leqslant \delta_2$. Therefore $\rho \leqslant \delta$ and the proof is complete.

An obvious corollary of the above result is that if $f \in \mathscr{A}(D)$ then f is a power series function when restricted to any open disc that is contained in D. For a less trivial application consider the real function tan. It is well known that *tan* is infinitely differentiable on, $\left(-\frac{\pi}{2}, \frac{\pi}{2}\right)$. It may be established by a fairly difficult power series argument that tan is represented by its Taylor expansion at 0 on some interval about 0. There is no simple power series technique that establishes that tan is represented by its Taylor expansion at 0 on the whole interval $\left(-\frac{\pi}{2}, \frac{\pi}{2}\right)$. On the other hand we know from the above proposition that the complex function tan is represented on all of $\Delta\left(0, \frac{\pi}{2}\right)$ by its Taylor expansion at 0. Moreover the Taylor coefficients at 0 of the real and complex functions tan are precisely the same. Therefore the real function tan is represented by its Taylor expansion at 0 on the whole interval $\left(-\frac{\pi}{2}, \frac{\pi}{2}\right)$.

The above example raises the following question. Suppose that $f: D \to \mathbf{C}$ is analytic on $\Delta(\alpha, R) \subset D$ and has a non-removable singularity on $C(\alpha, R)$. Does it follow that R is the radius of convergence of the Taylor expansion of f at α? We shall see in the next section that the answer is in the affirmative, but the student must not be deceived into thinking that the proof is obvious from Proposition 8.2. All we know at the present is that the Taylor expansion of f at α agrees with f on $\Delta(\alpha, R)$; on the face of it, the Taylor expansion of f at α might converge outside $\Delta(\alpha, R)$ to some function other than f.

8.2 Properties of zeros

In this section we derive some preliminary results on the nature of zeros of functions analytic on a domain. We shall return to this topic in §10 of this chapter. For the present we show that the zeros are all of finite order and form a countable subset of the domain that has no cluster points in the domain. This leads to the remarkable result that if $f, g \in \mathscr{A}(D)$ agree on a subset X of D with a cluster point in D, then $f = g$. A further corollary is that the algebra $\mathscr{A}(D)$ is in fact an integral domain.

Given $f: D \to \mathbf{C}$ we write Z_f for the set of zeros of f, so that

$$Z_f = \{z: z \in D, f(z) = 0\}.$$

Theorem 8.3. *If $f \in \mathscr{A}(D), f \neq 0$, then*

(i) *the zeros of f (if any) are of finite order;*

(ii) *the set Z_f has no cluster points in D;*

(iii) *if K is any compact subset of D then $K \cap Z_f$ is a finite set;*

(iv) *the set Z_f is at most countable.*

Proof. (i) We show that f has no zeros of infinite order. Let

$$E = \{\alpha: \alpha \in Z_f, f^{(n)}(\alpha) = 0 (n \in \mathbf{P})\}.$$

Since f is continuous on D, Z_f is closed in D. Similarly for each $n \in \mathbf{P}$, $\{\alpha: \alpha \in \mathbf{D}, f^{(n)}(\alpha) = 0\}$ is closed in D. Therefore E is the intersection of closed sets and so is closed in D. We show next that E is open. Given $\alpha \in E$, we have $\alpha \in D$ and so $\Delta(\alpha, R) \subset D$ for some $R > 0$. Since f is analytic on $\Delta(\alpha, R)$, Theorem 7.6 gives

$$f(z) = f(\alpha) + \sum_{n=1}^{\infty} \frac{f^{(n)}(\alpha)}{n!} (z - \alpha)^n = 0 \qquad (z \in \Delta(\alpha, R)).$$

Therefore $f^{(n)}(z) = 0$ $(z \in \Delta(\alpha, R), n \in \mathbf{P})$ and so $\Delta(\alpha, R) \subset E$. This shows that E is open. Since D is connected we now have either $E = D$ or $E = \varnothing$. Since $f \neq 0$, $E \neq D$ and so $E = \varnothing$, i.e. there are no zeros of infinite order.

(ii) Let $\alpha \in Z_f$ and let α be a zero of order n. By the corollary to Theorem 8.1 there is $g \in \mathscr{A}(D)$ such that $g(\alpha) \neq 0$ and $f(z) = (z - \alpha)^n g(z)$ $(z \in D)$. Since g is continuous at α there is $R > 0$ such that $\Delta(\alpha, R) \subset D$ and $g(z) \neq 0$ $(z \in \Delta(\alpha, R))$. Therefore $f(z) \neq 0$ $(z \in \Delta'(\alpha, R))$. This says that the zeros of f are isolated. If β is a cluster point of Z_f in D then $\beta \in Z_f$ since Z_f is closed. This is impossible since the zeros are isolated.

(iii) Let K be a compact subset of D and let $Q = Z_f \cap K$. If Q is infinite then by the remark on page 23 Q has a cluster point in K. This implies that Z_f has a cluster point in $K, K \subset D$. This contradicts part (ii) and hence Q must be finite.

(iv) By Proposition 2.7 we may choose a sequence $\{K_n\}$ of compact subsets of D such that

$$K_n \subset K_{n+1} \qquad (n \in \mathbf{P}), \qquad \bigcup \{K_n: n \in \mathbf{P}\} = D.$$

For each $n \in \mathbf{P}$, $Z_f \cap K_n$ is finite by part (iii). Since

$$Z_f = \bigcup \{Z_f \cap K_n : n \in \mathbf{P}\}$$

it follows that Z_f is at most countable.

The student should observe that although Z_f has no cluster points in D it may well happen that Z_f has cluster points in $b(D)$. For example if D is bounded and Z_f is infinite, then Z_f will certainly have at least one cluster point in $b(D)$ (see Problem 8.1). The above theorem provides another sharp contrast between real and complex analysis. To see this let f be defined on $\mathbf{R}$ by

$$f(x) = \begin{cases} e^{1x^2} & (x > 0) \\ 0 & (x \leqslant 0). \end{cases}$$

It is readily verified (from Problem 7.5) that f is infinitely differentiable on $\mathbf{R}$ and that $f^{(n)}(0) = 0$ $(n \in \mathbf{P})$. Thus 0 is a zero of infinite order, and Z_f has cluster points in $\mathbf{R}$ and is uncountable.

The following important application of Theorem 8.3 is sometimes called the *identity theorem.*

Theorem 8.4. *Let $f, g \in \mathscr{A}(D)$, and let X be a subset of D with a cluster point in D. If $f(z) = g(z)$ $(z \in X)$, then $f = g$.*

Proof. Let $h = f - g$, so that $h \in \mathscr{A}(D)$. Since $X \subset Z_h$ it follows from Theorem 8.3 that $h = 0$.

Corollary. *If $f, g \in \mathscr{A}(D)$ agree on a line segment or a non-empty open set, then $f = g$.*

Proof. Suppose $[\alpha, \beta] \subset D$ and $f(z) = g(z)$ $(z \in [\alpha, \beta])$. Take $z_n = \alpha + \dfrac{1}{n}(\beta - \alpha)$ $(n \in \mathbf{P})$ and we see that α is a cluster point of $[\alpha, \beta]$. Since $\alpha \in D$ it follows from the above theorem that $f = g$. If f and g agree on some non-empty open set then they agree on some open disc and therefore on some line segment.

The identity theorem is an extremely useful tool. For example if f and g are entire functions, then in order to show that $f = g$ it is sufficient to show that f and g agree on some interval of the real axis. The result is also very useful in evaluating integrals involving a complex parameter (see Problems 8.4, 8.5). Moreover the identity theorem is the starting point for the topic of analytic continuation to be considered in Chapter 10. We give below two further applications, the first algebraic and the second topological.

Theorem 8.5. *For any domain D, $\mathscr{A}(D)$ is an integral domain.*

Proof. In view of Proposition 3.7 it will be sufficient to show that if $f, g \in \mathscr{A}(D)$ and $fg = 0$, then $f = 0$ or $g = 0$. Suppose that $f \neq 0$. Since f is continuous the set

$$U = \{z: z \in D, f(z) \neq 0\} = f^{-1}(\mathbf{C} \setminus \{0\})$$

is a non-empty open subset of D. Since $fg = 0$ we must have $g(z) = 0$ $(z \in U)$ (since $\mathbf{C}$ is an integral domain!). Thus g agrees with the zero function on the non-empty open subset U of D. By the Corollary to Theorem 8.4 we have $g = 0$ and hence the proof is complete.

Proposition 8.6. *Let $f: D \to \mathbf{C}$ be analytic on $\Delta(\alpha, R) \subset D$, and let f have a non-removable isolated singularity at $\beta \in C(\alpha, R)$. Then R is the radius of convergence of the Taylor expansion of f at α.*

Proof. Let ρ be the radius of convergence of the Taylor expansion of f at α and let g be the associated power series function on $\Delta(\alpha, \rho)$. We have $\rho \geqslant R$ by Proposition 8.2. Suppose that $\rho > R$. Since β is an isolated singularity of f we may choose $\delta > 0$ such that $\bar{\Delta}(\beta, \delta) \subset \Delta(\alpha, R)$ and f is analytic on $\Delta'(\beta, \delta)$ (see Figure 8.1). By Proposition 1.23, $X = \Delta(\alpha, R) \cup \Delta'(\beta, \delta)$ is a domain. Then f and g are analytic on X and agree on $\Delta(\alpha, R)$. By the Corollary to Theorem 8.4, f and g therefore agree on $\Delta'(\beta, \delta)$. Since $\bar{\Delta}(\beta, \delta)$ is a compact subset of $\Delta(\alpha, \rho)$ and g is continuous on $\Delta(\alpha, \rho)$, it follows that g is bounded on $\bar{\Delta}(\beta, \delta)$. Therefore f is bounded on $\Delta'(\beta, \delta)$ and so has a removable singularity at β by Proposition 7.9. This contradiction shows that $\rho = R$ as required.

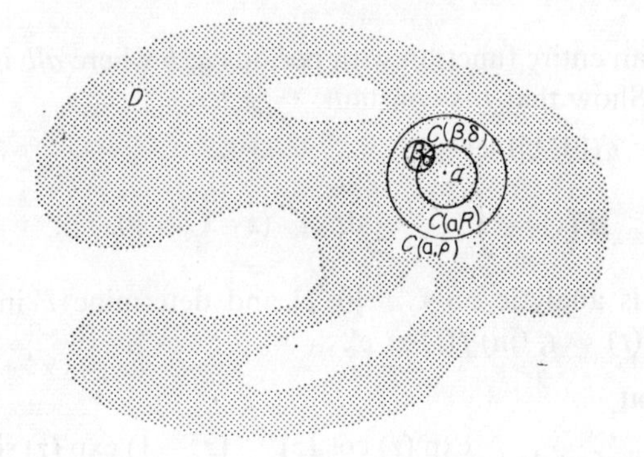

Figure 8.1

Consider now the real function f defined by $f(x) = \dfrac{1}{1 + x^2}$ $(x \in \mathbf{R})$. It is well known that f is infinitely differentiable on R and yet the Taylor expansion of f at 0, namely $\sum_{n=0}^{\infty} (-1)^n x^{2n}$, represents f only on $(-1, 1)$. The situation is now illumined by complex analysis. In fact if $g(z) = \dfrac{1}{1 + z^2}$ $(z \in \mathbf{C})$ then g is analytic on $\mathbf{C} \setminus \{i, -i\}$ and the points i, $-i$ are simple poles of g. By Proposition 8.2, g is represented on $\Delta(0, 1)$ by its Taylor expansion at 0. But the Taylor coefficients of f and g at 0 are exactly the same. Therefore f is represented on $(-1, 1)$ by its Taylor expansion at 0. By Proposition 8.6, 1 is the radius of convergence of the Taylor expansion of g at 0, and so also of the Taylor expansion of f at 0. In particular f is not represented by its Taylor expansion at 0 on any open interval larger than $(-1, 1)$. In this simple case it was of course obvious that 1 was the radius of convergence of the Taylor expansion of f at 0. If we now take $h(x) = \dfrac{\tan x}{1 + x^2}\left(x \in \left(-\dfrac{\pi}{2}, \dfrac{\pi}{2}\right)\right)$ it is by no means obvious from real analysis that the Taylor expansion of h at 0 has radius of convergence 1. On the other hand the result is obvious from the argument using complex analysis.

PROBLEMS 8

1. Let f be defined on $\Delta'(0, 1)$ by

$$f(z) = \sin\left(\frac{1}{z}\right)\sin\left(\frac{1}{1 - z}\right) \qquad (z \in \Delta'(0, 1)).$$

Determine Z_f and its cluster points.

2. Give an example of a function in $\mathscr{A}(\Delta(0, 1))$ with an infinite number of zeroes.

3. Let f be an entire function with periods a, b where a/b is an irrational real number. Show that f is constant.

4. Given $f \in \mathscr{C}([0, 1])$ let

$$F(z) = \int_0^1 \frac{f(t)}{t - z}\, dt \qquad (z \in \mathbf{C} \setminus [0, 1]).$$

Show that F is analytic on $\mathbf{C} \setminus [0, 1]$ and determine F in the cases (i) $f(t) = 1$, (ii) $f(t) = t$, (iii) $f(t) = e^t$.

5. Prove that

$$\int_0^1 t \exp(zt) \cos(zt)\, dt = \frac{\exp(z)\cos(z)}{2z} + \frac{(z - 1)\exp(z)\sin(z)}{2z^2} \qquad (z \in \mathbf{C}).$$

6. Determine the radius of convergence of each of the power series given in Problems 7.10, 7.24, 7.25, 7.26, 7.27. Use Problem 7.25 (ii) to obtain an estimate of the size of $|B_{2n}|$.

7. Let $\alpha \in \mathbf{C} \setminus N$. Show that the largest open disc on which log is represented by its Taylor expansion at α is $\Delta(\alpha, R)$ where $R = \text{dist}(\alpha, N)$. Show that the radius of convergence of the Taylor expansion at α is in fact $|\alpha|$. Compare this example with Proposition 8.6.

8.3 Entire Functions

In this section we derive some of the simple properties of entire functions. We give first an algebraic picture of the set of entire functions in terms of a certain convolution algebra of sequences. We then prove Liouville's theorem, that every bounded entire function is constant. This leads to an elegant proof of the fundamental theorem of algebra, namely that every complex polynomial equation has a root. Finally we describe the nature of an entire function with a finite number of zeros. This last result will be generalized to an arbitrary entire function in §6.

Let f be an entire function and let $\alpha \in \mathbf{C}$. By the corollary to Theorem 7.6 we have

$$f(z) = f(\alpha) + \sum_{n=1}^{\infty} \frac{f^{(n)}(\alpha)}{n!}(z - \alpha)^n \qquad (z \in \mathbf{C}).$$

We wish to investigate the relationship between f and its Taylor coefficients at α.

Let $\mathscr{S}$ denote the set of all complex sequences defined on the non-negative integers. It is easily verified that $\mathscr{S}$ is a complex linear algebra with respect to the operations defined by

$$\begin{aligned} \{\alpha_n\} + \{\beta_n\} &= \{\alpha_n + \beta_n\} \\ \lambda\{\alpha_n\} &= \{\lambda\alpha_n\} \qquad (\lambda \in \mathbf{C}) \\ \{\alpha_n\} * \{\beta_n\} &= \{\gamma_n\} \end{aligned}$$

where $\{\gamma_n\}$ is the convolution of $\{\alpha_n\}$ and $\{\beta_n\}$, i.e.

$$\gamma_n = \sum_{j=0}^{n} \alpha_j\beta_{n-j} \qquad (n = 0, 1, 2, \ldots).$$

We call $\mathscr{S}$ the *convolution algebra* of complex sequences. Let $\mathscr{E}$ denote the subset of $\mathscr{S}$ consisting of those sequences $\{\alpha_n\}$ such that $\lim_{n\to\infty} |\alpha_n|^{1/n} = 0$. It is a non-trivial exercise on sequences to show that $\mathscr{E}$ is a subalgebra of $\mathscr{S}$. On the other hand we may give an elegant proof via complex function theory.

Define $\psi: \mathscr{A}(\mathbf{C}) \to \mathscr{S}$ by $\psi(f) = \{\alpha_n\}$ where

$$\alpha_0 = f(\alpha), \qquad \alpha_n = \frac{1}{n!} f^{(n)}(\alpha) \qquad (n \in \mathbf{P}).$$

Since the Taylor expansion of f at α has infinite radius of convergence it follows from the remarks after Proposition 4.7 that $\overline{\lim_{n\to\infty}} |\alpha_n|^{1/n} = 0$ and therefore $\lim_{n\to\infty} |\alpha_n|^{1/n} = 0$. This shows that ψ maps $\mathscr{A}(\mathbf{C})$ into $\mathscr{E}$. Conversely, given $\{\alpha_n\} \in \mathscr{E}$ let

$$f(z) = \sum_{n=0}^{\infty} \alpha_n (z - \alpha)^n \qquad (z \in \mathbf{C}).$$

Then Theorem 4.10 gives $f \in \mathscr{A}(\mathbf{C})$ and $\psi(f) = \{\alpha_n\}$. This shows that ψ actually maps $\mathscr{A}(\mathbf{C})$ one-to-one onto $\mathscr{E}$. Moreover it is a straightforward exercise on Proposition 3.7 to show that

$$\begin{aligned} \psi(f + g) &= \psi(f) + \psi(g) \\ \psi(\lambda f) &= \lambda \psi(f) \qquad (\lambda \in \mathbf{C}) \\ \psi(fg) &= \psi(f) * \psi(g) \end{aligned}$$

for arbitrary f, g in $\mathscr{A}(\mathbf{C})$. It follows that $\mathscr{E}$ is a subalgebra of $\mathscr{S}$ and that ψ is an algebraic isomorphism of $\mathscr{A}(\mathbf{C})$ with $\mathscr{E}$.

We can now give an alternative proof that $\mathscr{A}(\mathbf{C})$ is an integral domain. In fact, $\mathscr{S}$ is easily shown to be an integral domain (use induction) and hence $\mathscr{E}$ is an integral domain. Since $\mathscr{A}(\mathbf{C})$ is isomorphic with $\mathscr{E}$ it follows that $\mathscr{A}(\mathbf{C})$ is an integral domain.

The above work is a simple example of what we might call 'mathematical trade' between algebra and analysis. In particular the student should note that it is both interesting and useful to consider mathematical problems from as many viewpoints as possible.

Our next investigation is concerned with the 'size' of entire functions. Recall that there are many examples of infinitely differentiable real functions that are bounded on $\mathbf{R}$. For example we may take $f(x) = e^{-x^2}$ $(x \in \mathbf{R})$, and in this case, even all the derivatives of f are bounded on $\mathbf{R}$. The result below gives the startling contrast for complex analysis; a simple generalization is given in Problem 8.9.

Theorem 8.7 (Liouville). *Every bounded entire function is constant.*

Proof. Let f be an entire function with $|f| \leqslant M$. Given $\alpha \in \mathbf{C}$, $R > 0$ it follows from Theorem 7.3 that

$$f'(\alpha) = \frac{1}{2\pi i} \int_{C(\alpha, R)} \frac{f(w)}{(w - \alpha)^2} \, dw.$$

Therefore

$$|f'(\alpha)| \leqslant \frac{1}{2\pi} 2\pi R \frac{M}{R^2} = \frac{M}{R}.$$

Since R was arbitrary this shows that $f'(\alpha) = 0$. Since α was arbitrary we conclude that f is constant.

Corollary. *$\mathscr{A}(\mathbf{C}^\infty)$ consists of the constant functions.*

Proof. If $f \in \mathscr{A}(\mathbf{C}^\infty)$ then f is continuous on the compact metric space $\mathbf{C}^\infty$ and so is bounded. By Liouville's theorem f is therefore constant on $\mathbf{C}$. Since f is continuous on $\mathbf{C}^\infty$, f must be constant on $\mathbf{C}^\infty$. Conversely it is clear that every constant function belongs to $\mathscr{A}(\mathbf{C}^\infty)$.

Liouville's theorem leads to an elegant proof of the fundamental theorem of algebra below. This is a non-trivial case of mathematical trading since algebraic proofs of the theorem (e.g. via the theory of field extensions) are rather complicated. On the other hand it should be realized that we have been involved in a good deal of work in arriving at Liouville's theorem.

Theorem 8.8. *Let p be a polynomial function of degree n. Then there exist $k \in \mathbf{C}$, $\alpha_1, \alpha_2, \ldots, \alpha_n \in \mathbf{C}$ (not necessarily distinct) such that*

$$p(z) = k(z - \alpha_1)(z - \alpha_2)\ldots(z - \alpha_n) \qquad (z \in \mathbf{C}).$$

Proof. Suppose that

$$p(z) = \beta_0 z^n + \beta_1 z^{n-1} + \cdots + \beta_n$$

where $\beta_0 \neq 0$. We easily see that

$$\lim_{z\to\infty} \frac{p(z)}{z^n} = \beta_0.$$

Hence there is $R > 0$ such that

$$\left|\frac{p(z)}{z^n}\right| > \tfrac{1}{2}|\beta_0| \qquad (z \in \nabla(0, R)) \qquad (*)$$

and so

$$\left|\frac{1}{p(z)}\right| < \frac{2}{|\beta_0|\,|z^n|} < \frac{2}{|\beta_0| R^n} \qquad (z \in \nabla(0, R)).$$

Suppose that p has no zeros in $\mathbf{C}$. Then $1/p$ is an entire function. In particular $1/p$ is continuous and so bounded on the compact set $\bar{\Delta}(0, R)$. Therefore $1/p$ is a bounded entire function and so constant by Liouville's theorem. We now have that p is constant and $\lim_{z\to\infty} p(z) = \infty$. This is impossible and so p must have a zero.

Let α_1 be a zero of p. By elementary algebra there is a polynomial function p_1 of degree $n - 1$ such that

$$p(z) = (z - \alpha_1)p_1(z) \qquad (z \in \mathbf{C}).$$

If $n = 1$ the proof is complete. Otherwise we may choose a zero α_2 of p_1 and then obtain a polynomial function p_2 of degree $n - 2$ such that

$$p(z) = (z - \alpha_1)(z - \alpha_2)p_2(z) \qquad (z \in \mathbf{C}).$$

We proceed in this fashion and the result follows after a finite number of steps.

The above theorem represents a polynomial function as the product of factors $z - \alpha_j$ where α_j is a zero of the function. In §6 we shall consider a generalization of this result that holds for arbitrary entire functions. For the present it is instructive to consider the simple case in which the entire function has only a finite number of zeros.

Theorem 8.9. *Let f be an entire function with no zeros. Then there is an entire function g such that $f = \exp \circ g$.*

Proof. If f is entire so also is f'. Since $\mathbf{C}$ is starlike and $0 \notin f(\mathbf{C})$ it follows from Proposition 6.4 that f has an analytic logarithm g on $\mathbf{C}$, i.e. there exists an entire function g such that $\exp \circ g = f$.

Theorem 8.10. *Let f be an entire function with zeros $\alpha_1, \alpha_2, \ldots, \alpha_n$, where α_j is a zero of order m_j. Then there is an entire function g such that*

$$f(z) = (z - \alpha_1)^{m_1}(z - \alpha_2)^{m_2}\ldots(z - \alpha_n)^{m_n} \exp(g(z)) \qquad (z \in \mathbf{C}).$$

Proof. By the corollary to Theorem 8.1 there is an entire function f_1 such that $f_1(\alpha_1) \neq 0$ and

$$f(z) = (z - \alpha_1)^{m_1}f_1(z) \qquad (z \in \mathbf{C}).$$

Clearly f_1 has zeros $\alpha_2, \ldots, \alpha_n$, α_j being a zero of order m_j. Applying the same argument to f_1 we obtain an entire function f_2 such that $Z_{f_2} = \{\alpha_3, \ldots, \alpha_n\}$ and

$$f_1(z) = (z - \alpha_2)^{m_2}f_2(z) \qquad (z \in \mathbf{C})$$

and so

$$f(z) = (z - \alpha_1)^{m_1}(z - \alpha_2)^{m_2}f_2(z) \qquad (z \in \mathbf{C}).$$

Proceeding in this manner we obtain an entire function h with no zeros such that

$$f(z) = (z - \alpha_1)^{m_1}(z - \alpha_2)^{m_2}\ldots(z - \alpha_n)^{m_n}h(z) \qquad (z \in \mathbf{C}).$$

The result now follows from Theorem 8.9.

PROBLEMS 8

8. Given $M \geqslant 0$ let $\mathscr{E}_M$ be the set of all those sequences $\{\alpha_n\}$ in $\mathscr{S}$ such that

$$\overline{\lim_{n\to\infty}}\ |\alpha_n|^{1/n} \leqslant M.$$

Show by two different methods that $\mathscr{E}_M$ is a subalgebra of $\mathscr{S}$.

9. Let f be an entire function such that

$$|f(z)| \leqslant M(1 + |z|^{\lambda}) \qquad (z \in \mathbf{C})$$

for some $M > 0$, $\lambda > 0$. Show that f is a polynomial function of degree at most n, where n is the largest integer such that $n \leqslant \lambda$.

10. (i) If f is an entire function such that $|f| \geqslant 1$, show that f is constant.

(ii) If f is an entire function such that $\operatorname{Re} f \leqslant 0$ or $\operatorname{Im} f \leqslant 0$, show that f is constant. (*Hint.* Consider $\exp \circ f$.)

(iii) If f is an entire function such that $\operatorname{Re} f$ or $\operatorname{Im} f$ has no zeros, show that f is constant.

11. Let $\alpha_j, \beta_j \in \mathbf{C}$ $(j = 0, 1, \ldots, n)$, where $\alpha_j \neq \alpha_k$ $(j \neq k)$. Show that there is exactly one polynomial function p of degree n such that $p(\alpha_j) = \beta_j$ $(j = 0, 1, \ldots, n)$. Show that

$$p(z) = \sum_{j=0}^{n} \beta_j \frac{p_j(z)}{p_j(\alpha_j)} \qquad (z \in \mathbf{C})$$

where

$$p_j(z) = (z - \alpha_0)\ldots(z - \alpha_{j-1})(z - \alpha_{j+1})\ldots(z - \alpha_n) \qquad (z \in \mathbf{C}).$$

8.4 Meromorphic functions

Given $f: D \to \mathbf{C}$ we say that f is *meromorphic* on D if the singularities of f are all poles. We denote by $\mathscr{M}(D)$ the set of all functions meromorphic on D. In this section we investigate some of the elementary properties of such functions. In particular we show that the set of poles of a meromorphic function has the same properties as the set of zeroes of a function analytic on D. On the algebraic side we show that $\mathscr{M}(D)$ is a field. Finally, the field $\mathscr{M}(\mathbf{C}^\infty)$ is shown to be precisely the field of rational functions.

Given $f \in \mathscr{M}(D)$ we write P_f for the set of poles of f. It is natural to ask if the set P_f has the same properties as those of Z_f, given by Theorem 8.3. It follows by definition that each pole is of finite order and is also an isolated singularity. But it is not directly obvious that P_f has no cluster point in D; on the face of it, f might be differentiable at a cluster point of P_f in D. We show below that this cannot occur.

Theorem 8.11. *If* $f \in \mathcal{M}(D)$ *then*

(i) *the set* P_f *has no cluster points in* D;
(ii) *if* K *is any compact subset of* P_f, *then* $K \cap P_f$ *is a finite set*;
(iii) *the set* P_f *is at most countable.*

Proof. (i) Let α be a cluster point of P_f with $\alpha \in D$. Choose $\{\alpha_n\} \subset P_f$ such that $\lim_{n\to\infty} \alpha_n = \alpha$. Since each α_n is a pole of f, Theorem 7.11 (iii) gives

$$\lim_{z\to\alpha_n} f(z) = \infty \qquad (n \in \mathbf{P}).$$

For each $n \in \mathbf{P}$ we may thus choose β_n such that $|\alpha_n - \beta_n| < 1/n$ and $|f(\beta_n)| > n$. Then $\lim_{n\to\infty} \beta_n = \alpha$ and $\lim_{n\to\infty} f(\beta_n) = \infty$. This shows that f cannot be differentiable at α, so that α is a singularity of f. Then α is a non-isolated singularity of f. This contradicts the hypothesis that $f \in \mathcal{M}(D)$ and so P_f has no cluster points in D.

(ii), (iii) These parts follow exactly as in Theorem 8.3.

Recall that whenever we wish to assign values of a function at its poles we adopt the convention

$$f(\alpha) = \infty \qquad (\alpha \in P_f).$$

This means that meromorphic functions are continuous as functions from D to $\mathbf{C}^\infty$.

We are already familiar with one class of meromorphic functions. If D is any domain in $\mathbf{C}$ or if $D = \mathbf{C}^\infty$ then each rational function belongs to $\mathcal{M}(D)$. This is clear from the fundamental theorem of algebra. Indeed if q is a rational function then there exist $k \in \mathbf{C}$, $\alpha_1, \ldots, \alpha_m \in \mathbf{C}$ (not necessarily distinct), and $\beta_1, \ldots, \beta_n \in \mathbf{C}$ (not necessarily distinct) such that

$$q(z) = k\,\frac{(z-\alpha_1)\ldots(z-\alpha_m)}{(z-\beta_1)\ldots(z-\beta_n)}$$

where $\{\alpha_1, \ldots, \alpha_m\} \cap \{\beta_1, \ldots, \beta_n\} = \varnothing$. It is a well-known fact of algebra that the set of rational functions forms a field. The same is true of the meromorphic functions, but care is needed in defining the sum and product of two meromorphic functions.

Let $f \in \mathcal{M}(D)$ and let $\alpha \in D$. By arguing as in Theorem 8.1 we easily see that there is $m \in \mathbf{Z}$ and $f_1 \in \mathcal{M}(D)$ such that

$$f(z) = (z-\alpha)^m f_1(z) \qquad (z \in D)$$

where f_1 is regular at α with $f_1(\alpha) \neq 0$. Similarly, if $g \in \mathcal{M}(D)$ there is $n \in \mathbf{Z}$ and $g_1 \in \mathcal{M}(D)$ such that

$$g(z) = (z-\alpha)^n g_1(z) \qquad (z \in D)$$

where g_1 is regular at α with $g_1(\alpha) \neq 0$. It follows readily that we may now define $f + g$ and fg by

$$(f + g)(\alpha) = \lim_{z \to \alpha} (f(z) + g(z))$$

$$(fg)(\alpha) = \lim_{z \to \alpha} (f(z)g(z)).$$

If $\alpha \notin P_f \cup P_g$ the above definitions agree with the usual pointwise definitions. It is now routine (though tedious) to verify that $\mathcal{M}(D)$ is a complex linear algebra.

Theorem 8.12. *$\mathcal{M}(D)$ is a field.*

Proof. In view of the above remarks it will be sufficient to show that $f \in \mathcal{M}(D)$ implies $1/f \in \mathcal{M}(D)$. Given $f \in \mathcal{M}(D)$ it is clear that $1/f$ is analytic on $D \setminus \{P_f \cup Z_f\}$. By Theorem 7.11 (i) $1/f$ is differentiable at each point of P_f (and has a zero there). By Theorem 7.11 (ii) $1/f$ has a pole at each point of Z_f. Thus the only singularities of $1/f$ are poles and so $1/f \in \mathcal{M}(D)$ as required.

The student may recall that the field of rational functions is the field of quotients obtained from the integral domain of polynomial functions. This is simply to say that the rational functions are of the form p_1/p_2 where p_1, p_2 are polynomial functions. (We have of course the convention that $p_1/p_2 = p_3/p_4$ if $p_1p_4 = p_2p_3$; we may thus suppose, as in our original definition of rational function, that p_1, p_2 have no common zero.) If D is any domain in $\mathbf{C}$ it is clear that

$$\{f/g: f, g \in \mathcal{A}(D)\} \subset \mathcal{M}(D).$$

In fact the above sets coincide so that $\mathcal{M}(D)$ is the field of quotients obtained from the integral domain $\mathcal{A}(D)$. We shall prove this result in §6 for the case $D = \mathbf{C}$. The result fails spectacularly for the case $D = \mathbf{C}^\infty$. Indeed we have seen from Liouville's theorem that $\mathcal{A}(\mathbf{C}^\infty)$ consists of the constant functions. This means that $\mathcal{A}(\mathbf{C}^\infty)$ is its own field of quotients. On the other hand we have already remarked that $\mathcal{M}(\mathbf{C}^\infty)$ contains all the rational functions. We show below that $\mathcal{M}(\mathbf{C}^\infty)$ actually consists of the rational functions. We need first a useful technical lemma.

Recall that if $f: D \to \mathbf{C}$ has an isolated singularity at α, then the principal part of f at α is denoted by $S(f; \alpha)$.

Lemma 8.13. *Let $f \in \mathcal{M}(D)$ and let $\alpha_1, \ldots, \alpha_n \in P_f$. Then $f - \sum_{j=1}^{n} S(f; \alpha_j)$ is regular at each α_j ($j = 1, \ldots, n$).*

Proof. Recall from Chapter 7 that each $S(f; \alpha_j)$ is analytic on $\mathbf{C} \setminus \{\alpha_j\}$, and $f - S(f; \alpha_j)$ is regular at α_j. This proves the result if $n = 1$. If $n > 1$

and $f_1 = f - S(f; \alpha_1)$ then f_1 is regular at α_1. Since $S(f; \alpha_1)$ is regular at α_2 it follows that $S(f_1; \alpha_2) = S(f; \alpha_2)$ and so

$$f_1 - S(f_1; \alpha_1) = f - S(f; \alpha_1) - S(f; \alpha_2).$$

Since $f_1 - S(f_1; \alpha_2)$ is regular at α_1 and α_2 this proves the result if $n = 2$. We continue in the fashion and the proof is completed in a finite number of steps.

Observe in passing that if $P_f = \{\alpha_1, \ldots, \alpha_n\}$ then there is $g \in \mathscr{A}(D)$ such that

$$f = g + \sum_{j=1}^{n} S(f; \alpha_j).$$

This remark will be generalized in §9.

Theorem 8.14. *$\mathscr{M}(\mathbf{C}^\infty)$ consists of the rational functions.*

Proof. It is clearly sufficient to show that each f in $\mathscr{M}(\mathbf{C}^\infty)$ is a rational function. Either f is regular at ∞ or f has a pole at ∞. In either case there is $R > 0$ such that f is analytic on $\nabla(0, R)$. By Theorem 8.11 (ii) f has a finite number of poles in the compact set $\bar{\Delta}(0, R)$, say $\alpha_1, \ldots, \alpha_n$. Let $g = f - \sum_{j=1}^{n} S(f; \alpha_j)$, so that g is an entire function by Lemma 8.13. Moreover $g - S(g; \infty)$ is regular at ∞. Since $S(g; \infty)$ is a polynomial function we have $g - S(g; \infty) \in \mathscr{A}(\mathbf{C}^\infty)$ and so $g - S(g; \infty)$ is constant by the corollary to Liouville's theorem. It follows that f is a rational function as required.

It is convenient at this point to remark on an obvious analogue of Theorem 8.10 for meromorphic functions on $\mathbf{C}$. Suppose that $f \in \mathscr{M}(\mathbf{C})$ with $Z_f = \{\alpha_1, \ldots, \alpha_m\}$, $P_f = \{\beta_1, \ldots, \beta_n\}$ where α_j is a zero of order μ_j and β_j is a pole of order ν_j. Then there is an entire function g such that

$$f(z) = \frac{(z - \alpha_1)^{\mu_1} \ldots (z - \alpha_m)^{\mu_m}}{(z - \beta_1)^{\nu_1} \ldots (z - \beta_n)^{\nu_n}} \exp(g(z)) \qquad (z \in \mathbf{C}).$$

The proof proceeds by the technique of Theorem 8.10. This result also will be generalized in §6.

Some of the results of this section (e.g. Lemma 8.13) apply equally well to functions all of whose singularities are isolated. We shall concentrate our attention in the text on the more interesting class of meromorphic functions and relegate such extensions to the problems.

PROBLEMS 8

12. (i) If $f(z) = \sin\left(\frac{i}{z}\right) \operatorname{cosec}\left(\frac{1}{z}\right)$ $(z \in \mathbf{C})$ determine Z_f, P_f and their sets of cluster points.

(ii) If $f(z) = \operatorname{cosec}\left(\frac{1}{1-z}\right)$ $(z \in \mathbf{C})$ show that f is meromorphic on $\Delta(0, 1)$ but not on $\mathbf{C}$.

13. Let $f: D \to \mathbf{C}$ have isolated singularities at each point of $E \subset D$ and let α be a cluster point of E in D.

(i) If each point of E is a removable singularity show that f may be differentiable at α.

(ii) If each point of E is an essential singularity show that f cannot be differentiable at α.

14. Let $\mathscr{I}(D)$ denote the set of functions $f: D \to \mathbf{C}$ all of whose singularities are non-removable isolated singularities.

(i) State and prove the analogue of Theorem 8.11 for $\mathscr{I}(D)$.

(ii) Show that $\mathscr{I}(D)$ is not a field.

(iii) State and prove the analogue of Lemma 8.13 for $\mathscr{I}(D)$.

(iv) Describe a general element of $\mathscr{I}(\mathbf{C}^\infty)$.

15. Let f be a rational function whose zeros and poles are all of even order. Show that there is a rational function g such that $f = g^2$.

16. (i) Let f be a rational function whose zeros and poles are all simple. Suppose the zeros are $\alpha_1, \ldots, \alpha_n$ and the poles are $\frac{1}{\alpha_1}, \ldots, \frac{1}{\alpha_n}$, where $|\alpha_j| \neq 1$ $(j = 1, \ldots, n)$. Show that $|f|$ is constant on $C(0, 1)$.

(ii) Given any rational function f with no poles on $C(0, 1)$, show that there is a rational function g with no poles on $\Delta(0, 1)$ such that $|g| = |f|$ on $C(0, 1)$.

8.5 Convergence in $\mathscr{A}(D)$

In this section we shall construct a metric on the set $\mathscr{A}(D)$. It turns out to be natural to define the metric on the larger set $\mathscr{C}(D)$ and then induce the metric on $\mathscr{A}(D)$. We show that convergence in this metric is equivalent to uniform convergence on the compact subsets of D. We show further that the metric space $\mathscr{C}(D)$ is complete. It then follows via a converse of Cauchy's theorem that $\mathscr{A}(D)$ is a closed subset of the metric space $\mathscr{C}(D)$. This means that if $f_n \in \mathscr{A}(D)$ $(n \in \mathbf{P})$ and if $f_n \to f$ uniformly on each compact subset of D then $f \in \mathscr{A}(D)$. Moreover for each $k \in \mathbf{P}$ we even have

$f_n^{(k)} \to f^{(k)}$ uniformly on each compact subset of D. This result is extremely important in complex analysis; two simple applications are given in this section and further significant applications appear in §6 and §9. At the end of this section we discuss (without proof) some of the deeper topological properties of the metric space $\mathscr{A}(D)$.

Let K be any compact subset of $\mathbf{C}$. If

$$d(f, g) = \sup\{|f(z) - g(z)| : z \in K\} \qquad (f, g \in \mathscr{C}(K))$$

recall that d is a metric (the usual metric) on $\mathscr{C}(K)$. The compactness of K ensures that the functions in $\mathscr{C}(K)$ are bounded and hence d is finite valued. Now let D be any domain in $\mathbf{C}$. The functions in $\mathscr{C}(D)$ are not all bounded (proof?) and a little care is needed to construct a useful metric on $\mathscr{C}(D)$. Let $\{K_n\}$ be a fixed compact exhaustion of D (cf. Proposition 2.7). Thus each K_n is compact, $D = \bigcup\{K_n : n \in \mathbf{P}\}$ and given any compact subset K of D there is $n \in \mathbf{P}$ such that $K \subset K_n$. Let d_n denote the usual metric on $\mathscr{C}(K_n)$. Given $f, g \in \mathscr{C}(D)$, $n \in \mathbf{P}$ let

$$\sigma_n(f, g) = \frac{d_n(f, g)}{1 + d_n(f, g)}$$

where, for simplicity of notation, we do not distinguish f, g from $f|_{K_n}$, $g|_{K_n}$ respectively. Observe that $0 \leqslant \sigma_n(f, g) \leqslant 1$ $(n \in \mathbf{P})$ so that the series $\sum 2^{-n}\sigma_n(f, g)$ converges by the comparison test. We may thus define σ on $\mathscr{C}(D) \times \mathscr{C}(D)$ by

$$\sigma(f, g) = \sum_{n=1}^{\infty} 2^{-n}\sigma_n(f, g).$$

Proposition 8.15. *($\mathscr{C}(D)$, σ) is a metric space.*

Proof. Given $f, g \in \mathscr{C}(D)$ it follows easily from the definition of σ that $0 \leqslant \sigma(f, g) < +\infty$, $\sigma(f, g) = \sigma(g, f)$ and $\sigma(f, f) = 0$. If $\sigma(f, g) = 0$ then $\sigma_n(f, g) = 0$ $(n \in \mathbf{P})$, and so $f(z) = g(z)$ $(z \in K_n, n \in \mathbf{P})$. Since $D = \bigcup\{K_n : n \in \mathbf{P}\}$ it follows that $f = g$. Finally, suppose $f, g, h \in \mathscr{C}(D)$. Since d_n is a metric on $\mathscr{C}(K_n)$ we have

$$d_n(f, g) \leqslant d_n(f, h) + d_n(h, g) \qquad (n \in \mathbf{P}).$$

Given $r, s, t \geqslant 0$ with $t \leqslant r + s$ it is easy to check that

$$\frac{t}{1+t} \leqslant \frac{r}{1+r} + \frac{s}{1+s}.$$

It follows that

$$\sigma_n(f, g) \leqslant \sigma_n(f, h) + \sigma_n(h, g)$$

and hence

$$\sigma(f, g) \leqslant \sigma(f, h) + \sigma(h, g).$$

Therefore σ is a metric on $\mathscr{C}(D)$.

Corollary. *$\mathscr{A}(D)$ is a metric space with the relative metric σ.*

The above metric σ is rather complicated but convergence in $(\mathscr{C}(D), \sigma)$ has a simple characterization. Recall that convergence in $(\mathscr{C}(K), d)$ is equivalent to uniform convergence on K.

Proposition 8.16. *Let $f_n \in \mathscr{C}(D)$ $(n \in \mathbf{P})$, $f \in \mathscr{C}(D)$. Then the following statements are equivalent.*

(i) $\lim_{n \to \infty} \sigma(f_n, f) = 0.$

(ii) *For each compact subset K of D, $f_n \to f$ uniformly on K.*

Proof. (i) $\Rightarrow$ (ii). Let $\sigma(f_n, f) \to 0$ and let K be any compact subset of D. Choose $m \in \mathbf{P}$ such that $K \subset K_m$. Since $\sigma(f, f_n) \to 0$ it follows that $\sigma_m(f, f_n) \to 0$ and hence $d_m(f, f_n) \to 0$. Since $K \subset K_m$ we conclude that $f_n \to f$ uniformly on K.

(ii) $\Rightarrow$ (i). Let $f_n \to f$ uniformly on each compact subset of D. Since each K_m is compact we deduce in particular that $d_m(f, f_n) \to 0$ and so $\sigma_m(f, f_n) \to 0$ for each $m \in \mathbf{P}$. Let $\epsilon > 0$. Since $\sigma_m \leqslant 1$ $(m \in \mathbf{P})$ we may choose $m_0 \in \mathbf{P}$ such that

$$\sum_{m=m_0+1}^{\infty} 2^{-m}\sigma_m(f, f_n) < \tfrac{1}{2}\epsilon \qquad (n \in \mathbf{P}).$$

Since $\sigma_m(f, f_n) \to 0$ for $m = 1, 2, \ldots, m_0$, we may now choose $n_0 \in \mathbf{P}$ such that

$$\sigma_m(f, f_n) < \tfrac{1}{2}\epsilon \qquad (n > n_0,\ m = 1, 2, \ldots, m_0).$$

For $n > n_0$ we now have

$$\begin{aligned}\sigma(f, f_n) &< \sum_{m=1}^{m_0} 2^{-m}\tfrac{1}{2}\epsilon + \tfrac{1}{2}\epsilon \\ &< \epsilon.\end{aligned}$$

This shows that $\sigma(f, f_n) \to 0$ as required.

The metric σ was defined in terms of a particular compact exhaustion of D, but we can now see that the choice of compact exhaustion is relatively unimportant. Suppose in fact that $\{K_n'\}$ is any other compact exhaustion of D and that τ is the corresponding metric on $\mathscr{C}(D)$ (defined in the same manner as σ). In general the metrics σ and τ will not be the same (example?). However it is clear from the proof of the above proposition that $\tau(f, f_n) \to 0$ if an only if $f_n \to f$ uniformly on each compact subset of D. Thus the metric spaces $(\mathscr{C}(D), \sigma)$ and $(\mathscr{C}(D), \tau)$ have precisely the same convergent sequences and hence the same closed sets. Equivalently, the identity mapping is a homeomorphism of $(\mathscr{C}(D), \sigma)$ with $(\mathscr{C}(D), \tau)$.

We show below that $(\mathscr{C}(D), \sigma)$ is a complete metric space. This will facilitate the subsequent proof that $(\mathscr{A}(D), \sigma)$ is a complete metric space.

Proposition 8.17.
(i) $(\mathscr{C}(K), d)$ *is complete for any compact metric space* K.
(ii) $(\mathscr{C}(D), \sigma)$ *is complete for any domain* D *in* $\mathbf{C}$.

Proof. (i) Let $\{f_n\}$ be a Cauchy sequence in $(\mathscr{C}(K), d)$. Since

$$|f_m(z) - f_n(z)| \leqslant d(f_m, f_n) \qquad (m, n \in \mathbf{P}, z \in K)$$

it follows that $\{f_n(z)\}$ is a Cauchy sequence in $\mathbf{C}$ for each $z \in K$. Since $\mathbf{C}$ is complete we may now define

$$f(z) = \lim_{n \to \infty} f_n(z) \qquad (z \in K).$$

Let $\epsilon > 0$. Since $\{f_n\}$ is a Cauchy sequence in $(\mathscr{C}(K), d)$ there is $n_0 \in \mathbf{P}$ such that

$$|f_m(z) - f_n(z)| < \tfrac{1}{2}\epsilon \qquad (m, n > n_0, z \in K).$$

For each $z \in K$ we may choose $m_z > n_0$ such that

$$|f(z) - f_{m_z}(z)| < \tfrac{1}{2}\epsilon.$$

For each $z \in K$, $n > n_0$ we now have

$$|f(z) - f_n(z)| \leqslant |f(z) - f_{m_z}(z)| + |f_{m_z}(z) - f_n(z)| < \epsilon.$$

This shows that $f_n \to f$ uniformly on K. It follows from Proposition 1.14 that $f \in \mathscr{C}(K)$. Finally, it is clear from above that $d(f, f_n) \to 0$, and so the proof is complete.

(ii) Let $\{f_n\}$ be a Cauchy sequence in $(\mathscr{C}(D), \sigma)$. Arguing as above we see that $\{f_n(z)\}$ is a Cauchy sequence in $\mathbf{C}$ for each $z \in D$ and so we may define

$$f(z) = \lim_{n \to \infty} f_n(z) \qquad (z \in D).$$

Moreover for each $m \in \mathbf{P}$ we may check that $\{f_n\}$ is a Cauchy sequence in $(\mathscr{C}(K_m), d_m)$. By part (i) above there is $g_m \in \mathscr{C}(K_m)$ such that $d_m(g_m, f_n) \to 0$ $(m \in \mathbf{P})$. By the uniqueness of limits of convergent sequences we have $g_m = f|_{K_m}$ $(m \in \mathbf{P})$. Therefore $f_n \to f$ uniformly on each K_m and hence uniformly on each compact subset of D. Given $z \in D$ there is $r > 0$ such that $\bar{\Delta}(z, r) \subset D$. Since $f_n \to f$ uniformly on $\Delta(z, r)$ it follows from Propositions 1.13, 1.14 that f is continuous at z. Thus $f \in \mathscr{C}(D)$. Since $f_n \to f$ uniformly on each compact subset of D we conclude from Proposition 8.16 that $\sigma(f, f_n) \to 0$ as required.

In preparation for the main result of this section we now prove a converse of Cauchy's theorem.

Theorem 8.18 (Morera). *Let $f \in \mathscr{C}(D)$ and suppose that for each $\alpha \in D$ there is $\Delta(\alpha, R) \subset D$ such that $\int_\Gamma f = 0$ for every simple closed contour Γ in $\Delta(\alpha, R)$. Then $f \in \mathscr{A}(D)$.*

Proof. Given $\alpha \in D$ and the corresponding $\Delta(\alpha, R)$ we have $\int_\Gamma f = 0$ for every triangular contour in $\Delta(\alpha, R)$. It follows from the argument employed in Theorem 6.2 that f has a primitive on $\Delta(\alpha, R)$, say g. Since g is analytic on $\Delta(\alpha, R)$ it follows from Theorem 7.3 that $f = g'$ is analytic on $\Delta(\alpha, R)$. Since α was any point of D we conclude that $f \in \mathscr{A}(D)$.

Theorem 8.19. *Let $f_n \in \mathscr{A}(D)$ $(n \in \mathbf{P})$ be such that $f_n \to f$ uniformly on each compact subset of D. Then $f \in \mathscr{A}(D)$ and for each $k \in \mathbf{P}$, $f_n^{(k)} \to f^{(k)}$ uniformly on each compact subset of D.*

Proof. It follows as in Proposition 8.17 that $f \in \mathscr{C}(D)$. Given $\alpha \in D$ there is $R > 0$ such that $\bar{\Delta}(\alpha, R) \subset D$. Let Γ be any simple closed contour in $\Delta(\alpha, R)$. Since f_n is analytic on $\Delta(\alpha, R)$, Cauchy's theorem gives $\int_\Gamma f_n = 0$ $(n \in \mathbf{P})$. Since Γ is a compact subset of D we have $f_n \to f$ uniformly on Γ and then **CI 7** gives

$$\int_\Gamma f = \lim_{n\to\infty} \int_\Gamma f_n = 0.$$

Therefore $f \in \mathscr{A}(D)$ by Theorem 8.18.

Let $k \in \mathbf{P}$ and let $r = \frac{1}{2}R$. For each $z \in \Delta(\alpha, R)$ Theorem 7.3 gives

$$f^{(k)}(z) = \frac{k!}{2\pi i} \int_{C(\alpha,R)} \frac{f(w)}{(w-z)^{k+1}}\,\mathrm{d}w$$

$$f_n^{(k)}(z) = \frac{k!}{2\pi i} \int_{C(\alpha,R)} \frac{f_n(w)}{(w-z)^{k+1}}\,\mathrm{d}w.$$

For each $z \in \bar{\Delta}(\alpha, r)$ we now have

$$|f^{(k)}(z) - f_n^{(k)}(z)| = \left|\frac{k!}{2\pi i} \int_{C(\alpha,R)} \frac{f(w) - f_n(w)}{(w-z)^{k+1}}\,\mathrm{d}w\right|$$

$$\leqslant \frac{k!}{2\pi}\, 2\pi R \left(\frac{2}{R}\right)^{k+1} \sup_{C(\alpha,R)} |f - f_n|.$$

Since $f_n \to f$ uniformly on $C(\alpha, R)$ it follows that $f_n^{(k)} \to f^{(k)}$ uniformly on $\bar{\Delta}(\alpha, r)$. If K is any compact subset of D then $\{\Delta(\alpha, r) : \alpha \in K\}$ is an open cover of K and so has a finite subcover, say $\{\Delta(\alpha_j, r_j) : j = 1, \ldots, m\}$. Since $f_n^{(k)} \to f^{(k)}$ uniformly on each $\Delta(\alpha_j, r_j)$, it follows easily that $f_n^{(k)} \to f^{(k)}$ uniformly on K. This completes the proof.

Corollary. *$(\mathscr{A}(D), \sigma)$ is a complete metric space.*

Proof. Let $\{f_n\}$ be Cauchy in $(\mathscr{A}(D), \sigma)$. Then $\{f_n\}$ is Cauchy in $(\mathscr{C}(D), \sigma)$ and so by Proposition 8.17 there is $f \in \mathscr{C}(D)$ such that $\sigma(f, f_n) \to 0$. By Proposition 8.16 and the above theorem $f \in \mathscr{A}(D)$. Since $\sigma(f, f_n) \to 0$ the proof is complete.

The above theorem will be a key tool in the next section. For the present we consider applications to the Riemann zeta function and the infinite Laplace transform. Recall that $n^{-z} = \exp(-z \log n)$ so that $|n^{-z}| = n^{-x}$ where $x = \operatorname{Re} z$. It follows from the comparison test that the series $\sum n^{-z}$ converges absolutely for $\operatorname{Re} z > 1$. The *Riemann zeta function* is defined by

$$\zeta(z) = \sum_{n=1}^{\infty} n^{-z} \qquad (\operatorname{Re} z > 1).$$

Example 8.20. *The Riemann zeta function is analytic on $\{z: \operatorname{Re} z > 1\}$.*

Proof. Given $n \in \mathbf{P}$ we define f_n on $\mathbf{C}$ by

$$f_n(z) = \sum_{j=1}^{n} j^{-z} \qquad (z \in \mathbf{C}).$$

It is then clear that each f_n is an entire function. Given $\delta > 0$ and z such that $\operatorname{Re} z \geqslant 1 + \delta$, we have

$$|\zeta(z) - f_n(z)| = \left| \sum_{j=n+1}^{\infty} j^{-z} \right| \leqslant \sum_{j=n+1}^{\infty} j^{-1-\delta}$$

and so $f_n \to \zeta$ uniformly on $\{z: \operatorname{Re} z \geqslant 1 + \delta\}$. Since δ was arbitrary it follows easily that $f_n \to \zeta$ uniformly on every compact subset of the domain $\{z: \operatorname{Re} z > 1\}$. By Theorem 8.19 ζ is analytic on $\{z: \operatorname{Re} z > 1\}$. Moreover the derivatives of ζ are given by

$$\zeta^{(k)}(z) = \sum_{n=1}^{\infty} (-1)^k \frac{(\log n)^k}{n^z} \qquad (\operatorname{Re} z > 1, k \in \mathbf{P}).$$

We shall show in §10.4 how the Riemann zeta function may be extended to a meromorphic function on $\mathbf{C}$.

Let f be a complex function defined on $\{t: t \geqslant 0\}$ and let f be Riemann integrable on any bounded subinterval of $\{t: t \geqslant 0\}$. We say that f is of *exponential type* with *index* λ if there is $M > 0$ such that

$$|f(t)| \leqslant M e^{\lambda t} \qquad (t \geqslant 0).$$

Given such f and $\operatorname{Re} z > \lambda$, the infinite integral

$$\int_0^{\infty} e^{-zt} f(t)\, dt$$

converges absolutely, i.e. $\int_0^R |e^{-zt}f(t)|\,dt$ converges as $R \to +\infty$. The (one-sided) *infinite Laplace transform* of f is defined by

$$F(z) = \int_0^\infty e^{-zt}f(t)\,dt \qquad (\operatorname{Re} z > \lambda).$$

The Laplace transform is a useful tool in the theory of differential equations and elsewhere.

Example 8.21. *If f is of exponential type with index λ, the Laplace transform of f is analytic on $\{z: \operatorname{Re} z > \lambda\}$.*

Proof. For each $n \in \mathbf{P}$ let

$$F_n(z) = \int_0^n e^{-zt}f(t)\,dt \qquad (z \in \mathbf{C}).$$

By Example 7.5 each F_n is an entire function. Given $\delta > 0$ and z such that $\operatorname{Re} z \geqslant \lambda + \delta$ we have

$$\begin{aligned}|F(z) - F_n(z)| &= \left|\int_n^\infty e^{-zt}f(t)\,dt\right| \\ &\leqslant \int_n^\infty e^{-(\lambda+\delta)t}M\,e^{\lambda t}\,dt \\ &= \frac{M}{\delta}\,e^{-\delta n}.\end{aligned}$$

Therefore $F_n \to F$ uniformly on $\{z: \operatorname{Re} z \geqslant \lambda + \delta\}$ and it follows as in the last example that F is analytic on $\{z: \operatorname{Re} z > \lambda\}$.

The student may observe that Theorem 8.19 leads to an alternative proof that a power series function is analytic on its open disc of convergence. In fact we need only recall from Proposition 4.8 that a power series converges uniformly on closed subdiscs of the disc of convergence. Since polynomial functions are entire, the result follows from Theorem 8.19. The result may be expressed in different terms as follows. Given any open disc Δ the polynomial functions are dense in $(\mathscr{A}(\Delta), \sigma)$. This leads us to ask for which domains D the polynomial functions are dense in $(\mathscr{A}(D), \sigma)$. A famous theorem of Runge asserts that the polynomial functions are dense in $(\mathscr{A}(D), \sigma)$ provided that D is *simply connected*, i.e. $\mathbf{C}^\infty \setminus D$ is connected. Recall from Problem 2.27 that every starlike domain is simply connected. Runge's theorem leads very easily to an extension of Cauchy's theorem to the case of a simply connected domain D. In fact given $f \in \mathscr{A}(D)$ and any simple closed contour $\Gamma \subset D$ we may choose a sequence of polynomial functions $\{p_n\}$ that converge to f uniformly on Γ. It follows from **CI 6, 7** that

$$\int_\Gamma f = \lim_{n\to\infty} \int_\Gamma p_n = 0.$$

(A proof of Runge's theorem is given in Saks and Zygmund, *Analytic Functions* (second English edition) Warsaw 1965.)

Finally we mention one further deep result concerning the metric space $(\mathscr{A}(D), \sigma)$. If $\{f_n\}$ is a sequence in $\mathscr{A}(D)$ that is uniformly bounded on each compact subset of D, then some subsequence of $\{f_n\}$ converges in $(\mathscr{A}(D), \sigma)$. (Observe the analogy with the result that every bounded real sequence has a convergent subsequence.) This remarkable result is called Montel's theorem. It is a key tool in the standard proofs of the Riemann mapping theorem (cf. page 243).

PROBLEMS 8

17. (i) Let D be any domain in $\mathbf{C}$ and let $H^\infty(D)$ denote the set of bounded functions that are analytic on D. Show that $H^\infty(D)$ is a complex Banach algebra with the supremum norm (cf. Problem 2.6).

(ii) Let D be any bounded domain in $\mathbf{C}$ and let A_D denote the set of functions in $\mathscr{C}(D^-)$ that are analytic on D. Show that A_D is a complex Banach algebra with the supremum norm. ($A_{\Delta(0,1)}$ is usually called the *disc algebra*. For further remarks on $H^\infty(\Delta(0, 1))$ and $A_{\Delta(0,1)}$ see page 303.)

18. Let $D = \{z: -1 < \operatorname{Im} z < 1\}$ and let

$$f(z) = \sum_{n=1}^{\infty} e^{-n} \sin(nz) \qquad (z \in D).$$

Show that $f \in \mathscr{A}(D)$ and find a 'closed expression' for f.

19. For each $n \in P$ let

$$f_n(z) = \sum_{j=1}^{\infty} \frac{nz^j}{nj(j+1)+1} \qquad (z \in \Delta(0, 1)).$$

Show that $f_n \in \mathscr{A}(\Delta(0, 1))$ $(n \in \mathbf{P})$ and that $\{f_n\}$ converges uniformly on $\Delta(0, 1)$ while $\{f_n'''\}$ does not converge uniformly on $\Delta(0, 1)$. Compare this example with Theorem 8.19. Also, find a closed expression for the limit of $\{f_n\}$.

20. For each $n \in \mathbf{P}$ let

$$f_n(z) = \frac{1}{n} \sin(nz) \qquad (z \in \mathbf{C}).$$

Show that $f_n \to 0$ uniformly on $\mathbf{R}$ while $\{f_n'\}$ converges only at the points $(n + \frac{1}{2})\pi$ $(n \in \mathbf{Z})$. Give two different proofs that $\{f_n\}$ fails to converge uniformly on any domain of $\mathbf{C}$.

21. Let $f \in \mathscr{C}(\mathbf{R})$ be such that there is $M > 0$ with

$$\int_{-m}^{n} |f(t)| \, \mathrm{d}t \leqslant M \qquad (m, n \in \mathbf{P}).$$

If

$$F(z) = \frac{1}{2\pi i} \int_{-\infty}^{\infty} \frac{f(t)}{t - z} \, \mathrm{d}t \qquad (z \in \mathbf{C} \setminus \mathbf{R})$$

show that F is analytic on $\mathbf{C} \setminus \mathbf{R}$ and that

$$\lim_{y \to 0+} \{F(x + iy) - F(x - iy)\} = f(x) \qquad (x \in \mathbf{R}).$$

Evaluate F in the case in which $f(t) = \dfrac{1}{1 + t^2}$.

22. Recall that $\mathscr{H}(D)$ denotes the set of harmonic functions on D. Show that $\mathscr{H}(D)$ is a closed subset of $(\mathscr{C}(D), \sigma)$.

23. Show that

$$(1 - 2^{1-z})\zeta(z) = \sum_{n=1}^{\infty} \frac{(-1)^{n+1}}{n^z} \qquad (\operatorname{Re} z > 1).$$

24. Let f be of exponential type with index λ and let F be the Laplace transform of f. Show that

$$F^{(n)}(z) = \int_0^{\infty} (-t)^n \, \mathrm{e}^{-zt} f(t) \, \mathrm{d}t \qquad (\operatorname{Re} z > \lambda).$$

If f is such that f' is of exponential type with index λ, find the Laplace transform of f' in terms of F.

25. Find the Laplace transforms of the following functions.

(i) $f(t) = \mathrm{e}^{\alpha t}$ $\quad (t \geqslant 0, \alpha \in \mathbf{C})$

(ii) $f(t) = t^n$ $\quad (t \geqslant 0, n = 0, 1, 2, \ldots)$

(iii) $f(t) = \sin(t + a)$ $\quad (t \geqslant 0, a \in \mathbf{R})$.

26. Show that the polynomial functions are not dense in $(\mathscr{A}(\Delta'(0, 1)), \sigma)$. Generalize this result.

8.6 Weierstrass expansions

In this section we derive the Weierstrass expansions of entire functions and meromorphic functions on $\mathbf{C}$. These expansions are given in terms of infinite products in which the zeros and poles appear in the expected manner. The crux of the theory rests in constructing an entire function with arbitrarily prescribed zeros. As an application of the results we establish that the set of meromorphic functions on **C** is the field of quotients obtained from the integral domain of entire functions. In order to keep the exposition brief we refrain from developing the general theory of infinite products.*

It is trivial to construct entire functions with a given finite number of zeros. In fact

$$f(z) = (z - \alpha_1)^{\mu_1}(z - \alpha_2)^{\mu_2}\ldots(z - \alpha_n)^{\mu_n} \qquad (z \in \mathbf{C})$$

gives an entire function with $Z_f = \{\alpha_1, \ldots, \alpha_n\}$, α_j being a zero of order μ_j. In order to extend this to deal with an infinite number of zeros it is natural to investigate infinite products. A moment's thought will convince the student that it is better to consider products of the form

$$\left(1 - \frac{z}{\alpha_1}\right)^{\mu_1}\left(1 - \frac{z}{\alpha_2}\right)^{\mu_2}\ldots\left(1 - \frac{z}{\alpha_n}\right)^{\mu_n}.$$

It is convenient to work up to the statement of our first result on the convergence of infinite products.

Let D be any domain in **C** and let $\{u_n\}$ be a sequence in $\mathscr{A}(D)$ such that $\sum|u_n|$ converges uniformly on each compact subset of D. For $n \in \mathbf{P}$ let

$$H_n(z) = (1 + u_1(z))(1 + u_2(z))\ldots(1 + u_n(z)).$$

We then have

$$H_{n+1}(z) - H_n(z) = u_{n+1}(z)H_n(z).$$

Given $m > n$ it follows that

$$\begin{aligned} |H_m(z) - H_n(z)| &= \left|\sum_{j=n+1}^{m} u_{j+1}(z)H_j(z)\right| \\ &\leqslant \sum_{j=n+1}^{m} |u_{j+1}(z)| \exp(|u_1(z)|)\ldots\exp(|u_j(z)|) \\ &\leqslant \exp\left(\sum_{j=1}^{\infty} |u_j(z)|\right) \sum_{j=n+1}^{m} |u_{j+1}(z)|. \end{aligned}$$

Let K be a compact subset of D and let $\epsilon > 0$. Since $\sum |u_n|$ converges uniformly on K there is $M > 0$ such that $\exp\left(\sum_{j=1}^{\infty} |u_j(z)|\right) \leqslant M$ $(z \in K)$. It follows that there is $n_0 \in \mathbf{P}$ such that

$$|H_m(z) - H_n(z)| < M\epsilon \qquad (m, n > n_0,\ z \in K)$$

* See for example Hille, *Analytic Function Theory*, vol. 1, Blaisdell 1963.

and thus $\{H_n\}$ is a Cauchy sequence in $(\mathscr{A}(D), \sigma)$. By the Corollary to Theorem 8.19 $\{H_n\}$ converges in $(\mathscr{A}(D), \sigma)$. We write

$$\lim_{n\to\infty} H_n(z) = \prod_{n=1}^{\infty} (1 + u_n(z)).$$

Proposition 8.22. *Let $\{u_n\} \subset \mathscr{A}(D)$ be such that $\sum |u_n|$ converges uniformly on each compact subset of D and let*

$$f(z) = \prod_{n=1}^{\infty} (1 + u_n(z)) \qquad (z \in D).$$

Then $f \in \mathscr{A}(D)$ and $f(z) = 0$ if and only if $1 + u_n(z) = 0$ for some $n \in \mathbf{P}$.

Proof. We have established above that $f \in \mathscr{A}(D)$. If $1 + u_m(z) = 0$ then $H_n(z) = 0$ $(n \geqslant m)$ and so $f(z) = 0$. Suppose conversely that $1 + u_n(z) \neq 0$ $(n \in \mathbf{P})$. Then $H_n(z) \neq 0$ $(n \in \mathbf{P})$ and we easily verify for $m > n$ that

$$\left|\frac{H_m(z)}{H_n(z)} - 1\right| \leqslant (1 + |u_{n+1}(z)|)\ldots(1 + |u_n(z)|) - 1$$

$$\leqslant \exp\left(\sum_{j=n+1}^{m} |u_j(z)|\right) - 1.$$

Since $\sum |u_j(z)|$ converges it follows that there is $n_0 \in \mathbf{P}$ such that

$$\exp\left(\sum_{j=n+1}^{m} |u_j(z)|\right) - 1 < \tfrac{1}{2} \qquad (m > n \geqslant n_0).$$

We thus have

$$\left|\frac{H_m(z)}{H_{n_0}(z)} - 1\right| < \frac{1}{2} \qquad (m > n_0)$$

and so

$$|H_m(z)| \geqslant \tfrac{1}{2}|H_{n_0}(z)| \qquad (m > n_0).$$

It follows that $|f(z)| \geqslant \frac{1}{2}|H_{n_0}(z)|$ so that $f(z) \neq 0$ as required.

The infinite products to be considered below are constructed from the Weierstrass *primary factors*. These latter are defined by

$$E_0(z) = 1 - z$$

$$E_n(z) = (1 - z)\exp\left(z + \frac{z^2}{2} + \ldots + \frac{z^n}{n}\right) \qquad (n \in \mathbf{P}).$$

We need first two simple lemmas.

Lemma 8.23. $|E_n(z) - 1| \leqslant 3|z|^{n+1}$ $(z \in \bar{\Delta}(0, \frac{1}{2}), n \in \mathbf{P})$.

Proof. Given $z \in \Delta(0, 1)$ we have

$$1 - z = \exp(\log(1 - z)) = \exp\left(-\sum_{n=1}^{\infty} \frac{z^n}{n}\right)$$

and thus

$$E_n(z) = \exp\left(-\sum_{j=n+1}^{\infty} \frac{z^j}{j}\right).$$

If $|z| \leqslant \frac{1}{2}$ then

$$\sum_{j=n+1}^{\infty} \frac{|z|^j}{j} \leqslant \frac{1}{2} \sum_{j=n+1}^{\infty} |z|^j \leqslant |z|^{n+1}.$$

It is elementary to verify that

$$|\exp(w) - 1| \leqslant |w| \exp(|w|) \qquad (w \in \mathbf{C})$$

and it now follows that for $z \in \bar{\Delta}(0, \frac{1}{2})$

$$|E_n(z) - 1| \leqslant |z|^{n+1} \exp(|z|^{n+1}) \leqslant 3|z|^{n+1}.$$

Lemma 8.24. *Given $\{\mu_n\} \subset \mathbf{P}$ there exists $\{p_n\} \subset \mathbf{P}$ such that $\sum \mu_n(\frac{1}{2})^{p_n}$ converges.*

Proof. Choose $p_n > \dfrac{1}{\log 2}(n + \log \mu_n)$ $(n \in \mathbf{P})$ and it follows that

$$\log(\mu_n) + p_n \log(\tfrac{1}{2}) \leqslant n \log(\tfrac{1}{2}).$$

Therefore $\mu_n(\frac{1}{2})^{p_n} \leqslant (\frac{1}{2})^n$ $(n \in \mathbf{P})$ and the result follows from the comparison test.

We are now ready for the key result of this section. Recall from Theorem 8.3 that the zeros of an entire function are at most countable. If there are an infinite number of zeros, say $\{\alpha_n\}$, we then have $\lim_{n \to \infty} \alpha_n = \infty$, since the set of zeros has no cluster point in $\mathbf{C}$. We may thus assume that the zeros are indexed such that $|\alpha_{n+1}| \geqslant |\alpha_n|$ $(n \in \mathbf{P})$. We show now that any such sequence can be the set of zeros of an entire function.

Theorem 8.25 (Weierstrass). *Let $\{\alpha_n\} \subset \mathbf{C} \setminus \{0\}$ be such that*

$$|\alpha_{n+1}| \geqslant |\alpha_n| \qquad (n \in \mathbf{P})$$

and $\lim_{n \to \infty} \alpha_n = \infty$, and let $\{\mu_n\}$ be any sequence of positive integers. Then there is an entire function g such that $Z_g = \{\alpha_n\}$, where α_n is a zero of order μ_n.

Proof. Let K be any compact subset of $\mathbf{C}$. Since $\alpha_n \to \infty$ we may choose $n_0 \in \mathbf{P}$ such that

$$\left|\frac{z}{\alpha_n}\right| \leqslant \tfrac{1}{2} \qquad (n > n_0, z \in K).$$

Let p_n be as in Lemma 8.24. We have by Lemma 8.23

$$\left|E_{p_n}\left(\frac{z}{\alpha_n}\right) - 1\right| \leqslant 3\left|\frac{z}{\alpha_n}\right|^{p_n+1} \leqslant \frac{3}{2}\left(\frac{1}{2}\right)^{p_n} \qquad (n > n_0, z \in K).$$

Using Lemma 8.24 we now see that $\sum \mu_n \left|E_{p_n}\left(\frac{z}{\alpha_n}\right) - 1\right|$ converges uniformly on K. Since K was any compact subset of $\mathbf{C}$ it follows from Proposition 8.22 that if

$$g(z) = \prod_{n=1}^{\infty} \left\{E_{p_n}\left(\frac{z}{\alpha_n}\right)\right\}^{\mu_n} \qquad (z \in \mathbf{C})$$

then g is an entire function with $Z_g = \{\alpha_n\}$. It is also clear from Proposition 8.22 that α_n is a zero of order μ_n.

Remark. If we wish also to have a zero of order k at 0 we simply take $g_1(z) = z^k g(z)$ $(z \in \mathbf{C})$.

Theorem 8.26 (Weierstrass). *If f is an entire function, there is an entire function h such that*

$$f(z) = z^k \exp(h(z)) \prod_{n=1}^{\infty} \left\{E_{p_n}\left(\frac{z}{\alpha_n}\right)\right\}^{\mu_n} \qquad (z \in \mathbf{C})$$

where $k \geqslant 0$, $\{\alpha_n\} = Z_f \setminus \{0\}$, α_n being a zero of order μ_n, and p_n is chosen as in Lemma 8.24.

Proof. By Theorem 8.25 and the remark following, we may choose an entire function g such that $Z_g = Z_f$, each zero of g having the same order as the corresponding zero of f. It follows that f/g is an entire function with no zeros. By Theorem 8.9 there is an entire function h such that $f/g = \exp \circ h$. With g as in Theorem 8.25 the result now follows.

Theorem 8.27 (Weierstrass). *If $f \in \mathscr{M}(\mathbf{C})$ there exist entire functions g, h such that $f = g/h$.*

Proof. Since $f \in \mathscr{M}(\mathbf{C})$ it follows from Theorem 8.11 that the set P_f is at most countable and has no cluster points in $\mathbf{C}$. By Theorem 8.25 we may choose an entire function h such that $Z_h = P_f$, each zero of h being of the same order as the corresponding pole of f. It follows that hf is an entire function, say g. Then $f = g/h$ as required.

Observe that the above entire function g has the property that $Z_g = Z_f$, corresponding zeros being of the same order. Note also from Theorem 8.25 that each $f \in \mathscr{M}(\mathbf{C})$ can be represented as the quotient of two infinite product expansions.

It should be understood that the results of this section are essentially existence theorems. Given an entire function f with $Z_f = \{\alpha_n\}$ we may readily obtain the Weierstrass expansion of f *modulo* an entire function with no zeros. The determination of the exact expansion is generally quite difficult. One such example will be given later in Example 8.38.

PROBLEMS 8

27. Given $\{\alpha_n\} \subset \mathbf{C} \setminus \{0\}$ let $p_n = \alpha_1 \alpha_2 \ldots \alpha_n$ $(n \in \mathbf{P})$. Show that $\{p_n\}$ converges iff $\sum \log \alpha_n$ converges, in which case $\lim_{n\to\infty} p_n = \exp\left(\sum_{n=1}^{\infty} \log \alpha_n\right)$.

28. For which $z \in \mathbf{C}$ do the following infinite products converge to a non-zero complex number?

(i) $\prod_{n=1}^{\infty}\left(1 + \frac{z}{n}\right)$ (ii) $\prod_{n=1}^{\infty}\left(1 + \frac{z^2}{n^2}\right)$ (iii) $\prod_{n=1}^{\infty} n \sin\left(\frac{z}{n}\right)$.

29. Let $\{u_n\} \subset \mathscr{A}(D)$ be such that $\sum |u_n|$ converges uniformly on each compact subset of D and $1 + u_n(z) \neq 0$ $(z \in D)$. If

$$f(z) = \prod_{n=1}^{\infty} (1 + u_n(z)) \qquad (z \in D)$$

show that

$$\frac{f'(z)}{f(z)} = \sum_{n=1}^{\infty} \frac{u_n'(z)}{1 + u_n(z)} \qquad (z \in D)$$

the series converging uniformly on each compact subset of D. What modifications are required if $Z_f \neq \varnothing$?

30. Let $\{\alpha_n\}, \{\beta_n\} \subset \mathbf{C}$ with $\lim_{n\to\infty} \alpha_n = \lim_{n\to\infty} \beta_n = \infty$ and $\alpha_m \neq \beta_n$ $(m, n \in \mathbf{P})$, and let $\{\mu_n\}, \{\nu_n\} \subset \mathbf{P}$. Construct $f \in \mathscr{M}(\mathbf{C})$ with $Z_f = \{\alpha_n\}$, $P_f = \{\beta_n\}$, α_n being a zero of order μ_n, β_n being a pole of order ν_n. Comment on the case when $\lim_{n\to\infty} (\alpha_n - \beta_n) = 0$.

31. Given $0 < r < 1$ show that there is $M_r > 0$ with

$$|E_n(w) - 1| \leqslant M_r |w|^{n+1} \qquad (w \in \Delta(0, r)).$$

Derive the analogues of Theorems 8.25, 26, 27 when $\mathbf{C}$ is replaced by $\Delta(0, 1)$. Can you generalize the argument to an arbitrary domain D of $\mathbf{C}$?

8.7 Topological index

In this section we define the topological index of a simple closed contour with respect to a point not on the contour. Roughly speaking the topological index specifies how often a curve winds round a point and in which direction. In fact the topological index leads to a very useful definition of the positive direction on a starlike simple closed contour. As such the topological index simplifies the statement of several subsequent theorems in this chapter. We consider first a weakened concept of the primitive of a function analytic on a domain.

Let Γ denote, as usual, an oriented piecewise smooth arc or a simple closed contour with representative γ (γ being suitably differentiable). Let $f \in \mathscr{A}(D)$ where $\Gamma \subset D$. We say that $g \in \mathscr{C}([0, 1])$ is a *primitive of f along* Γ if for each $\tau \in [0, 1]$ there is a primitive F_τ of f in some neighbourhood of $\gamma(\tau)$ such that $g(t) = F_\tau(\gamma(t))$ for t in some neighbourhood of τ. If f has a primitive on D in the usual sense, i.e. if there is $F \in \mathscr{A}(D)$ with $F' = f$, then f has a primitive along Γ. In fact we simply take $g = F \circ \gamma$. The concept of a primitive along Γ is thus a weaker concept than the usual concept of a primitive. We know that a given $f \in \mathscr{A}(D)$ need not have a primitive on D (see Example 4.17 and Proposition 4.15). On the other hand f always has a primitive along any Γ in D.

Theorem 8.28. *If $f \in \mathscr{A}(D)$ and $\Gamma \subset D$, then*
(i) *f has a primitive along Γ, g say;*
(ii) *the set of all primitives of f along Γ is given by $\{g + c\mathbf{1}: c \in \mathbf{C}\}$;*
(iii) $\int_\Gamma f = g(1) - g(0)$.

Proof. Note that the case $D = \mathbf{C}$ is trivial since f then has a primitive on D by Theorem 6.2.

(i) Let $\epsilon = \operatorname{dist}(\Gamma, \mathbf{C} \setminus D)$ so that $\epsilon > 0$ by Proposition 1.21. Since γ is continuous on the compact set $[0, 1]$ it is uniformly continuous on $[0, 1]$ by Proposition 1.19. Hence there is $\delta > 0$ such that

$$x, y \in [0, 1], |x - y| < \delta \quad \Rightarrow \quad |\gamma(x) - \gamma(y)| < \epsilon.$$

Choose $n \in \mathbf{P}$ such that $\frac{1}{n} < \delta$ and let

$$t_j = \frac{j}{n} \qquad (j = 0, 1, \ldots, n).$$

Then

$$\gamma([t_j, t_{j+1}]) \subset \Delta(\gamma(t_j), \epsilon) \subset D \qquad (j = 0, 1, \ldots, n - 1)$$

(see Figure 8.2). Since

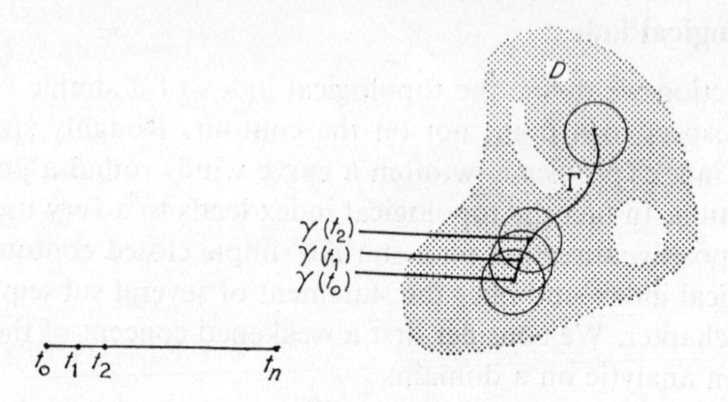

Figure 8.2

$$\gamma(t_{j+1}) \in \Delta(\gamma(t_j), \epsilon) \cup \Delta(\gamma(t_{j+1}), \epsilon)$$

it follows from Proposition 1.23 that

$$D_j = \Delta(\gamma(t_j), \epsilon) \cup \Delta(\gamma(t_{j+1}), \epsilon)$$

is a domain for $j = 0, 1, \ldots, n-1$. Since f is analytic on $\Delta(\gamma(t_0), \epsilon)$, f has a primitive F_0 on $\Delta(\gamma(t_0), \epsilon)$ by Theorem 6.2. Similarly f has a primitive G_1 on $\Delta(\gamma(t_1), \epsilon)$. Then F_0, G_1 are primitives of f on the domain D_0 and so they differ by a constant, say $k\mathbf{1}$. If $F_1 = G_1 + k\mathbf{1}$ then F_1 is a primitive of f on $\Delta(\gamma(t_1), \epsilon)$ and $F_1 = F_0$ on D_0. We proceed in this manner one step at a time and so obtain for each $j = 0, 1, \ldots, n$ a primitive F_j of f on $\Delta(\gamma(t_j), \epsilon)$ such that $F_{j+1} = F_j$ on D_j. Now define g on $[0, 1]$ by

$$g(t) = F_j(\gamma(t)) \qquad (t \in [t_j, t_{j+1}], j = 0, 1, \ldots, n).$$

Then g is well defined and is clearly continuous on each (t_j, t_{j+1}). Moreover, for $j = 1, \ldots, n-1$

$$\lim_{t \to t_j -} g(t) = F_j(\gamma(t_j)) = F_{j+1}(\gamma(t_j)) = \lim_{t \to t\ +} g(t).$$

It is also clear that g is continuous at 0 and 1. Therefore $g \in \mathscr{C}([0, 1])$ and we readily verify now that g is a primitive of f along Γ.

(ii) Given $c \in \mathbf{C}$ it is clear that $g + c\mathbf{1}$ is also a primitive of f along Γ. Conversely let h be any primitive of f along Γ. Given $\tau \in [0, 1]$ there is a neighbourhood of τ in which $h(t) = G_\tau(\gamma(t))$ where G_τ is a primitive of f on some neighbourhood of $\gamma(\tau)$. It follows that g and h differ by a constant on some neighbourhood of τ. Thus $g - h$ is a locally constant continuous function on the connected space $[0, 1]$. By Proposition 1.28, $g - h$ is constant on $[0, 1]$, as required.

(iii) We have

$$\int_\Gamma f = \int_0^1 (f \circ \gamma)\gamma'$$
$$= \sum_{j=0}^{n-1} \int_{t_j}^{t_{j+1}} (F_j' \circ \gamma)\gamma'$$
$$= \sum_{j=0}^{n-1} \int_{t_j}^{t_{j+1}} g^1$$
$$= g(1) - g(0).$$

Corollary. *If Γ is a simple closed contour and $p \in \mathbf{C} \setminus \Gamma$ then*

$$\frac{1}{2\pi i} \int_\Gamma \frac{\mathrm{d}z}{z - p} \in \mathbf{Z}.$$

Proof. In this case we may take the local primitives F_j of the form $z \to \log_{\theta_j}(z - p)$. The integral is then the difference between two branches-of-log at a point and so the corollary follows from Proposition 4.13.

Remark. The above definition of a primitive along an arc or contour, and parts (i) and (ii) of the theorem are readily generalized to the case in which γ is *any* continuous function on [0, 1] with range contained in D. This leads to a simple definition of the integral of $f \in \mathscr{A}(D)$ with respect to any such γ and thence to generalizations of Cauchy's theorem (see Problem 8.32).

Let Γ be any simple closed contour. Given $p \in \mathbf{C} \setminus \Gamma$ the *topological index* (or *winding number*) of Γ with respect to p is defined by

$$n(\Gamma; p) = \frac{1}{2\pi i} \int_\Gamma \frac{\mathrm{d}z}{z - p}.$$

It follows from the above corollary that $n(\Gamma; p)$ is an integer. We easily see by the method of Lemma 7.2 that the mapping $P \to n(\Gamma; p)$ is contınous on $\mathbf{C} \setminus \Gamma$. Since the mapping is integer valued it follows from Proposition 1.27 that the mapping is constant on each component of $\mathbf{C} \setminus \Gamma$. Without assuming the Jordan curve theorem we do not know in general how many components the set $\mathbf{C} \setminus \Gamma$ has. We may easily show that $\mathbf{C} \setminus \Gamma$ has exactly one unbounded component. Indeed since Γ is compact there is $R > 0$ such that $\Gamma \subset \Delta(0, R)$. Then $\nabla(0, R)$ is a connected subset of $\mathbf{C} \setminus \Gamma$ and so is contained in a component of $\mathbf{C} \setminus \Gamma$. Any other components of $\mathbf{C} \setminus \Gamma$ are evidently bounded. Moreover $n(\Gamma; p) = 0$ for p in the unbounded component of $\mathbf{C} \setminus \Gamma$. To see this it is sufficient to show that $n(\Gamma; p) = 0$ for some point in the unbounded component. Choose $p \in \nabla(0, R)$ and the mapping $z \to \dfrac{1}{z-p}$ is analytic on $\Delta(0, R)$. Since $\Gamma \subset \Delta(0, R)$ it follows from Theorem 6.3 that $n(\Gamma; p) = 0$.

To summarize, we have established that the mapping $p \to n(\Gamma; p)$ is integer valued, constant on each component of $\mathbf{C} \setminus \Gamma$, and zero on the unbounded component. For a general simple closed contour we cannot assert much more without recourse to the Jordan curve theorem; for starlike contours we can give much more detailed information.

Theorem 8.29. *If Γ is a starlike simple closed contour then*
(i) $n(\Gamma; p) = 0$ $(p \in \text{ext}\,(\Gamma))$;
(ii) *either* $n(\Gamma; p) = 1$ $(p \in \text{int}\,(\Gamma))$ *or* $n(\Gamma, p) = -1$ $(p \in \text{int}\,(\Gamma))$.

Proof. (i) This has been established above.

(ii) Let α be a star centre for Γ. Since the mapping $p \to n(\Gamma; p)$ is constant on int (Γ) it will be sufficient to show that $n(\Gamma; \alpha)$ is $+1$ or -1. Let the ray $\{\alpha + r : r \geqslant 0\}$ meet Γ at β_1 and let the ray $\{\alpha - r : r \geqslant 0\}$ meet Γ at β_2 (see Figure 8.3). We may choose a representative function γ for Γ such that $\beta_1 = \gamma(0)$. Suppose that $\beta_2 = \gamma(a)$, so that $0 < a < 1$. Let Γ_1, Γ_2 be the arcs represented by $\gamma|_{[0,a]}$, $\gamma|_{[a,1]}$ respectively, and let

$$U_\alpha = \{z : \text{Im}\, z \geqslant \text{Im}\, \alpha\}, \qquad L_\alpha = \{z : \text{Im}\, z \leqslant \text{Im}\, \alpha\}.$$

We assert that $\Gamma_1 \subset U_\alpha$ or $\Gamma_1 \subset L_\alpha$. To see this we first observe that there is exactly one point β_3 on Γ such that $\arg(\beta_3 - \alpha) = -\frac{1}{2}\pi$. Suppose that $\beta_3 \in \Gamma_2$ and then define χ on $[0, a]$ by

$$\chi(t) = \arg_{\pi/2}(\gamma(t) - \alpha) \qquad (t \in [0, a]).$$

Since $\beta_3 \notin \Gamma_1$, χ is continuous. Since Γ is starred with respect to α, χ is one-to-one. It follows that χ is monotonic on $[0, a]$ and has range $[0, \pi]$.

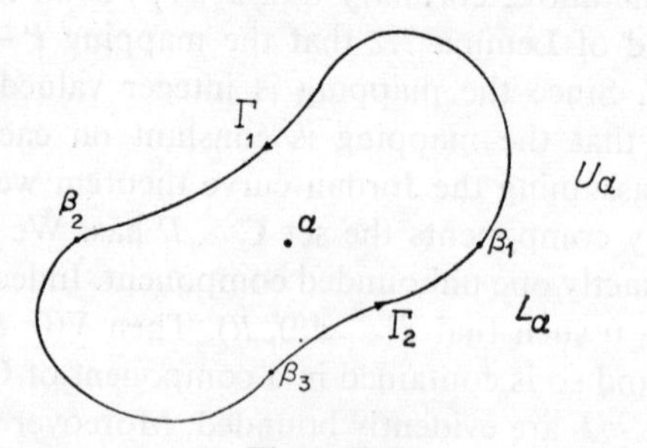

Figure 8.3

Therefore $\Gamma_1 \subset U_\alpha$ and so $\Gamma_2 \subset L_\alpha$. Similarly if $\beta_3 \in \Gamma_1$ we may show that $\Gamma_2 \subset U_\alpha$ and $\Gamma_1 \subset L_\alpha$.

Suppose that $\Gamma_1 \subset U_\alpha$ and define g on $[0, 1]$ by

$$g(t) = \begin{cases} \log_{\frac{\pi}{2}}(\gamma(t) - \alpha) & \text{if } 0 \leqslant t \leqslant a, \\ \log_{\frac{3\pi}{2}}(\gamma(t) - \alpha) & \text{if } a < t \leqslant 1. \end{cases}$$

It is clear that g is continuous except possibly at $t = a$. Since

$$\lim_{t\to a+} g(t) = \log_{\frac{3\pi}{2}}(\beta_2 - \alpha) = \log_{\frac{\pi}{2}}(\beta_2 - \alpha) = g(a)$$

it follows that $g \in \mathscr{C}([0, 1])$. It is now clear that g is a primitive along Γ of the function $z \to \dfrac{1}{z - \alpha}$. Theorem 8.28 (iii) now gives

$$\begin{aligned} n(\Gamma; \alpha) &= \frac{1}{2\pi i}\{g(1) - g(0)\} \\ &= \frac{1}{2\pi i}\{\log|\beta_1 - \alpha| + 2\pi i - \log|\beta_1 - \alpha|\} \\ &= 1. \end{aligned}$$

Similarly if $\Gamma_1 \subset L_\alpha$ we obtain $n(\Gamma; \alpha) = -1$. The proof is now complete.

Given any starlike simple closed contour Γ with star centre α we say that Γ has the *positive* (or *counter-clockwise*) *direction* if $n(\Gamma; \alpha) = +1$, and we say that Γ has the *negative* (or *clockwise*) *direction* if $n(\Gamma; \alpha) = -1$. This accords with our earlier convention for circles since $n(C(\alpha, R); \alpha) = +1$. There is a very simple geometrical characterization of the positive direction on Γ. It is clear from the proof of the above theorem that $n(\Gamma; \alpha) = 1$ if and only if there is $a \in (0, 1)$ such that $\gamma([0, a]) \subset U_\alpha$.

PROBLEMS 8

32. Let $h \in \mathscr{C}([0, 1])$ and let D be a domain in $\mathbf{C}$. We say that h is a *path in* D if $h([0, 1]) \subset D$. Given $f \in \mathscr{A}(D)$ we define a *primitive* of f *along the path* h in analogous manner to the corresponding definition in the text for an arc or contour. Prove the analogues of Theorems 8.28 (i), (ii). Define the *integral* of f *along the path* h by

$$\int_h f = g(1) - g(0)$$

where g is any primitive of f along the path h.

(i) Discuss the analogues of **CI 1–7**.

(ii) We say that h is a *closed path* if $h(0) = h(1)$. If D is starlike show that $\int_h f = 0$ for $f \in \mathscr{A}(D)$ and any closed path h in D.

(iii) Given a domain D in $\mathbf{C}$ show that $f \in \mathscr{A}(D)$ has a primitive on D iff $\int_h f = 0$ for every closed path h in D.

(iv) Define the *topological index* of a *closed path* with respect to a point not in the range of the path. Give an example of a closed path h such that $\mathbf{C} \setminus h([0, 1])$ has three components and the values of the topological index on these components are 0, 1, 2.

33. Let Γ_1, Γ_2 be simple closed contours represented by γ_1, γ_2, respectively.

(i) If $0 \in (\mathbf{C} \setminus \Gamma_1) \cap (\mathbf{C} \setminus \Gamma_2)$ and $\gamma_1\gamma_2$ represents a simple closed contour Γ_3, show that

$$n(\Gamma_3; 0) = n(\Gamma_1; 0) + n(\Gamma_2; 0).$$

(ii) If $0 \in \mathbf{C} \setminus \Gamma_1$, $|\gamma_2| < |\gamma_1|$, and $\gamma_1 + \gamma_2$ represents a simple closed contour Γ_4 show that

$$n(\Gamma_4; 0) = n(\Gamma_1; 0).$$

(*Hint.* Use $\gamma_1 + \gamma_2 = \gamma_1\left(1 + \dfrac{\gamma_2}{\gamma_1}\right)$ and part (i).)

34. Let Γ be a starlike simple closed contour with star centre α. Let $z_1 \in \Gamma$ with $\operatorname{Im} z_1 < \operatorname{Im} \alpha$ and let $z_2 \in \Gamma$ with $\operatorname{Im} z_2 > \operatorname{Im} \alpha$. Let $\{(\Gamma_1, z_1), (\Gamma_2, z_2)\}$ be the associated decomposition of Γ. If Γ_1 intersects the ray $\{\alpha + r : r \geqslant 0\}$ show that $n(\Gamma; \alpha) = 1$. (*Hint.* Use Figure 8.4.)

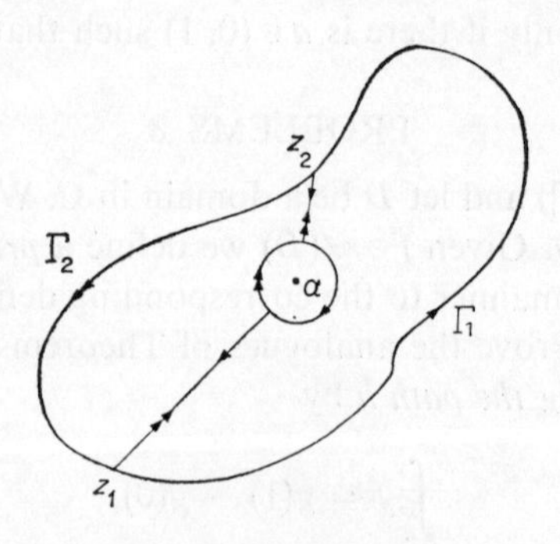

Figure 8.4

8.8 Cauchy's residue theorem

In this section we prove the famous Cauchy residue theorem for the case of starlike simple closed contours. This case is adequate for most applications. Occasionally it is necessary to derive a slight extension beyond the starlike case, and we illustrate the technique in an example. Roughly speaking the residue theorem evaluates a contour integral in terms of the residues at the singularities inside the contour. For the student with no previous experience of complex analysis one of the most exciting applications of the residue theorem concerns the evaluation of certain types of definite integrals. We illustrate the general technique in the text and give additional ideas in the problems. Further applications of the residue theorem will appear in the next two sections.

Theorem 8.30 (Cauchy). *Let $f \in \mathscr{M}(D)$ and let Γ be a starlike simple closed contour in D, with the positive direction, such that $\Gamma^* \subset D$. If f has no poles on Γ then f has a finite number of poles on int (Γ), say $\{\alpha_j : j = 1, \ldots, n\}$ and*

$$\int_\Gamma f = 2\pi i \sum_{j=1}^{n} \operatorname{Res}(f; \alpha_j).$$

Proof. Since Γ^* $(= \text{int}(\Gamma)^-)$ is compact, f can have only a finite number of poles on Γ^*, α_j $(j = 1, \ldots, n)$ say, and by hypothesis none of these lie on Γ. Let $E = P_f \setminus \{\alpha_j : j = 1, \ldots, n\}$ and suppose that $E \neq \varnothing$. Since P_f has no cluster points in D we must have dist $(z, E) > 0$ $(z \in \Gamma)$ and since Γ is compact we must have $\epsilon =$ dist $(\Gamma, E) > 0$ by Proposition 1.21. Let D_ϵ be the starlike domain given in the corollary to Theorem 5.8 so that $\Gamma^* \subset D_\epsilon$ and $P_f \cap D_\epsilon = \{\alpha_j : j = 1, \ldots, n\}$. If $E = \varnothing$ we take $\epsilon =$ dist $(\Gamma, C \setminus D)$ and the same assertion holds for D_ϵ.

Recall that $S(f; \alpha_j)$ denotes the principal part of f at α_j. If

$$g = f - \sum_{j=1}^{n} S(f; \alpha_j)$$

it follows from Lemma 8.13 that $g \in \mathscr{A}(D_\epsilon)$. Since $\Gamma \subset D_\epsilon$, Theorem 6.3 gives $\int_\Gamma g = 0$. Therefore

$$\begin{aligned}\int_\Gamma f &= \sum_{j=1}^{n} \int_\Gamma S(f; \alpha_j)\\ &= \sum_{j=1}^{n} \operatorname{Res}(f; \alpha_j)\, n(\Gamma; \alpha_j)\\ &= 2\pi i \sum_{j=1}^{n} \operatorname{Res}(f; \alpha_j).\end{aligned}$$

The example below gives a simple illustration of how the theorem may be extended to non-starlike contours.

Example 8.31. Let $f \in \mathcal{M}(\mathbf{C})$ with $P_f = \{0, 1 + i, -1 + i\}$, and let Γ be the simple closed contour indicated in Figure 8.5, where $0 < r < 1$, $R > 2$. Then

$$\int_\Gamma f = 2\pi i\{\text{Res}\,(f;\ 1 + i) + \text{Res}\,(f;\ -1 + i)\}.$$

Proof. Join ir, iR to form two simple closed contours Γ_1, Γ_2 as in Figure 8.5. By the argument used in Proposition 6.7 we see that Γ_1 and Γ_2 are starlike. It follows from the remark at the end of the last section

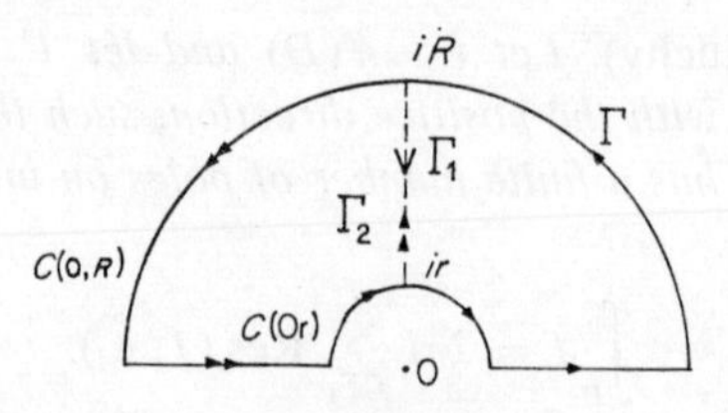

Figure 8.5

that Γ_1, Γ_2 both have the positive direction. The residue theorem now gives

$$\int_{\Gamma_1} f = 2\pi i\,\text{Res}\,(f;\ 1 + i), \qquad \int_{\Gamma_2} f = 2\pi i\,\text{Res}\,(f;\ -1 + i).$$

It is clear that

$$\int_\Gamma f = \int_{\Gamma_1} f + \int_{\Gamma_2} f$$

and so the proof is complete.

We shall now see how the residue theorem enables us to evaluate certain definite integrals without performing any integration. One of the standard techniques is as follows. Suppose we wish to evaluate $\int_{-\infty}^{\infty} f(x)\,dx$ (assuming that the infinite integral converges, i.e. $\lim\limits_{R,S\to+\infty} \int_{-S}^{R} f(x)\,dx$ exists). Suppose that f admits an extension to a function g in $\mathcal{M}(\mathbf{C})$ with only a finite number of poles in the upper half plane and none on the real line. We form a semi-circular contour $\Gamma(R)$ in the upper half plane that has diameter $[-R, R]$ and contains all the poles of g in the upper half plane. We now apply the residue theorem to $\int_{\Gamma(R)} g$ and then let $R \to +\infty$. The residues can usually be evaluated by the techniques of Proposition 7.12. The success of the technique thus depends on being able to deal with the 'extraneous' part of the contour, i.e. the circular part of the contour in the present case. In most cases a straight application of **CI 5** is sufficient, but occasionally a little more finesse is required (see Example 8.33).

Proposition 8.32. *Let f be a rational function with* $\deg(f) \leqslant -2$. *If f has no poles on* $\mathbf{R}$ *then*

$$\int_{-\infty}^{\infty} f(x)\,\mathrm{d}x = 2\pi i \sum_{j=1}^{n} \operatorname{Res}(f; \alpha_j)$$

where $\{\alpha_j : j = 1, \ldots, n\}$ *is the set of poles of f in the upper half plane.*

Proof. It is straightforward to show that there is $M > 0$ with

$$|(1 + x^2)f(x)| \leqslant M \qquad (x \in \mathbf{R}).$$

Since f is continuous on $\mathbf{R}$ it follows by the comparison test that f is integrable on R and so

$$\int_{-\infty}^{\infty} f(x)\,\mathrm{d}x = \lim_{R \to +\infty} \int_{-R}^{R} f(x)\,\mathrm{d}x.$$

Let $\Gamma(R)$ be the semi-circular contour described above, with the positive direction. If all the poles of f in the upper half plane lie inside $\Gamma(R_0)$ then

$$\int_{\Gamma(R)} f = 2\pi i \sum_{j=1}^{n} \operatorname{Res}(f; \alpha_j) \qquad (R > R_0).$$

Since $\deg(f) \leqslant -2$ there is $K > 0$ and $R_1 > R_0$ with

$$|f(z)| \leqslant \frac{K}{|z|^2} \qquad (|z| \geqslant R_1).$$

If $\Gamma_1(R)$ denotes the arc $\{R\,\mathrm{e}^{i\theta} : \theta \in [0, \pi]\}$ it follows from **CI 5** that

$$\left|\int_{\Gamma_1(R)} f\right| \leqslant \pi R \frac{K}{R^2} = \frac{\pi K}{R} \qquad (R > R_1)$$

and so

$$\lim_{R \to +\infty} \int_{\Gamma_1(R)} f = 0.$$

We now conclude that

$$\int_{-\infty}^{\infty} f(x)\,\mathrm{d}x = \lim_{R \to +\infty} \int_{\Gamma(R)} f = 2\pi i \sum_{j=1}^{n} \operatorname{Res}(f; \alpha_j).$$

Example 8.33. *Given $a > 0$ we have*

$$\int_0^\infty \frac{x \sin x}{x^2 + a^2}\, dx = \frac{\pi}{2} e^{-a}.$$

Proof. Let

$$g(z) = \frac{z \exp(iz)}{z^2 + a^2} \qquad (z \in \mathbf{C})$$

so that $g \in \mathscr{M}(\mathbf{C})$ with $P_g = \{ia, -ia\}$. The pole ia in the upper half plane is simple (Theorem 7.11) and so by Proposition 7.12 we have

$$\operatorname{Res}(g: ia) = \frac{ia \exp(i^2 a)}{2ia} = \tfrac{1}{2} e^{-a}.$$

Integrating g round the contour $\Gamma(R)$ of the above proposition we obtain

$$\int_{\Gamma(R)} g = \pi i\, e^{-a} \qquad (R > a).$$

For $R > 2a$ we have

$$\begin{aligned}
\left|\int_{\Gamma_1(R)} g\right| &\leqslant \frac{R}{R^2 - a^2}\int_0^\pi |\exp(iR e^{i\theta}) iR e^{i\theta}|\, d\theta \\
&= \frac{R^2}{R^2 - a^2}\int_0^\pi e^{-R\sin\theta}\, d\theta \\
&\leqslant 6\int_0^{\pi/2} e^{-R\sin\theta}\, d\theta \\
&\leqslant 6\int_0^{\pi/2} e^{-(2R/\pi)\theta}\, d\theta \\
&= \frac{3\pi}{R}(1 - e^{-R})
\end{aligned}$$

so that

$$\lim_{R\to+\infty} \int_{\Gamma_1(R)} g = 0.$$

It follows that

$$\lim_{R\to+\infty} \int_{-R}^{R} \frac{x e^{ix}}{x^2 + a^2}\, dx = \pi i\, e^{-a}.$$

On taking imaginary parts we conclude that

$$\int_0^\infty \frac{x \sin x}{x^2 + a^2}\, dx = \tfrac{1}{2} \lim_{R\to+\infty} \int_{-R}^{R} \frac{x \sin x}{x^2 + a^2}\, dx = \frac{\pi}{2} e^{-a}.$$

Further examples are given in the problems below.

PROBLEMS 8

35. Let Γ be a starlike simple closed contour, with the positive direction, and let $f \in \mathscr{C}(\Gamma^*)$ be analytic on int (Γ). Show that

$$f(z) = \frac{1}{2\pi i}\int_\Gamma \frac{f(w)}{w-z}\,\mathrm{d}w \qquad (z \in \text{int}(\Gamma)).$$

36. Extend the residue theorem to the case in which $f \in \mathscr{I}(D)$ (see Problem 8.14).

37. Transform the integral $\int_0^{2\pi} f(\cos\theta, \sin\theta)\,\mathrm{d}\theta$ to one of the form $\int_{C(0,1)} g(z)\,\mathrm{d}z$ and hence show that

(i) $\displaystyle\int_0^{2\pi} \frac{\mathrm{d}\theta}{a+\cos\theta} = \frac{2\pi}{(a^2-1)^{1/2}} \qquad (a > 1),$

(ii) $\displaystyle\int_0^{2\pi} \frac{\cos^2 3\theta}{1-2a\cos 2\theta + a^2}\,\mathrm{d}\theta = \pi\,\frac{1-a+a^2}{1-a} \qquad (0 < a < 1),$

(iii) $\displaystyle\int_0^{2\pi} \mathrm{e}^{2\cos\theta}\,\mathrm{d}\theta = 2\pi\sum_{n=0}^{\infty}\frac{1}{(n!)^2}.$

38. Show that

(i) $\displaystyle\int_{-\infty}^{\infty} \frac{\mathrm{d}x}{1+x^6} = \frac{2\pi}{3},$

(ii) $\displaystyle\int_0^{\infty} \frac{\cos(\lambda x)}{x^2+a^2}\,\mathrm{d}x = \frac{\pi}{2a}\,\mathrm{e}^{-\lambda a} \qquad (\lambda \in \mathbf{R},\ a > 0).$

39. Integrals of the form $\int_0^\infty t^{a-1}f(t)\,\mathrm{d}t$ may be evaluated as follows. Let $t = \mathrm{e}^x$ and the integral becomes $\int_{-\infty}^{\infty} \mathrm{e}^{ax}f(\mathrm{e}^x)\,\mathrm{d}x$. Now let $g(z) = \exp(az)f(\mathrm{e}^z)$ $(z \in \mathbf{C})$. Integrate g round the boundary of the rectangle with vertices S, $S + 2\pi i$, $-T + 2\pi i$, $-T$, and take limits as $S, T \to +\infty$. Use this method to show that

(i) $\displaystyle\int_0^{\infty} \frac{x^{a-1}}{1+x^b}\,\mathrm{d}x = \frac{\pi}{b\sin\left(\frac{\pi a}{b}\right)} \qquad (0 < a < b),$

(ii) $\displaystyle\int_0^{\infty} \frac{x^a}{1+2x\cos\theta + x^2}\,\mathrm{d}x = \frac{\pi}{\sin \pi a}\,\frac{\sin a\theta}{\sin\theta} \quad (-1 < a < 1,\ -\pi < \theta < \pi).$

40. Let f have a simple pole at α and let $\Gamma(r)$ be the arc represented by

$$\gamma(t) = \alpha + r\,\mathrm{e}^{it} \qquad (t \in [\theta_1, \theta_2]).$$

Show that

$$\lim_{r\to 0+}\int_{\Gamma(r)} f = i(\theta_2 - \theta_1)\,\text{Res}\,(f;\alpha).$$

41. Let $f(z) = \frac{1}{z}\exp(iz)$ ($z \in \mathbf{C}$). Integrate f round the contour given in Example 8.31. Let $r \to 0+$, $R \to +\infty$ and deduce that

$$\int_0^\infty \frac{\sin x}{x}\,\mathrm{d}x = \frac{\pi}{2}.$$

Use a similar argument to show that

$$\int_0^\infty \frac{\sin x}{x(x^2 + a^2)}\,\mathrm{d}x = \frac{\pi}{2a^2}(1 - \mathrm{e}^{-a}) \qquad (a > 0).$$

42. Establish the following results.

(i) $\displaystyle\int_0^\infty \frac{\log x}{1 + x^2}\,\mathrm{d}x = 0.$

(ii) $\displaystyle\int_0^\infty \frac{\log x}{(1 + x^2)^2}\,\mathrm{d}x = -\frac{\pi}{4}.$

(iii) $\displaystyle\int_0^\infty \frac{\sinh ax}{\sinh \pi x}\,\mathrm{d}x = \tan(\tfrac{1}{2}a) \qquad (-\pi < a < \pi).$

8.9 Mittag-Leffler expansions

In this section we derive the Mittag-Leffler expansion of a function f that is meromorphic on $\mathbf{C}$. This expansion represents f as a series involving the principal parts of f at each pole of f. Recall that the Weierstrass expansion represents f as the quotient of two infinite products, one involving the zeros of f and the other involving the poles of f. As such, the principal parts at the poles of f play no significant role in the Weierstrass expansion. The Mittag-Leffler expansion emphasizes the principal parts at the poles of f, but gives no information about the zeros of f. The Mittag-Leffler expansion, like the Weierstrass expansion, is essentially an, existence theorem, but we shall give a method, due to Cauchy, for determining the exact expansion in certain cases. A related technique leads to the evaluation of sums of the form $\sum_{-\infty}^{\infty} f(n)$ and $\sum_{-\infty}^{\infty} (-1)^n f(n)$. This enables us to sum suitable series of partial fractions and thus to recover (in special cases) the closed expression for a meromorphic function on $\mathbf{C}$ from its Mittag-Leffler expansion.

Recall from the proof of Theorem 8.14 that if f is meromorphic on $\mathbf{C}^\infty$ then f is a rational function and f is the sum of the principal parts at its poles, i.e.

$$f(z) = p_0(z) + \sum_{j=1}^{n} p_j\left(\frac{1}{z - \alpha_j}\right) \qquad (z \in \mathbf{C})$$

where f has poles at $\alpha_1, \ldots, \alpha_n$ and (possibly) ∞, and $p_0, p_1, \ldots, p_n$ are polynomial functions. This is none other than the usual expansion of f in partial fractions. If we now suppose that f is meromorphic only on $\mathbf{C}$ then f may have an infinite number of poles and the series of principal parts at the poles of f, $\sum p_n\left(\frac{1}{z-\alpha_n}\right)$, need not converge. It is thus necessary to modify the terms to produce a convergent series.

Theorem 8.34 (Mittag-Leffler). *Let $\alpha_0 = 0$ and let α_n be non-zero complex numbers with $\lim_{n\to\infty} \alpha_n = \infty$. Let $\{p_n\}$ be any sequence of polynomial functions such that $p_n(0) = 0$. Then there exists $g \in \mathscr{M}(\mathbf{C})$ such that $P_g = \{\alpha_n : n = 0, 1, \ldots\}$ and $S(g; \alpha_n)(z) = p_n\left(\frac{1}{z-\alpha_n}\right)$.*

Proof. We may suppose that $\{\alpha_n\}$ is indexed so that $|\alpha_{n+1}| \geqslant |\alpha_n|$ $(n \in \mathbf{P})$. Given $n \in \mathbf{P}$ the function $z \to p_n\left(\frac{1}{z-\alpha_n}\right)$ is regular at 0 and its Taylor expansion converges on $\Delta(0, |\alpha_n|)$ (Proposition 8.2) and so converges uniformly on $\bar{\Delta}(0, \frac{1}{2}|\alpha_n|)$. Choose a partial sum h_n of the Taylor expansion such that

$$\left|p_n\left(\frac{1}{z-\alpha_n}\right) - h_n(z)\right| \leqslant 2^{-n} \quad (z \in \bar{\Delta}(0, \tfrac{1}{2}|\alpha_n|)).$$

Let K be any compact subset of $\mathbf{C}$. Since $\lim_{n\to\infty} \alpha_n = \infty$ we may choose $n_0 \in \mathbf{P}$ such that $K \subset \bar{\Delta}(0, \frac{1}{2}|\alpha_n|)$ for $n \geqslant n_0$. It follows from above that the series

$$\sum_{n=n_0}^{\infty} \left\{p_n\left(\frac{1}{z-\alpha_n}\right) - h_n(z)\right\}$$

converges uniformly on K. Since K was arbitrary we may now define g on $\mathbf{C}$ by

$$g(z) = p_0\left(\frac{1}{z}\right) + \sum_{n=1}^{\infty} \left\{p_n\left(\frac{1}{z-\alpha_n}\right) - h_n(z)\right\} \quad (z \in \mathbf{C}).$$

It follows from Theorem 8.19 that g is analytic on $\mathbf{C} \setminus \{\alpha_n : n = 0, 1, \ldots\}$. It is also clear that each α_n is a pole of g, so that $g \in \mathscr{M}(\mathbf{C})$ and $P_g = \{\alpha_n : n = 0, 1, \ldots\}$. Finally we have

$$S(g; \alpha_n)(z) = p_n\left(\frac{1}{z-\alpha_n}\right)$$

since the function $z \to g(z) - p_n\left(\frac{1}{z-\alpha_n}\right)$ is regular at α_n.

The above theorem constructs a meromorphic function on **C** with assigned poles (cf. Theorem 8.11) and assigned principal parts at these poles. It is the analogue of Theorem 8.25.

Theorem 8.35 (Mittag-Leffler). *If f is meromorphic on* **C** *there is an entire function h and polynomial functions h_n such that*

$$f(z) = h(z) + p_0\left(\frac{1}{z}\right) + \sum_{n=1}^{\infty}\left\{p_n\left(\frac{1}{z-\alpha_n}\right) - h_n(z)\right\} \qquad (z \in \mathbf{C})$$

where f has poles at $\alpha_0 = 0$ (possibly) and $\alpha_n \in \mathbf{C} \setminus \{0\}$, and $p_n\left(\dfrac{1}{z-\alpha_n}\right) = S(f; \alpha_n)(z)$.

Proof. Choose g as in Theorem 8.34 so that $P_g = P_f$ and g, f have the same principal parts at their poles. By our convention for subtracting meromorphic functions we have that $f - g$ is an entire function, h say. Then $f = h + g$ and the result follows from Theorem 8.34.

Remark. Note that the Mittag-Leffler expansion converges uniformly on compact subsets of **C** provided that for each compact subset K of **C** we ignore the finite number of terms with poles on K.

Given $f \in \mathscr{M}(\mathbf{C})$ we shall always assume that $P_f = \{\alpha_n\}$ is indexed so that $\{|\alpha_n|\}$ is non-decreasing. To simplify the details of the Cauchy technique for determining the exact expansion of f we shall suppose that all the poles are simple. The essential tool in the argument is the residue theorem.

Theorem 8.36 (Cauchy). *Let $f \in \mathscr{M}(\mathbf{C})$ with $P_f = \{\alpha_n\}$, each pole being simple, and let f be regular at* 0. *Suppose there is a sequence $\{\Gamma_n\}$ of starlike simple closed contours such that*

(i) $0 \in \operatorname{int}(\Gamma_n) \subset \operatorname{int}(\Gamma_{n+1})$ $(n \in \mathbf{P})$;

(ii) $\lim_{n\to\infty} \operatorname{dist}(0, \Gamma_n) = +\infty$;

(iii) *there is* $A > 0$ *with* $l_{\Gamma_n} \leqslant A \operatorname{dist}(0, \Gamma_n)$ $(n \in \mathbf{P})$;

(iv) *there is* $B > 0$ *with* $|f(z)| \leqslant B$ $(z \in \Gamma_n, n \in \mathbf{P})$;

(v) *there are $m(n)$ poles of f on* $\operatorname{int}(\Gamma_n)$.

Then

$$f(z) = f(0) + \lim_{n\to\infty} \sum_{j=1}^{m(n)} \operatorname{Res}(f; \alpha_j)\left(\frac{1}{z-\alpha_j} + \frac{1}{\alpha_j}\right) \qquad (z \in \mathbf{C})$$

the series converging uniformly on compact subsets of $\mathbf{C} \setminus P_f$.

Proof. Observe that condition (iv) implies that f has no poles on any of the contours Γ_n. Let K be any compact subset of $\mathbf{C} \setminus P_f$. Since K is bounded, say $K \subset \Delta(0, R)$, it follows from condition (ii) that there is $n_0 \in \mathbf{P}$ such that $K \subset \operatorname{int}(\Gamma_{n_0})$. For $n > n_0$ define F_n on $K \cup \{0\}$ by

$$F_n(z) = \frac{1}{2\pi i} \int_{\Gamma_n} \frac{f(w)}{w - z} \, \mathrm{d}w \qquad (z \in K \cup \{0\}).$$

The function $w \to \dfrac{f(w)}{w - z}$ has simple poles on $\operatorname{int}(\Gamma_n)$ at $\alpha_1, \alpha_2, \ldots, \alpha_{m(n)}$ and z. The residue at α_j is given by

$$\lim_{w \to \alpha_j} \frac{(w - \alpha_j) f(w)}{w - z} = \frac{\operatorname{Res}(f; \alpha_j)}{\alpha_j - z}$$

and the residue at z is clearly $f(z)$. If each Γ_n has the positive direction the residue theorem gives

$$F_n(z) = f(z) + \sum_{j=1}^{m(n)} \frac{\operatorname{Res}(f; \alpha_j)}{\alpha_j - z} \qquad (z \in K \cup \{0\}).$$

In particular

$$F_n(0) = f(0) + \sum_{j=1}^{m(n)} \frac{\operatorname{Res}(f; \alpha_j)}{\alpha_j}$$

and therefore for $z \in K$

$$f(z) = f(0) + \sum_{j=1}^{m(n)} \operatorname{Res}(f; \alpha_j) \left\{ \frac{1}{z - \alpha_j} + \frac{1}{\alpha_j} \right\} + F_n(z) - F_n(0).$$

For $n > n_0$, $z \in K$ we have

$$\begin{aligned} |F_n(z) - F_n(0)| &= \left| \frac{1}{2\pi i} \int_{\Gamma_n} \frac{z f(w)}{w(w - z)} \, \mathrm{d}w \right| \\ &\leqslant \frac{1}{2\pi} l_{\Gamma_n} \frac{RB}{\operatorname{dist}(0, \Gamma_n) \operatorname{dist}(z, \Gamma_n)} \\ &\leqslant \frac{ARB}{2\pi \operatorname{dist}(K, \Gamma_n)} \end{aligned}$$

and thus $\{F_n(z) - F_n(0)\}$ converges to 0 uniformly on K. Since K was arbitrary the proof is now complete.

In most applications Γ_n is taken to be a circle or the boundary of a rectangle. Although the theorem has a large number of conditions, in practice only condition (iv) presents any real difficulty. It often occurs that the meromorphic function we wish to expand has a pole at 0, so that the above theorem is not directly applicable. We need only subtract the principal part at 0 and then apply the theorem. This is the case in the example below.

Example 8.37.

$$\pi \cot(\pi z) = \frac{1}{z} + \sum_{n=1}^{\infty} \frac{2z}{z^2 - n^2} \qquad (z \in \mathbf{C} \setminus \mathbf{Z}).$$

Proof. Let $f(z) = \pi \cot(\pi z) - \dfrac{1}{z}$ $(z \in \mathbf{C})$ so that f is meromorphic on $\mathbf{C}$. We easily see that f is regular at 0 with $f(0) = 0$ and that $P_f = \{\pm n : n \in \mathbf{P}\}$, each pole being simple. Moreover

$$\operatorname{Res}(f; n) = \frac{\pi \cos(\pi n)}{\pi \cos(\pi n)} = 1 \qquad (n \in \mathbf{Z} \setminus \{0\}).$$

Let Γ_n be the boundary of the square with vertices $\pm(n + \frac{1}{2}) \pm i(n + \frac{1}{2})$. Evidently conditions (i) and (ii) of the theorem are satisfied and also $l_{\Gamma_n} = 8 \operatorname{dist}(0, \Gamma_n)$ $(n \in \mathbf{P})$. Recall that if $z = x + iy$ then

$$|\cos^2(\pi z)| = \cos^2(\pi x) + \sinh^2(\pi y), \; |\sin^2(\pi z)| = \sin^2(\pi x) + \sinh^2(\pi y).$$

Thus if $z = \pm(n + \frac{1}{2}) + iy$ then

$$|f(z)| \leqslant \pi \frac{\sinh \pi y}{\cosh \pi y} + \frac{1}{n + \frac{1}{2}} \leqslant \pi + 1 \qquad (n \in \mathbf{P}).$$

If $z = x \pm i(n + \frac{1}{2})$ then

$$|f(z)| \leqslant \pi \frac{\cosh \pi(n + \frac{1}{2})}{\{\cosh^2 \pi (n + \frac{1}{2}) - 1\}^{1/2}} + \frac{1}{n + \frac{1}{2}} \leqslant 2\pi + 1 \quad (n \in \mathbf{P}).$$

It now follows from Theorem 8.36 that

$$f(z) = \lim_{n \to \infty} \sum_{j=1}^{n} \left\{ \frac{1}{z - j} + \frac{1}{j} + \frac{1}{z + j} - \frac{1}{j} \right\}$$

and therefore

$$\pi \cot(\pi z) = \frac{1}{z} + \sum_{n=1}^{\infty} \frac{2z}{z^2 - n^2} \qquad (z \in \mathbf{C} \setminus \mathbf{Z}).$$

We shall now use the above example to obtain the Weierstrass expansion of the entire function $z \to \sin(\pi z)$.

Example 8.38.

$$\sin(\pi z) = \pi z \prod_{n=1}^{\infty} \left(1 - \frac{z^2}{n^2}\right) \qquad (z \in \mathbf{C}).$$

Proof. It is clear that $\sum \dfrac{|z|^2}{n^2}$ converges uniformly on compact subsets of

C. Thus if

$$g(z) = \pi z \prod_{n=1}^{\infty} \left(1 - \frac{z^2}{n^2}\right) \qquad (z \in \mathbf{C})$$

it follows from Proposition 8.22 that g is an entire function. Observe that $g(x) > 0$ $(0 < x < 1)$ and so

$$\log g(x) = \log(\pi x) + \sum_{n=1}^{\infty} \log\left(1 - \frac{x^2}{n^2}\right) \qquad (0 < x < 1).$$

We now have

$$\begin{aligned} \frac{\mathrm{d}}{\mathrm{d}x} \log g(x) &= \frac{1}{x} + \sum_{n=1}^{\infty} \frac{2x}{x^2 - n^2} \\ &= \cot(\pi x) \\ &= \frac{\mathrm{d}}{\mathrm{d}x} \log \sin(\pi x) \qquad (0 < x < 1) \end{aligned}$$

using the uniform convergence of the derived series and Example 8.37. It follows that there is $k \in \mathbf{R}$ with

$$\sin(\pi x) = k g(x) \qquad (0 < x < 1)$$

and so

$$\frac{\sin(\pi x)}{\pi x} = k \prod_{n=1}^{\infty} \left(1 - \frac{x^2}{n^2}\right) \qquad (0 < x < 1).$$

Taking limits as $x \to 0+$ we obtain $k = 1$, so that

$$\sin(\pi x) = g(x) \qquad (0 < x < 1).$$

The result now follows from Theorem 8.4.

Recall from Example 8.37 that the function $z \to \pi \cot(\pi z)$ has a simple pole at each $n \in \mathbf{Z}$ with residue 1, and is uniformly bounded on Γ_n. A similar argument shows that the function $z \to \pi \operatorname{cosec}(\pi z)$ has a simple pole at each $n \in \mathbf{Z}$ with residue $(-1)^n$, and is uniformly bounded on Γ_n. We thus obtain a similar expansion for $\pi \operatorname{cosec}(\pi z)$. For convenience we give both expansions below.

$$\pi \cot(\pi z) = \frac{1}{z} + \sum_{n=1}^{\infty} \frac{2z}{z^2 - n^2}$$

$$\pi \operatorname{cosec}(\pi z) = \frac{1}{z} + \sum_{n=1}^{\infty} \frac{(-1)^n 2z}{z^2 - n^2}.$$

With the aid of these two standard meromorphic functions we can evaluate the sums of the form $\sum_{-\infty}^{\infty} f(n)$, $\sum_{-\infty}^{\infty} (-1)^n f(n)$ for certain meromorphic functions f.

Let $f \in \mathcal{M}(\mathbf{C})$ be such that f has no poles and no zeros on $\mathbf{Z}$ and let

$$g(z) = \pi \cot(\pi z) f(z) \qquad (z \in \mathbf{C}).$$

Then $f \in \mathcal{M}(\mathbf{C})$ and $P_g = P_f \cup \mathbf{Z}$. We readily see that $\operatorname{Res}(g; n) = f(n)$ for each $n \in \mathbf{Z}$. Let Γ_n be as in Example 8.37 and suppose f has $m(n)$ poles inside Γ_n at α_j $(j=1,\ldots, m(n))$ and no poles on Γ_n. The residue theorem now gives

$$\frac{1}{2\pi i}\int_{\Gamma_n} g = \sum_{j=-n}^{n} f(j) + \sum_{j=1}^{m(n)} \operatorname{Res}(g; \alpha_j).$$

If g satisfies appropriate conditions we may let $n \to \infty$ and thus evaluate $\sum_{-\infty}^{\infty} f(j)$. In similar fashion we may attempt to evaluate $\sum_{-\infty}^{\infty} (-1)^j f(j)$ by considering the meromorphic function defined by

$$h(z) = \pi \operatorname{cosec}(\pi z) f(z) \qquad (z \in \mathbf{C}).$$

In particular if f is a rational function with $\deg(f) \leqslant -2$ then we can evaluate the sums.

Proposition 8.39. *Let f be a rational function with no poles and no zeros on $\mathbf{Z}$ and with* $\deg(f) \leqslant -2$. *Let g, h be defined as above. Then*

$$\sum_{-\infty}^{\infty} f(n) = -\sum_{j=1}^{m} \operatorname{Res}(g; \alpha_j)$$

$$\sum_{-\infty}^{\infty} (-1)^n f(n) = -\sum_{j=1}^{m} \operatorname{Res}(h; \alpha_j)$$

where $P_f = \{\alpha_j : j = 1, \ldots, m\}$.

Proof. We use the notation and method introduced before the proposition. We may choose $n_0 \in \mathbf{P}$ such that $P_f \subset \operatorname{int}(\Gamma_n)$ $(n > n_0)$. For $n > n_0$ we then have

$$\frac{1}{2\pi i}\int_{\Gamma_n} g = \sum_{j=-n}^{n} f(j) + \sum_{j=1}^{m} \operatorname{Res}(g; \alpha_j).$$

Since $\deg(f) \leqslant -2$, it follows that the double sequence $\sum_{-\infty}^{\infty} f(j)$ converges. It is thus sufficient to show that

$$\lim_{n\to\infty} \int_{\Gamma_n} g = 0.$$

We have already remarked that the function $z \to \pi \cot(\pi z)$ is uniformly bounded on the contours Γ_n. Since $\deg(f) \leqslant -2$ it follows that there is $A > 0$ with

$$\left|\int_{\Gamma_n} g\right| \leqslant l_{\Gamma_n} \frac{A}{(n+\frac{1}{2})^2} \leqslant \frac{8A}{n+1} \qquad (n > n_0).$$

This establishes the result for the case $\sum_{-\infty}^{\infty} f(n)$. The case $\sum_{-\infty}^{\infty} (-1)^n f(n)$ is clearly similar.

If the poles α_j of f are simple then it is easy to calculate the residues of g, h at α_j. In fact we have

$$\begin{aligned}\operatorname{Res}(g; \alpha_j) &= \pi \cot(\pi\alpha_j) \operatorname{Res}(f; \alpha_j)\\ \operatorname{Res}(h; \alpha_j) &= \pi \operatorname{cosec}(\pi\alpha_j) \operatorname{Res}(f; \alpha_j).\end{aligned}$$

Example 8.40. *Given* $\alpha \in \mathbf{C} \setminus i\mathbf{Z}$ *we have*

$$\sum_{n=1}^{\infty} \frac{1}{n^2 + \alpha^2} = \frac{\pi}{2\alpha} \coth(\pi\alpha) - \frac{1}{2\alpha^2} \qquad (z \in \mathbf{C} \setminus i\mathbf{Z}).$$

Proof. If $f(z) = \dfrac{1}{z^2 + \alpha^2}$ $(z \in \mathbf{C})$, then f satisfies the conditions of the above proposition. Since f has simple poles at $i\alpha$, $-i\alpha$ we have

$$\sum_{-\infty}^{\infty} \frac{1}{n^2 + \alpha^2} = -\pi \cot(\pi i\alpha) \frac{1}{2i\alpha} + \pi \cot(-\pi i\alpha) \frac{1}{2i\alpha} = \frac{\pi}{\alpha} \coth(\pi\alpha).$$

Therefore

$$\frac{1}{\alpha^2} + 2 \sum_{n=1}^{\infty} \frac{1}{n^2 + \alpha^2} = \frac{\pi}{\alpha} \coth(\pi\alpha)$$

from which the result follows.

As a corollary of the above example we may obtain the expansion for the function $z \to \pi \coth(\pi z)$. Moreover if we let $\alpha \to 0$ we obtain the delightful formula

$$\sum_{n=1}^{\infty} \frac{1}{n^2} = \frac{\pi^2}{6}.$$

A generalization of this formula is given in Problem 8.47.

PROBLEMS 8

43. Generalize the Mittag-Leffler theorems to $\Delta(0, 1)$ and then to any domain in **C**.

44. Verify the expansion given for $\pi \operatorname{cosec}(\pi z)$.

45. Determine the Mittag-Leffler expansions of the following functions in $\mathscr{M}(\mathbf{C})$.

(i) $f(z) = \dfrac{1}{\exp(z) - 1}$

(ii) $f(z) = \pi \operatorname{sech}(\pi z)$

(iii) $f(z) = \pi \tan(\pi z)$.

(*Hint.* The mappings $z \to iz$, $z \to z - \dfrac{\pi}{2}$ may be helpful.)

46. Use Theorem 8.19 to derive the Mittag-Leffler expansions of the following functions in $\mathscr{M}(\mathbf{C})$.

(i) $f(z) = \dfrac{\exp(z)}{\{\exp(z) - 1\}^2}$

(ii) $f(z) = \pi^2 \operatorname{cosec}^2(\pi z)$

(iii) $f(z) = \pi^2 \tanh^2(\pi z)$.

47. Given $n \in \mathbf{P}$ let

$$f_{2n}(z) = \frac{1}{z^{2n}\{\exp(z) - 1\}} \qquad (z \in \mathbf{C}).$$

Given $m \in \mathbf{P}$ let Γ_m be the boundary of the square with vertices

$$\pm(2m+1)\pi \pm (2m+1)\pi i.$$

Show that

$$|f_{2n}(z)| \leqslant \frac{2}{\{(2m+1)\pi\}^{2n}} \qquad (z \in \Gamma_m,\ m, n \in \mathbf{P})$$

and deduce that

$$\sum_{j=1}^{\infty} \frac{1}{j^{2n}} = \frac{(-1)^{n+1}(2\pi)^{2n} B_{2n}}{2.(2n)!} \qquad (n \in \mathbf{P})$$

where B_{2n} is the *2n-th* Bernoulli number. (*Note.* This shows that the numbers B_{2n} alternate in sign—a fact that was not obvious from Example 7.13.)

48. Sum the following series.

(i) $\displaystyle\sum_{n=1}^{\infty} \frac{(-1)^{n+1}}{n^2}$

(ii) $\displaystyle\sum_{n=0}^{\infty} \frac{(-1)^n}{(2n+1)^3}$

(iii) $\displaystyle\sum_{n=1}^{\infty} \frac{n^2}{n^4 + a^4} \qquad (a > 0)$

(iv) $\displaystyle\sum_{n=1}^{\infty} (-1)^n \frac{n \sin(na)}{z^2 - n^2} \qquad (-\pi < a < \pi)$.

8.10 Zeros and poles revisited

In this section we obtain more detailed information about the zeros and poles of meromorphic functions. In particular we are concerned to count the number inside a given starlike contour. This has an obvious application in estimating the zeros of a given function. We give two other applications. In the first we derive some of the results of Hurwitz concerning the zeros of uniform limits of analytic functions. In the second we derive the Jensen formula for a function meromorphic on a disc. This formula is the starting point for the classical treatment of some of the deeper properties of meromorphic functions.

Suppose that $f \in \mathcal{M}(D)$ has a zero of order k at $\alpha \in D$. For the purposes of this section we shall regard α as contributing k zeros to f. This convention is not unfamiliar in elementary algebra where we sometimes say that an equation has k equal roots. When we say that the zeros of f are counted *according to their multiplicity* we mean that a zero of order k counts as k zeros. A similar convention applies for poles.

Theorem 8.41. *Let $f \in \mathcal{M}(D)$ and let Γ be a starlike simple closed contour with positive direction such that $\Gamma^* \subset D$. If f has no poles and no zeros on Γ then*

$$\frac{1}{2\pi i}\int_\Gamma \frac{f'}{f} = N - P$$

where N is the number of zeros of f on int (Γ) *and P is the number of poles on* int (Γ), *each counted according to their multiplicity.*

Proof. We know from Theorems 8.3 and 8.11 that N and P are both finite. It is clear that the singularities of f'/f can occur only at the zeros and poles of f. If f has a zero of order k at α, there is $g \in \mathcal{M}(D)$ such that

$$f(z) = (z - \alpha)^k g(z) \qquad (z \in D)$$

where g is regular at α with $g(\alpha) \neq 0$. Then g'/g is regular in some neighbourhood of α, and since

$$\frac{f'(z)}{f(z)} = \frac{k}{z - \alpha} + \frac{g'(z)}{g(z)}$$

it follows that f'/f has a simple pole at α with residue k. If f has a pole of order m at α, a similar argument shows that f'/f has a simple pole at α with residue $-m$. The result is now immediate from the residue theorem.

Corollary. *If $f \in \mathcal{A}(D)$ then* $\dfrac{1}{2\pi i}\displaystyle\int_\Gamma \frac{f'}{f} = N.$

The above result is sometimes called the *principle of the argument*. To see why, observe that there is a domain D_1 with $\Gamma \subset D_1 \subset D$ such that $f'/f \in \mathcal{A}(D_1)$. By Theorem 8.28 there is a primitive of f'/f along Γ. The local primitives are analytic logarithms of f and it follows that the integral is the difference between two analytic logarithms of f at a point. Intuitively speaking this represents the continuous change in the argument of $f(z)$ as z moves round Γ.

In general the principle of the argument does not enable us to determine N and P separately unless we already know one of them. Moreover the result is clearly most satisfactory for the case in which the zeros and poles are all simple. In most examples the above integral is not easy to evaluate. A much more useful theorem is given below.

Theorem 8.42 (Rouché). *Let $f, g \in \mathscr{M}(D)$ and let Γ be a starlike simple closed contour with $\Gamma^* \subset D$ such that*

$$|g(z)| < |f(z)| < +\infty \qquad (z \in \Gamma).$$

Then the number $N-P$ of the last theorem is the same for the functions f and $f+g$.

Proof. It is clear from the condition on f that f has no zeros and no poles on Γ. Moreover

$$0 < |g(z)| - |f(z)| \leqslant |(f+g)(z)| < 2|f(z)| < +\infty \qquad (z \in \Gamma)$$

so that $f+g$ has no zeros and no poles on Γ. By Theorem 8.41 it is now sufficient to show that

$$\int_\Gamma \frac{(f+g)'}{f+g} = \int_\Gamma \frac{f'}{f}.$$

Let $F = \dfrac{f+g}{f}$ so that

$$\frac{F'}{F} = \frac{(f+g)'}{f+g} - \frac{f'}{f}.$$

It is now sufficient to show that $\displaystyle\int_\Gamma \frac{F'}{F} = 0$. Observe that

$$|F(z) - 1| = \left|\frac{g(z)}{f(z)}\right| < 1 \qquad (z \in \Gamma).$$

Let $\gamma \in \mathscr{C}([0, 1])$ represent Γ and let $h = \log \circ F \circ \gamma$. Since $F(\Gamma) \subset \Delta(1, 1)$ it follows that h is continuous and is a primitive of F'/F along Γ. Theorem 8.28 now gives

$$\int_\Gamma \frac{F'}{F} = h(1) - h(0) = 0.$$

Corollary. *If $f, g \in \mathscr{A}(D)$ then f and $f+g$ have the same number of zeros on* int (Γ), *counted according to their multiplicity.*

Example 8.43. *Given $|\alpha| > \mathrm{e}$, the equation*

$$\alpha z \exp(z) = 1$$

has exactly one root in $\Delta(0, 1)$. If $\alpha > \mathrm{e}$ this root is real and positive.

Proof. Let f, g be the entire functions defined by

$$f(z) = \alpha z, \qquad g(z) = -\exp(-z) \qquad (z \in \mathbf{C}).$$

Then f has exactly one zero, at 0, and it is simple. For $z \in C(0, 1)$ we have

$$|g(z)| = \mathrm{e}^{-\mathrm{Re}\, z} \leqslant \mathrm{e} < |\alpha| = |f(z)|.$$

By Rouché's theorem $f + g$ has thus exactly one zero on $\Delta(0, 1)$ and so the given equation has exactly one root in $\Delta(0, 1)$.

Suppose now that $\alpha > \mathrm{e}$ and let

$$h(x) = \alpha x \mathrm{e}^x - 1 \qquad (x \in [0, 1]).$$

Then $h \in \mathscr{C}_{\mathbf{R}}([0, 1])$, $h(0) = -1$, $h(1) = \mathrm{e} - 1 > 0$. By the intermediate value theorem there is $t \in (0, 1)$ such that $h(t) = 0$. This completes the proof.

Theorem 8.44 (Hurwitz). *Let $\{f_n\} \subset \mathscr{A}(D)$ converge to f uniformly on each compact subset of D, and let Γ be a starlike simple closed contour with $\Gamma^* \subset D$. If f has no zeros on Γ and k zeros on* int (Γ) *then, for some $n_0 \in \mathbf{P}$, f_n has k zeros on* int (Γ) $(n > n_0)$, *zeros being counted according to their multiplicity.*

Proof. Observe from Theorem 8.19 that $f \in \mathscr{A}(D)$. Since f has no zeros on the compact set Γ, it follows from Proposition 1.20 that

$$\epsilon = \inf\{|f(z)| : z \in \Gamma\} > 0.$$

Since $f_n \to f$ uniformly on Γ there is $n_0 \in \mathbf{P}$ such that

$$|f_n(z) - f(z)| < \epsilon \leqslant |f(z)| \qquad (z \in \Gamma, n > n_0).$$

The conditions of Rouché's theorem are now satisfied with $g = f_n - f$. Therefore f_n has k zeros on int (Γ) for $n > n_0$.

Corollary. *If f has no zeros on* int (Γ) *neither has f_n for $n > n_0$.*

The above Hurwitz theorem gives a good deal of information about the zeros of f in terms of the zeros of f_n. A little more analysis provides even more detailed information.

Theorem 8.45. *Let $\{f_n\}$, f be as in the above theorem with $f \neq 0$.*

(i) *If $\alpha_n \in Z_{f_n}$ with $\lim\limits_{n \to \infty} \alpha_n = \alpha \in D$, then $\alpha \in Z_f$.*

(ii) *Given $\alpha \in Z_f$ there is $\alpha_j \in Z_{f_{n_j}}$ such that $\lim\limits_{j \to \infty} \alpha_j = \alpha$.*

Proof. (i) (This part requires only that $f_n \in \mathscr{C}(D)$.) Choose $R > 0$ such that $\bar{\Delta}(\alpha, R) \subset D$. Since $\alpha_n \to \alpha$ there is $m_0 \in \mathbf{P}$ with $\alpha_n \in \bar{\Delta}(\alpha, R)$ $(n > m_0)$. Let $\epsilon > 0$. Since $f_n \to f$ uniformly on $\bar{\Delta}(\alpha, R)$ there is $n_0 > m_0$ such that

$$|f(z) - f_n(z)| < \tfrac{1}{2}\epsilon \qquad (z \in \bar{\Delta}(\alpha, R), n > n_0).$$

In particular

$$|f(\alpha_n) - f_n(\alpha_n)| < \tfrac{1}{2}\epsilon \qquad (n > n_0).$$

Since f is continuous at α there is $n_1 \in \mathbf{P}$ such that

$$|f(\alpha) - f(\alpha_n)| < \tfrac{1}{2}\epsilon \qquad (n > n_1).$$

For $n > \max(n_0, n_1)$ we now have

$$\begin{aligned} |f(\alpha)| &= |f(\alpha) - f_n(\alpha_n)| \\ &\leqslant |f(\alpha) - f(\alpha_n)| + |f(\alpha_n) - f_n(\alpha_n)| \\ &< \epsilon. \end{aligned}$$

Since ϵ was arbitrary this shows that $f(\alpha) = 0$ as required.

(ii) Choose $R > 0$ such that α is the only zero of f on $\bar{\Delta}(\alpha, R)$. Suppose α is a zero of order k. Given $0 < r < R$ it follows from Hurwitz's theorem that there is $n(r) \in \mathbf{P}$ such that f_n has k zeros on $\Delta(\alpha, r)$ for $n > n(r)$. In particular choose $r_j = R/j$ $(j \in \mathbf{P})$. For each $j \in \mathbf{P}$ we may choose some zero α_j of some f_{n_j} with $\alpha_j \in \Delta(\alpha, r_j)$. We then have $\lim_{j\to\infty} \alpha_j = \alpha$ as required.

The next result is a simple extension of the principle of the argument. If $f \in \mathscr{M}(D)$ has n zeros, counted according to multiplicity, on some compact subset of D, we shall here denote these zeros by $\alpha_1, \ldots, \alpha_n$. Thus if α is a zero of order k then α will appear k times in $\alpha_1, \ldots, \alpha_n$. A similar convention applies for poles.

Proposition 8.46. *Let $f \in \mathscr{M}(D)$, $g \in \mathscr{A}(D)$ and let Γ be a starlike simple closed contour, with positive direction, such that $\Gamma^* \subset D$. Suppose f has no zeros and no poles on Γ and has zeros on* int (Γ) *at $\alpha_1, \ldots, \alpha_n$ and poles on* int (Γ) *at $\beta_1, \ldots, \beta_m$. Then*

$$\frac{1}{2\pi i}\int_\Gamma g\frac{f'}{f} = \sum_{j=1}^{n} g(\alpha_j) - \sum_{j=1}^{m} g(\beta_j).$$

Proof. We employ the same method as that used for Theorem 8.41. It is only necessary to verify that

$$\operatorname{Res}\left(g\frac{f'}{f}; \alpha_j\right) = g(\alpha_j), \qquad \operatorname{Res}\left(g\frac{f'}{f}; \beta_j\right) = -g(\beta_j).$$

In general it is not possible to apply the above result directly to the case in which g is a logarithmic function. In the case when Γ is a circle such an application would lead formally to the Jensen formula (Theorem 8.48) which relates the size of $\log |f|$ on a circle to the zeros and poles inside the circle. We choose a different approach to derive the result, beginning with a simple lemma.

Lemma 8.47.

$$\int_0^{2\pi} \log |1 + \alpha\, e^{i\theta}|\, d\theta = 0 \qquad (\alpha \in \Delta(0, 1)).$$

Proof. Given $\alpha \in \Delta(0, 1)$ let

$$f(z) = \frac{1}{z} \log (1 + \alpha z) \qquad \left(z \in \Delta\left(0, \frac{1}{|\alpha|}\right)\right).$$

Since $\log 1 = 0$ it follows that $f \in \mathscr{A}\left(\Delta\left(0, \frac{1}{|\alpha|}\right)\right)$, and so Cauchy's theorem gives $\int_{C(0,1)} f = 0$. Therefore

$$\int_0^{2\pi} \frac{\log (1 + \alpha\, e^{i\theta})}{e^{i\theta}}\, i\, e^{i\theta}\, d\theta = 0$$

and so

$$\int_0^{2\pi} \log |1 + \alpha\, e^{i\theta}|\, d\theta = 0$$

as required.

Theorem 8.48 (Jensen). *Let $f \in \mathscr{M}(\Delta(0, R))$ be regular at* 0 *with $f(0) \neq 0$. Given $0 < r < R$ let f have no zeros and no poles on $C(0, r)$, and let the zeros of f on $\Delta(0, r)$ be $\alpha_1, \ldots, \alpha_n$, and the poles $\beta_1, \ldots, \beta_m$. Then*

$$\frac{1}{2\pi} \int_0^{2\pi} \log |f(re^{i\theta})|\, d\theta = \log |f(0)| + \sum_{j=1}^{n} \log \frac{r}{|\alpha_j|} - \sum_{j=1}^{m} \log \frac{r}{|\beta_j|}.$$

Proof. Define F on $\Delta(0, R)$ by

$$F(z) = \frac{\alpha_1}{\alpha_1 - z} \cdot \ldots \cdot \frac{\alpha_n}{\alpha_n - z} \cdot \frac{\beta_1 - z}{\beta_1} \cdot \ldots \cdot \frac{\beta_m - z}{\beta_m} \cdot \frac{f(z)}{f(0)}.$$

Clearly $F \in \mathscr{M}(\Delta(0, R))$ and F has no zeros and no poles on $\bar{\Delta}(0, r)$. Hence there is t such that $r < t < R$ and F is analytic on $\Delta(0, t)$ and has no zeros on $\Delta(0, t)$. By Proposition 6.4 there is $G \in \mathscr{A}(\Delta(0, t))$ such that

$$\exp (G(z)) = F(z) \qquad (z \in \Delta(0, t)).$$

Since $F(0) = 1$ we have $G(0) = 2k\pi i$ for some $k \in \mathbf{Z}$. By subtracting a constant from G if necessary we may suppose that $G(0) = 0$. It follows that the function $z \to \dfrac{G(z)}{z}$ is analytic on $\Delta(0, t)$ and Cauchy's theorem then gives

$$\int_{C(0,r)} \frac{G(z)}{z}\, dz = 0.$$

Thus

$$\int_0^{2\pi} G(re^{i\theta})\,d\theta = 0$$

and so

$$\int_0^{2\pi} \operatorname{Re} G(re^{i\theta})\,d\theta = 0.$$

Since $|F(z)| = \exp(\operatorname{Re} G(z))$ we have $\operatorname{Re} G(z) = \log |F(z)|$ and thus

$$\frac{1}{2\pi}\int_0^{2\pi} \log |f(re^{i\theta})|\,d\theta = \log |f(0)| + \sum_{j=1}^{m} \log |\beta_j| - \sum_{j=1}^{n} \log |\alpha_j| + \sum_{j=1}^{n} \frac{1}{2\pi}\int_0^{2\pi} \log |\alpha_j - re^{i\theta}|\,d\theta - \sum_{j=1}^{m} \frac{1}{2\pi}\int_0^{2\pi} \log |\beta_j - re^{i\theta}|\,d\theta.$$

It is clear that

$$\frac{1}{2\pi}\int_0^{2\pi} \log |\alpha_j - re^{i\theta}|\,d\theta = \log r + \frac{1}{2\pi}\int_0^{2\pi} \log \left|1 - \frac{\alpha_j}{r} e^{i\theta}\right| d\theta = \log r$$

by Lemma 8.47, and similarly

$$\frac{1}{2\pi}\int_0^{2\pi} \log |\beta_j - re^{i\theta}|\,d\theta = \log r.$$

The theorem now follows.

Corollary. *If $f \in \mathscr{A}(\Delta(0, R))$ then*

$$\log |f(0)| \leqslant \frac{1}{2\pi}\int_0^{2\pi} \log |f(re^{i\theta})|\,d\theta$$

with equality if and only if f has no zeros on $\Delta(0, r)$.

PROBLEMS 8

49. Use Rouché's theorem to prove the fundamental theorem of algebra.

50. Show that four of the roots of the equation $z^5 + 15z + 1 = 0$ belong to $A(0; 2, \frac{3}{2})$.

51. Let $f(z) = \tan z - \alpha z$ $(z \in \mathbf{C})$ and let Γ_n be the boundary of the square with vertices $\pm n\pi \pm n\pi i$. Show that f has $2n + 1$ zeros on int (Γ_n) provided that

$$\pi n|\alpha| > \frac{1 + e^{-2n\pi}}{1 - e^{-2n\pi}}.$$

Show that the zeros are real if α is real.

52. Show that the equation $z = \exp(z)$ has an infinite number of roots and that they belong to $\{z\colon \operatorname{Re} z < -1\}$.

53. Given $r > 0$ show that there is $n(r) \in \mathbf{P}$ such that all the roots of

$$1 + z + \frac{z^2}{2!} + \cdots + \frac{z^n}{n!} = 0$$

belong to $\nabla(0, r)$ for $n > n(r)$.

54. Let $f(z) = \sum_{n=0}^{\infty} \alpha_n z^n$ $(z \in \Delta(0, 1))$, let $0 < r < 1$, let $m(r)$ be the number of zeros of f on $\Delta(0, r)$, counted according to multiplicity and let

$$\mu = \inf \{|f(z)|\colon z \in C(0, r)\}.$$

Show that

$$\mu \leqslant |\alpha_0| + |\alpha_1| + \cdots + |\alpha_{m(r)}|.$$

55. Let p be a polynomial function with $\deg(p) = n$. Show that the equation $p(z) = \beta$ has n roots for each $\beta \in \mathbf{C}$.

56. Let q be a rational function with m poles on $\mathbf{C}^\infty$. Show that the equation $q(z) = \beta$ has m roots for each $\beta \in \mathbf{C}^\infty$. Deduce Problem 55 above as a special case.

57. (i) Let $f \in \mathscr{A}(D)$ and let Γ be a starlike simple closed contour with $\Gamma^* \subset D$ and $f(\Gamma) \subset \mathbf{R}$. Show that f is constant. Show that f need not be constant if $\operatorname{int}(\Gamma) \not\subset D$.

(ii) Let $f \in \mathscr{A}(D)$ and let Γ be a starlike simple closed contour such that $\Gamma^* \subset D$ and $|f|$ is constant on Γ. If f is not constant show that f has at least one zero on $\operatorname{int}(\Gamma)$.

58. Given $\{\alpha_n\} \subset \Delta'(0, 1)$ such that $\sum_{n=1}^{\infty} (1 - |\alpha_n|) < +\infty$, and $k \in \mathbf{P}$, let

$$B(z) = z^k \prod_{n=1}^{\infty} \frac{\alpha_n - z}{1 - \bar{\alpha}_n z} \frac{|\alpha_n|}{\alpha_n} \qquad (z \in \Delta(0, 1)).$$

Show that $B \in H^\infty(\Delta(0, 1))$ (see Problem 8.17) and that the zeros of B occur at α_n $(n \in \mathbf{P})$ and 0. (The function B is called a *Blaschke product*). Conversely, given $f \in H^\infty(\Delta(0, 1)), f \neq 0, Z_f = \{\alpha_n\}$, α_n being a zero of order μ_n, use Jensen's theorem to show that $\sum_{n=1}^{\infty} \mu_n(1 - |\alpha_n|) < +\infty$. This characterizes the zero sets of functions in $H^\infty(\Delta(0, 1))$.

8.11 The open mapping theorem

In this section we discuss a very important topological property of functions that are analytic on a domain, namely the fact that such functions map open sets onto open sets. This result has many consequences. It implies that if $f \in \mathscr{A}(D)$ then $f(D)$ is a domain. If, further, f is one-to-one on D then f^{-1} is analytic on $f(D)$. When combined with the principle of the argument the result also leads to a very useful method for determining the images of domains under suitable mappings. A further application will be given in the final section of this chapter.

Recall that $f\colon D \to \mathbf{C}$ is continuous if and only if $f^{-1}(U)$ is open in D whenever U is open in $\mathbf{C}$. In the opposite direction, f is said to be *open* if $f(U)$ is open in $\mathbf{C}$ whenever U is open in D. In general, a continuous function fails to be open. For example any constant function on D is continuous but clearly not open. Less trivially, any continuous real valued function on D fails to be open.

Theorem 8.49. *If $f \in \mathscr{A}(D)$ is not constant then f is open.*

Proof. Let U be open in D (so that U is simply an open subset of $\mathbf{C}$ contained in D) and let $\alpha \in U$. We shall show that there is $\delta > 0$ such that $\Delta(f(\alpha), \delta) \subset f(U)$. This will show that $f(U)$ is open and hence that f is an open mapping.

By Theorem 8.1 there is $k \geqslant 1$ and $g \in \mathscr{A}(D)$ such that

$$f(z) - f(\alpha) = (z - \alpha)^k g(z) \qquad (z \in D)$$

where $g(\alpha) \neq 0$. Choose $r > 0$ such that $\bar{\Delta}(\alpha, r) \subset U, g(z) \neq 0\ (z \in \bar{\Delta}(\alpha, r))$, and let

$$\delta = \inf\{|(z - \alpha)^k g(z)|: z \in C(\alpha, r)\}.$$

Since $C(\alpha, r)$ is compact it follows from Proposition 1.20 that $\delta > 0$. Given $\beta \in \Delta(f(\alpha), \delta)$ we have

$$f(z) - \beta = (z - \alpha)^k g(z) + f(\alpha) - \beta \qquad (z \in D).$$

Since

$$|f(\alpha) - \beta| < \delta \leqslant |(z - \alpha)^k g(z)| \qquad (z \in C(\alpha, r))$$

it follows from Rouché's theorem that the equation $f(z) = \beta$ has k roots on $\Delta(\alpha, \delta)$. Thus each point of $\Delta(f(\alpha), \delta)$ is the image under f of at least one point of $\Delta(\alpha, r) \subset U$, i.e. $\Delta(f(\alpha), \delta) \subset f(U)$ as required.

Corollary. *$f(D)$ is a domain.*

Proof. By the theorem $f(D)$ is open and by Proposition 1.27 $f(D)$ is also connected, i.e. $f(D)$ is a domain.

The above theorem is sometimes called the *open mapping theorem* or the *interior mapping principle.* Using the idea in the proof we can relate when a function is locally one-to-one to the behaviour of the derivative.

Proposition 8.50. *Given $f \in \mathscr{A}(D)$ and $\alpha \in D$, f is one-to-one on some neighbourhood of α if and only if $f'(\alpha) \neq 0$.*

Proof. Suppose that $f'(\alpha) \neq 0$. In the notation of the proof of the last theorem we have $k = 1$ so that f is one-to-one on $\Delta(\alpha, r)$.

Suppose now that f is one-to-one on $\Delta(\alpha, R) \subset D$ and that $f'(\alpha) = 0$. If $r = \frac{1}{2}R$ it follows from the proof of the last theorem that for each $\beta \in \Delta(f(\alpha), \delta)$ the roots of $f(z) = \beta$ on $\Delta(\alpha, r)$ are repeated. Thus

$$f'(z) = 0 \qquad (z \in f^{-1}(\Delta(f(\alpha), \delta))).$$

Since f is continuous there is t such that $0 < t \leqslant r$ and

$$f'(z) = 0 \qquad (z \in \Delta(\alpha, t))$$

and so f is constant on $\Delta(\alpha, t)$. This contradicts the hypothesis that f is one-to-one on $\Delta(\alpha, R)$ and so we must have $f'(\alpha) \neq 0$.

Observe that the above result is essentially a local result. For example if $D = \mathbf{C}$ and $f = \exp$ then f' has no zeros on D. Given $\alpha \in \mathbf{C}$, exp is one-to-one on $\Delta(\alpha, R)$ if and only if $0 < R \leqslant \pi$. In particular exp is certainly not one-to-one on D. On the other hand, for any $f \in \mathscr{A}(D)$, if f is one-to-one on D it does follow from the above proposition that $f'(z) \neq 0$ $(z \in D)$. This leads to a considerable strengthening of Proposition 3.8 (ii). The result below is sometimes called the *inverse mapping theorem.*

Theorem 8.51. *If $f \in \mathscr{A}(D)$ is one-to-one on D then $f^{-1} \in \mathscr{A}(f(D))$.*

Proof. Let U be open in D. By the open mapping theorem we have $f(U) = (f^{-1})^{-1}(U)$ open in $f(D)$. It follows from Proposition 1.11 that f^{-1} is continuous on $f(D)$. We also have $f'(z) \neq 0$ $(z \in D)$ by Proposition 8.50 and the result now follows from Proposition 3.8 (ii).

Given $f \in \mathscr{A}(D)$ it is often desirable to be able to determine explicitly what the domain $f(D)$ looks like. Up till now we have mainly used *ad hoc* techniques to determine $f(D)$. The result below allows for a more systematic method.

Theorem 8.52. *Let $f \in \mathscr{A}(D)$ and let γ represent a starlike simple closed contour Γ, with positive direction, such that $\Gamma^* \subset D$. If $f \circ \gamma$ represents a starlike simple closed contour Γ_1, with positive direction, then f maps* int (Γ) *one-to-one onto* int (Γ_1).

Proof. Given $\beta \in \mathbf{C} \setminus \Gamma_1$ the number of roots on int (Γ) of the equation $f(z) = \beta$ is given by

$$N(\beta) = \frac{1}{2\pi i}\int_\Gamma \frac{f'(z)}{f(z)-\beta}\,dz$$
$$= \frac{1}{2\pi i}\int_0^1 \frac{(f'\circ\gamma)\gamma'}{f\circ\gamma - \beta}$$
$$= \frac{1}{2\pi i}\int_{\Gamma_1} \frac{dw}{w-\beta}$$

Therefore

$$N(\beta) = \begin{cases} 1 & \text{if } \beta \in \text{int}(\Gamma_1), \\ 0 & \text{if } \beta \in \text{ext}(\Gamma_1). \end{cases}$$

It follows that

$$\text{int}(\Gamma_1) \subset f(\text{int}(\Gamma)) \subset \Gamma_1{}^*.$$

To complete the proof it is now sufficient to show that $f(\text{int}(\Gamma)) \subset \text{int}(\Gamma_1)$. Suppose conversely that there is $\alpha \in \text{int}(\Gamma)$ such that $f(\alpha) \in \Gamma_1$. Since int (Γ) is open and f is an open mapping, it follows that there is $r > 0$ such that $\Delta(f(\alpha), r) \subset f(\text{int}(\Gamma))$. But $f(\alpha) \in b(\text{ext}(\Gamma_1))$ and so $\Delta(f(\alpha), r)$ must contain points of ext (Γ_1). This contradicts the fact established above that $f(\text{int}(\Gamma)) \cap \text{ext}(\Gamma_1) = \varnothing$, and so the proof is complete.

The theorem may not lead us directly to $f(D)$. In some cases it is necessary to proceed via the 'squeezing principle' as we illustrate below. The theorem has extensions to the cases in which the contours need not be starlike, but the theorem in its present form suffices for most applications.

Example 8.53. *If* $D = \left\{z\colon -\frac{\pi}{2} < \text{Re}\, z < \frac{\pi}{2}\right\}$ *then*

$$\sin(D) = \mathbf{C} \setminus \{x\colon x \leqslant -1 \text{ or } x \geqslant 1\}.$$

Proof. Recall that sin is an entire function with $\sin(x+iy) = \sin x \cosh y + i \cos x \sinh y$. Let Γ_n be the boundary of the rectangle with vertices

$$\frac{\pi}{2},\quad \frac{\pi}{2} + in,\quad -\frac{\pi}{2} + in,\quad -\frac{\pi}{2}$$

(cf. Figure 8.6). Then sin is one-to-one on Γ_n and maps $[-\frac{1}{2}\pi, \frac{1}{2}\pi]$ onto $[-1, 1]$, $\left[\frac{\pi}{2}, \frac{\pi}{2} + in\right]$ onto $[1, \cosh n]$, $\left[\frac{\pi}{2} + in, -\frac{\pi}{2} + in\right]$ onto the arc of the ellipse

$$\frac{u^2}{\cosh^2 n} + \frac{v^2}{\sinh^2 n} = 1$$

in the upper half plane, and $\left[-\frac{\pi}{2} + in, -\frac{\pi}{2}\right]$ onto $[-\cosh n, -1]$.

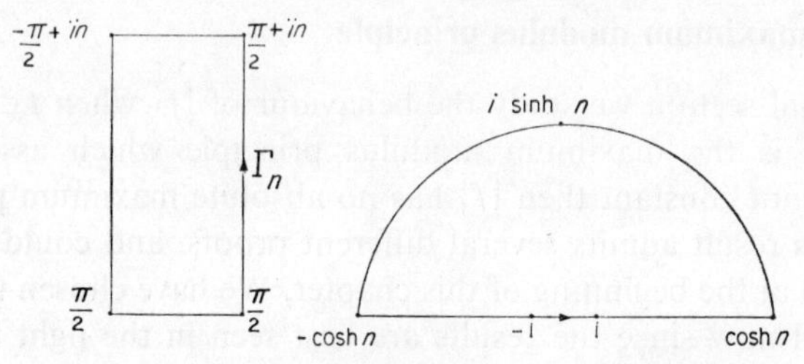

Figure 8.6

By Theorem 8.52, sin maps the inside of Γ_n one-to-one onto the inside of the semi-ellipse in Figure 8.6. Since n was arbitrary it follows that sin maps $\left\{z: -\frac{\pi}{2} < \operatorname{Re} z < \frac{\pi}{2}, \operatorname{Im} z > 0\right\}$ one-to-one onto $\{z: \operatorname{Im} z > 0\}$. A similar argument shows that sin maps $\left\{z: -\frac{\pi}{2} < \operatorname{Re} z < \frac{\pi}{2}, \operatorname{Im} z < 0\right\}$ one-to-one onto $\{z: \operatorname{Im} z < 0\}$. The required result follows.

PROBLEMS 8

59. Let $\mathcal{O}(D)$ denote the set of all open mapping from D into $\mathbf{C}$.
 (i) Show that $\mathcal{O}(D)$ is not closed under sums or products.
 (ii) Show that $f \in \mathcal{O}(D)$ implies $f^n \in \mathcal{O}(D)$ $(n \in \mathbf{P})$.
 (iii) If $f \in \mathcal{O}(D)$ has no zeros show that $1/f \in \mathcal{O}(D)$.
 (iv) Given $f \in \mathcal{O}(D_1)$, $g \in \mathcal{O}(D_2)$ with $g(D_2) \subset D_1$ show that $f \circ g \in \mathcal{O}(D_2)$.
 (v) Give an example of $f \in \mathcal{O}(D)$ that is not continuous on D.

60. Let $f \in \mathscr{A}(D)$ be one-to-one on D, and let Γ be a starlike contour in D with star centre α. If γ represents Γ with the positive direction show that $f \circ \gamma$ represents a simple closed contour Γ_1 such that $n(\Gamma_1; f(\alpha)) = 1$.

61. Determine $f(D)$ in each of the following cases.
 (i) $f(z) = \cosh(z)$, $D = \{z: 0 < \operatorname{Im} z < \pi\}$.
 (ii) $f(z) = \tan(z)$, $D = \left\{z: -\frac{\pi}{2} < \operatorname{Re} z < \frac{\pi}{2}\right\}$.
 (iii) $f(z) = z + \frac{1}{z}$, $D = \Delta'(0, 1)$.
 (iv) $f(z) = \log z$, $D = \Delta(1, 1)$.
 (v) $f(z) = \log(1 - z^2)$, $D = \{z: \operatorname{Im} z > 0\}$.

62. Let f be as in part (ii) above. Show that

$$f^{-1}(z) = \int_{[0,z]} \frac{dw}{1 + w^2} = \frac{1}{2i} \log \frac{1 + iz}{1 - iz}$$

Obtain analogous results for inverse functions of sin, cos.

8.12 The maximum modulus principle

In this final section we study the behaviour of $|f|$ when $f \in \mathscr{A}(D)$. The basic result is the maximum modulus principle which asserts that if $f \in \mathscr{A}(D)$ is not constant then $|f|$ has no absolute maximum point on D. This famous result admits several different proofs, and could easily have been proved at the beginning of this chapter. We have chosen to delay this section until now since the results are best seen in the light of the open mapping theorem.

We use the maximum modulus principle to derive some deeper results. The first concerns a sharpening of the bound of a function and the second concerns the behaviour on the boundary of D when D is an unbounded domain. As a final application we derive the corresponding maximum modulus principle for harmonic functions.

Theorem 8.54 (Maximum Modulus Principle). (i) *If $f \in \mathscr{A}(D)$ is not constant then $|f|$ has no absolute maximum points on D.*

(ii) *If D is bounded and $g \in \mathscr{C}(D^-)$ is analytic and not constant on D then the absolute maximum points of $|g|$ belong to $b(D)$.*

Proof. (i) Suppose that α is an absolute maximum point for $|f|$, i.e. $|f(z)| \leqslant |f(\alpha)| (z \in D)$. By the open mapping theorem $f(D)$ is open and so there is $r > 0$ such that $\Delta(f(\alpha), r) \subset f(D)$. We may clearly choose $\beta \in \Delta(f(\alpha), r)$ such that $|\beta| > |f(\alpha)|$. Since $\beta \in f(D)$ this contradicts the hypothesis that α is an absolute maximum point for $|f|$. This proves (i).

(ii) If D is bounded then D^- is closed and bounded, and so compact. Since $|g|$ is a continuous real function on D^- it has at least one absolute maximum point on D^-. Suppose α is an absolute maximum point for $|g|$ and $\alpha \in D$. We then have in particular

$$|g(z)| \leqslant |g(\alpha)| \qquad (z \in D)$$

so that α is an absolute maximum point for $|g|$ on D. This contradicts part (i) and so the absolute maximum points of $|g|$ must therefore belong to $D^- \times D = b(D)$.

Corollary. (i) *If f has no zeros on D then $|f|$ has no absolute minimum points on D.*

(ii) *If g has no zeros on D^- then the absolute minimum points of $|g|$ belong to $b(D)$.*

Proof. Apply the theorem to the functions $1/f$ and $1/g$.

The maximum modulus principle may be expressed in geometrical terms as follows. Given $f \in \mathscr{A}(D)$ the graph of $|f|$ is a subset of $\mathbf{R}^3$—it may be thought of as the landscape of $|f|$ over D. The first part of the theorem says that there are no mountain tops. In fact there are not even local hills for we would then have an absolute maximum point for $|f|$ restricted to some open disc. A zero of f is obviously an absolute minimum point for $|f|$, but if f has no zeros on D then the graph of f has no hollows over D. The second part of the theorem says that the highest part of the graph always occurs at the edge. Moreover if the function has no zeros then the lowest point also occurs at the edge. The student should illustrate these ideas by drawing the landscapes associated with exp, sin, and log.

It is worthwhile to consider one of the other standard proofs of the maximum modulus principle. The student should realize that it is sufficient to show that if $f \in \mathscr{A}(\Delta(\alpha, R))$ is such that $|f|$ has an absolute maximum point at α then f is constant. If $f(\alpha) = 0$, this last statement is trivial. Suppose that $f(\alpha) \neq 0$. By the Corollary to Theorem 7.1 we have

$$f(\alpha) = \frac{1}{2\pi} \int_0^{2\pi} f(\alpha + re^{i\theta})\, d\theta \qquad (0 < r < R).$$

Let

$$h(\theta) = \operatorname{Re}\left\{\frac{f(\alpha + re^{i\theta})}{f(\alpha)}\right\} \qquad (\theta \in [0, 2\pi]).$$

Then $h \in \mathscr{C}([0, 2\pi])$ and since α is an absolute maximum point for $|f|$ it follows that $h(\theta) \leqslant 1$ $(\theta \in [0, 2\pi])$. Since

$$\int_0^{2\pi} \{1 - h(\theta)\}\, d\theta = \operatorname{Re} \int_0^{2\pi} \left\{1 - \frac{f(\alpha + re^{i\theta})}{f(\alpha)}\right\} d\theta = 0$$

it follows by integration theory that $h(\theta) = 1$ $(\theta \in [0, 2\pi])$. Since r was any point of $(0, R)$ it follows that $\operatorname{Re} f$ is constant on $\Delta(\alpha, R)$. We conclude from Problem 3.20 that f is constant on $\Delta(\alpha, R)$ as required.

The maximum modulus principle leads to a study of the growth of $|f|$ on circles of increasing radius. More precisely suppose that $f \in \mathscr{A}(\Delta(\alpha, R))$ is not constant and define

$$M_\alpha(r) = \sup \{|f(z)| : z \in C(\alpha, r)\}.$$

Since f is continuous on the compact set $C(\alpha, r)$ we have

$$M_\alpha(r) = \max \{|f(z)| : z \in C(\alpha, r)\}.$$

Moreover by the maximum modulus principle we may equally write

$$M_\alpha(r) = \max \{|f(z)| : z \in \bar{\Delta}(\alpha, r)\}.$$

It follows that M_α is an increasing function. In fact (Problem 8.64) it is strictly increasing and continuous. Further properties of M_α are given in the problems at the end of the section. By way of illustration observe that if $f(z) = z^n$ then $M_\alpha(r) = (r + |\alpha|)^n$, and if $f(z) = \exp(z)$ then $M_\alpha(r) = \exp(r + \operatorname{Re}\alpha)$.

The next result is usually called Schwarz's Lemma but we have dignified it as a theorem. Observe that the result gives a sharpening for the bound of a function. We shall give a significant application of the result in Lemma 9.10.

Theorem 8.55 (Schwarz). *Let $f \in \mathscr{A}(\Delta(\alpha, R))$ be such that $f(\alpha) = 0$ and $|f(z)| \leqslant M$ $(z \in \Delta(\alpha, R))$. Then*

$$|f(z)| \leqslant \frac{M}{R}|z - \alpha| \qquad (z \in \Delta(\alpha, R))$$

with equality if and only if $f(z) = \lambda(z - \alpha)$ $(z \in \Delta(\alpha, R))$ where $|\lambda| = M/R$.

Proof. Since $f(\alpha) = 0$ there is $g \in \mathscr{A}(\Delta(\alpha, R))$ such that

$$f(z) = (z - \alpha)\, g(z) \qquad (z \in \Delta(\alpha, R)).$$

Given $0 < r < R$ we have

$$|g(z)| \leqslant \left|\frac{M}{z - \alpha}\right| = \frac{M}{r} \qquad (z \in C(\alpha, r)).$$

It follows from the maximum modulus principle that

$$|g(z)| \leqslant \frac{M}{r} \qquad (z \in \bar{\Delta}(\alpha, r)).$$

Given $\beta \in \Delta(\alpha, R)$ we have

$$|g(\beta)| \leqslant \frac{M}{r} \qquad (|\beta - \alpha| < r < R)$$

and therefore $|g(\beta)| \leqslant \frac{M}{R}$. This shows that

$$|f(z)| \leqslant \frac{M}{R}|z - \alpha| \qquad (z \in \Delta(\alpha, R)).$$

If equality holds at any point of $\Delta(\alpha, R)$ then $|g|$ has an absolute maximum point on $\Delta(\alpha, R)$ and so must be constant. The remaining details of the proof are trivial.

The second part of the maximum modulus principle may be phrased as follows: if D is a bounded domain and if $f \in \mathscr{C}(D^-)$ is analytic on D with $|f| \leqslant M$ on $b(D)$ then $|f| \leqslant M$ on D. We now consider the situation in which D is an unbounded domain. For this case the theorem fails in general as we see in Example 8.56 below, and it is necessary to impose some condition on the rate at which $|f|$ grows towards infinity. Several results of this nature were obtained by Phragmen and Lindelöf, but we shall simply give one illustrative theorem.

Example 8.56. *Let* $D = \left(z: -\frac{\pi}{2} < \operatorname{Im} z < \frac{\pi}{2}\right)$ *and let*

$$f(z) = \exp(\exp(z)) \ (z \in \mathbf{C}).$$

Then $|f| \leqslant 1$ *on* $b(D)$ *while* f *is unbounded on* D.

Proof. Clearly $b(D) = \left(x \pm i\frac{\pi}{2}: x \in \mathbf{R}\right)$ and

$$\left|f\left(x \pm i\frac{\pi}{2}\right)\right| = |\exp(\pm ix)| \leqslant 1 \qquad (x \in \mathbf{R}).$$

Thus $|f| \leqslant 1$ on $b(D)$ but it is obvious that f is not bounded on $\mathbf{R} \subset D$.

Theorem 8.57 (Phragmen-Lindelöf). *Let* $D = \left\{z: -\frac{\pi}{2} \leqslant \operatorname{Re} z < \frac{\pi}{2}\right\}$ and *let* $f \in \mathscr{C}(D^-)$ *be analytic on* D. *Suppose there exist* M, $a > 0$, $\lambda < 1$ *such that*

$$|f(z)| \leqslant M \ (z \in b(D)), \qquad |f(z)| \leqslant \exp(a\, e^{\lambda|\operatorname{Re} z|})(z \in D).$$

Then $|f(z)| \leqslant M \ (z \in D)$.

Proof. Choose $\mu > 0$ with $\lambda < \mu < 1$. Given $\epsilon > 0$ let

$$g_\epsilon(z) = \exp(-\epsilon \cosh(\mu z)) \qquad (z \in \mathbf{C})$$

so that each g_ϵ is an entire function. For $z = x + iy \in D^-$ we have

$$\operatorname{Re}(2\cosh(\mu z)) = \cosh(\mu x)\cos(\mu y) \geqslant \delta \cosh(\mu x)$$

where $\delta = \cos\left(\mu\frac{\pi}{2}\right) > 0$, since $0 < \mu < 1$. It follows that

$$|g_\epsilon(z)| \leqslant \exp(-\epsilon\delta\cosh(\mu x)) \leqslant 1 \qquad (z \in D^-).$$

Therefore $|fg_\epsilon| \leqslant M$ on $b(D)$ and

$$|f(z)g_\epsilon(z)| \leqslant \exp(a\, e^{\lambda|z|} - \epsilon\delta\cosh(\mu x)) \qquad (z \in D).$$

Since $\epsilon\delta > 0$ and $\lambda < \mu$ it follows that

$$\lim_{|x| \to +\infty} f(z)g_\epsilon(z) = 0.$$

Given $\alpha \in D$ we may choose $R > \operatorname{Re}\alpha$ such that

$$|f(z)g_\epsilon(z)| \leqslant M \qquad (|x| > R)$$

It follows that $|fg_\epsilon| \leqslant M$ on the boundary of the rectangle with vertices $\pm R \pm i\frac{\pi}{2}$. By the maximum modulus principle $|fg_\epsilon| \leqslant M$ on all of the rectangle and in particular $|f(\alpha)g_\epsilon(\alpha)| \leqslant M$. Since α was any point of D this shows that $|fg_\epsilon| \leqslant M$ on D. Since

$$\lim_{\epsilon \to 0+} g_\epsilon(z) = 1 \qquad (z \in D)$$

we conclude that $|f| \leqslant M$ on D as required.

Recall from Chapter 3 that there is a close relation between functions analytic on a domain and functions harmonic on a domain. We show finally how the maximum modulus principle for $\mathscr{A}(D)$ leads to a corresponding principle for $\mathscr{H}(D)$.

Theorem 8.58. (i) *If $u \in \mathscr{H}(D)$ is not constant then u has no absolute maximum nor minimum points on D.*

(ii) *If D is bounded and u is a continuous real function on D^-, harmonic on D, then the absolute maximum and minimum points for u belong to $b(D)$.*

Proof. In view of the method of proof for Theorem 8.53 it will be sufficient to show that if $u \in \mathscr{H}(D)$ is not constant then u is an open mapping from D into $\mathbf{R}$. Given $\alpha \in D$ there is $R > 0$ such that $\Delta(\alpha, R) \subset D$. By Proposition 3.13 there is $v \in \mathscr{H}(D)$ such that $f = u + iv$ is analytic on $\Delta(\alpha, R)$. By the open mapping theorem there is $\delta > 0$ such that

$$\Delta(f(\alpha), \delta) \subset f(\Delta(\alpha, R)).$$

It follows that

$$(u(\alpha) - \delta, u(\alpha) + \delta) \subset u(\Delta(\alpha, R)) \subset u(D).$$

This proves that u is an open mapping as required.

PROBLEMS 8

63. Let D be a bounded domain and let $f_n \in \mathscr{C}(D^-)$ be analytic on D for each $n \in \mathbf{P}$. If $\{f_n\}$ converges uniformly on $b(D)$ show that $\{f_n\}$ converges uniformly on D^-. (*Hint.* Use the maximum modulus principle to show that $\{f_n\}$ is Cauchy in the metric space $\mathscr{C}(D^-)$.)

64. Given $f \in \mathscr{A}(\Delta(\alpha, R))$ show that the function M_α is strictly increasing and continuous. (*Hint.* For the continuity part derive contradictions from the suppositions

$$\sup\{M_\alpha(r)\colon r < r_0\} < M_\alpha(r_0) < \inf\{M_\alpha(r)\colon r > r_0\}.)$$

65. (Hadamard's three circle theorem.) Given $f \in \mathscr{A}(\Delta'(\alpha, R))$ and $0 < r_1 \leqslant r \leqslant r_2 < R$ show that

$$M_\alpha(r)^{\log r_2 - \log r_1} \leqslant M_\alpha(r_1)^{\log r_2 - \log r} M_\alpha(r_2)^{\log r - \log r_1}.$$

(*Hint*. Apply the maximum modulus principle to the function $z \to z^p(f(z)^q)$ $(p, q \in \mathbf{Z}, q > 0)$. Choose $\lambda \in \mathbf{R}$ such that $r_1{}^\lambda M_\alpha(r_1) = r_2{}^\lambda M_\alpha(r_2)$ and then choose p_n, q_n so that $\lim\limits_{n \to \infty} \dfrac{p_n}{q_n} = \lambda$.) If $g(u) = \log M_\alpha(e^u)$ show that g is a *convex* function i.e.

$$g(tu_1 + (1 - t)u_2) \leqslant tg(u_1) + (1 - t)g(u_2) \qquad (t \in (0, 1)).$$

66. Given $f \in \mathscr{A}(\mathbf{C})$, $\lambda \in (0, 1)$ let

$$h(r) = \frac{M_0(\lambda r)}{M_0(r)} \qquad (r > 0).$$

Show that h is strictly decreasing and that $\lim\limits_{r \to +\infty} h(r) = 0$ unless f is a polynomial function. What is the limit if f is a polynomial function?

67. Given $f \in \mathscr{A}(\Delta(\alpha, R))$ with $f(\alpha) = 0$, use Schwarz's lemma to show that the function $r \to \dfrac{M_\alpha(r)}{r}$ is increasing. Show also that the function is strictly increasing if it is not constant.

68. Let $f \in \mathscr{A}(\Delta(0, 1))$ be such that $f(\Delta(0, 1)) \subset \Delta(0, 1)$. If $f(\alpha) = \alpha$, $f(\beta) = \beta$ where $\alpha, \beta \in \Delta(0, 1)$, $\alpha \neq \beta$, show that $f(z) = z$ $(z \in \Delta(0, 1))$, i.e. if f fixes two distinct points of $\Delta(0, 1)$ then it fixes all the points of $\Delta(0, 1)$. (*Hint*. Consider the following functions defined on $\Delta(0, 1$.)

$$h(z) = f\left(\frac{z + \alpha}{1 - \bar{\alpha} z}\right), \qquad g(z) = \frac{h(z) - \alpha}{1 - \bar{\alpha} h(z)}.$$

69. Let $f \in \mathscr{C}(\bar{\Delta}(0, R))$ be analytic on $\Delta(0, R)$ and let

$$A(r) = \max\{\operatorname{Re} f(z) : z \in C(0, r)\}.$$

(i) Show that A is strictly increasing and continuous.
(ii) If $f(0) = 0$ show that

$$M_0(r) \leqslant \frac{2r}{R - r} A(R) \qquad (0 < r < R).$$

(*Hint*. Let $g(z) = \dfrac{f(z)}{2A(R) - f(z)}$ $(z \in \Delta(0, R))$ and apply Schwarz's lemma.)

(iii) Show that

$$M_0(r) \leqslant \frac{2r}{R-r} A(R) + \frac{R+r}{R-r} |f(0)| \qquad (0 < r < R).$$

70. Give an alternative proof of Theorem 8.58 by considering the function $\exp \circ f$ where $f \in \mathscr{A}(\Delta(\alpha, R))$.

9

CONFORMAL MAPPING

The aim of the last chapter was to study the fundamental properties of functions that are analytic or meromorphic on a domain D. The aim of this chapter is to study the existence of functions in $\mathscr{A}(D)$ with special properties. In particular we are interested in functions in $\mathscr{A}(D)$ that are one-to-one. As a preliminary motivation we consider some ideas from applied mathematics.

The flow of air past an aerofoil (wing) is clearly of great importance in the design of an aircraft, but the associated problems are complicated because of the shape of the aerofoil. If the aerofoil has uniform cross-section then the original three-dimensional problem can be reduced to a problem in two real dimensions—or one complex dimension. The next step is to find a change of variable that will transform the aerofoil to a simpler shape —and in this context the simplest shape is a disc. In fact in this case it is more appropriate to transform the outside of the aerofoil (where the flow occurs) to the outside of a disc (see Problem 9.19 for an example). The original problem about the aerofoil may now be transformed to a corresponding problem about the disc and this latter problem is usually much more amenable. Having solved the problem for the disc we then transform back again to solve the problem for the aerofoil. We consider next the properties that should hold for this change of variable.

We wish to transform two real variables, x, y say, to two new real variables, u, v say; or equivalently we wish to transform one complex variable z to a new complex variable w, say by the relation $w = f(z)$. The mapping f should be one-to-one so that we can transform the problem back and forth between the aerofoil and the disc. Since differential equations figure prominently in the aerodynamics of the problem, both u and v ought to be suitably differentiable in terms of x and y (rather than merely continuous). The potential theory aspects of aerodynamics make it desirable that the mapping should preserve angles between curves; thus if two curves intersect at right angles in the x, y plane the transformed curves should intersect at right angles in the u, v plane. The appropriate condition for this property is that $f = u + iv$ should be analytic and such that f' has no zeros (see Problem 9.1). Recall from Proposition 8.50 that a necessary condition for f to be one-to-one is that f' have no zeros. Alternatively we may recall from real analysis that the mapping $(x, y) \to (u, v)$ is one-to-one on some neighbourhood of a point provided the Jacobian function

$$J = \begin{vmatrix} \dfrac{\partial u}{\partial x} & \dfrac{\partial v}{\partial x} \\ \\ \dfrac{\partial u}{\partial y} & \dfrac{\partial v}{\partial y} \end{vmatrix}$$

does not vanish at that point. If $f = u + iv$ is analytic then the Cauchy-Riemann equations give

$$J = \left(\frac{\partial u}{\partial x}\right)^2 + \left(\frac{\partial v}{\partial x}\right)^2 = |f'|^2$$

and we see again the necessity of the condition that f' have no zeros.

We thus see in the context of aerodynamics the relevance of investigating the existence of functions in $\mathscr{A}(D)$ that are one-to-one and have a specified domain as their range. It is now natural for the mathematician to pose several general questions about analytic mappings between domains. The first four questions below are suggested by the above example. The final question arises in connection with potential theory and the existence of harmonic functions with specified behaviour on the boundary of a domain; it is also suggested by Cauchy's integral formula.

1. Given domains D, D_1, does there exist $f \in \mathscr{A}(D)$ which maps D one-to-one onto D_1?
2. If so, what is the set of all such mappings?
3. What special features arise in the case $D_1 = D$?
4. Can the above functions be extended to be continuous on D^- and one-to-one on $b(D)$?
5. Given a continuous function g on $b(D)$, when can g be extended to a continuous function on D^- that is analytic on D?

In their full generality the above questions are beyond the scope of this book. In §1 we discuss, without proof, part of the answer to question 1. In the process we discuss an equivalence relation on the set of all domains that is *finer* (i.e. induces more equivalence classes) than the equivalence relation induced by homeomorphism. Questions 2 and 3 are answered in §2 and some part answers to questions 4 and 5 are given in §3. In the final section we give some illustrative mappings.

9.1 Discussion of the Riemann mapping theorem

The classical definition of a conformal mapping on a domain D is that of a function $f \in \mathscr{A}(D)$ whose derivative never vanishes, and the classical property of such functions is that they preserve angles between curves. Thus conformal mappings transform two families of mutually orthogonal curves into two other families of mutually orthogonal curves. This fact is significant for the application of conformal mapping to classical potential theory. We have chosen to relegate this material to Problem 9.1 since we wish to emphasize in our discussion the algebraic aspects of the theory of conformal mapping.

Let D, D_1 be two domains in $\mathbf{C}$. An *isomorphism* of D with D_1 is a homeomorphism f of D with D_1 such that $f \in \mathscr{A}(D)$ and $f^{-1} \in \mathscr{A}(D_1)$. (An isomorphism is more commonly called a *conformal equivalence*.) If such a mapping exists we say that D is *isomorphic* with D_1, and we write $D \approx D_1$. The set of all isomorphisms of D with D_1 is denoted by $I(D, D_1)$. It is clear from the inverse mapping theorem (Theorem 8.51) that if $f \in \mathscr{A}(D)$ and if f maps D one-to-one onto D_1, then f is an isomorphism of D with D_1. An *automorphism* of a domain D is an isomorphism of D with itself. The set of all automorphisms of D is denoted by $A(D)$. Observe that $A(D)$ is non empty for any domain D since it always contains the identity mapping.

Remark. Some may object to the use of the standard algebraic terms isomorphism and automorphism in the present context. In fact we shall see that the above isomorphisms and automorphisms are intimately related to certain isomorphisms and automorphisms in the usual algebraic sense.

Proposition 9.1. *The relation* $\approx$ *is an equivalence relation on the set* $\mathscr{D}$ *of all domains in* $\mathbf{C}$.

Proof. Every domain D of $\mathbf{C}$ is isomorphic with itself under the identity mapping, i.e. $D \approx D$. If $D_1 \approx D_2$ there exists some f in $I(D_1, D_2)$. It follows from the inverse mapping theorem that $f^{-1} \in I(D_2, D_1)$, i.e. $D_2 \approx D_1$. Suppose now that $D_1 \approx D_2$, $D_2 \approx D_3$ and that $f \in I(D_1, D_2)$, $g \in I(D_2, D_3)$. Then $g \circ f$ is a homeomorphism of D_1 with D_3 and is analytic on D_1. Thus $g \circ f \in I(D_1, D_3)$ so that $D_1 \approx D_3$. This completes the proof.

Question 1 of the introduction may be reworded by asking which domains of $\mathbf{C}$ are isomorphic. Moreover it is also natural to ask how many distinct equivalence classes are induced on D by the equivalence relation $\approx$.

In the first place, intuition suggests that isomorphic domains should at least have the same number of 'holes'. It is easy to give a mathematical definition of what we mean by a domain D having n holes. In fact we have already discussed the case $n = 0$ in connection with Runge's theorem (see page 195). Recall that a domain D is *simply connected* if $\mathbf{C}^\infty \setminus D$ is connected, and recall that any starlike domain is simply connected. A domain D is said to be *doubly connected* if $\mathbf{C}^\infty \setminus D$ has 2 components. This corresponds to the domain D having one hole; for example $D = A(0; 2, 1)$ or $D = \Delta'(0, 1)$. More generally, a domain D is said to have *connectivity* n if $\mathbf{C}^\infty \setminus D$ has n components. Note that we have not yet exhausted all the possibilities since $\mathbf{C}^\infty \setminus D$ may have an infinite number of components. The connectivity of a domain is preserved under homeomorphisms and in fact any two domains of connectivity n are homeomorphic; but we shall not prove these statements.

The most important case is the simply connected case (which includes in particular the starlike case). Although all simply connected domains are homeomorphic they are not all isomorphic. In fact $\mathbf{C}$ is not isomorphic with $\Delta(0, 1)$. Otherwise we should have an entire function f such that $|f| < 1$ on $\mathbf{C}$ and f is one-to-one on $\mathbf{C}$. This contradicts Liouville's theorem and so there are at least two equivalence classes amongst the simply connected domains. It is a surprising fact that there are only two such equivalence classes. This is a simple consequence of the famous Riemann mapping theorem stated below.

Every simply connected domain D of $\mathbf{C}$, *other than* $\mathbf{C}$ *itself, is isomorphic with* $\Delta(0, 1)$.

More precisely the theorem asserts that given $\alpha \in D$ there is exactly one mapping $f \in I(D, \Delta(0, 1))$ such that $f(\alpha) = 0$ and $f'(\alpha) > 0$. We shall not attempt to prove the theorem since it involves more topological machinery than we have developed. (The proof would be well within our grasp had we first proved Runge's theorem and Montel's theorem (see page 196).)

The Riemann mapping theorem implies that, up to isomorphism, there are only two simply connected domains in $\mathbf{C}$, namely $\mathbf{C}$ and $\Delta(0, 1)$. For many problems this means that if we can deal with the cases $D = \mathbf{C}$ and $D = \Delta(0, 1)$ then we can (at least in theory) deal with the case of any simply connected domain. For this reason particular attention is given in advanced texts to the case $D = \Delta(0, 1)$. We shall return to this point in a moment.

It must be clearly understood that the Riemann mapping theorem is an existence theorem and gives little clue as to how we might determine an actual isomorphism of $\Delta(0, 1)$ with a given simply connected domain other than $\mathbf{C}$. In practice one usually consults a 'dictionary' of conformal mappings, or resorts to numerical techniques.

The results for doubly connected domains are more complicated. Every doubly connected domain is isomorphic with some annulus centred on the origin; but $A(0; R_2, R_1)$ is isomorphic with $A(0; r_2, r_1)$ if and only if $R_2/R_1 = r_2/r_1$. It follows that the equivalence classes for doubly connected domains are in one-to-one correspondence with the members of $[0, 1)$. The results for higher connectivity are even more complicated.

We turn now to more algebraic considerations.

Proposition 9.2. *Given $\varphi \in I(D_1, D_2)$ the mapping $f \to f \circ \varphi$ is an algebraic isomorphism of $\mathscr{A}(D_2)$ with $\mathscr{A}(D_1)$.*

Proof. Let $Tf = f \circ \varphi \; (f \in \mathscr{A}(D_2))$. Then $Tf = Tg$ implies $f = Tf \circ \varphi^{-1} = Tg \circ \varphi^{-1} = g$, so that T is one-to-one. Given $h \in \mathscr{A}(D_1)$ we have $h \circ \varphi^{-1} \in \mathscr{A}(D_2)$ and $T(h \circ \varphi^{-1}) = h$, so that T maps $\mathscr{A}(D_2)$ onto $\mathscr{A}(D_1)$. Given $f, g \in \mathscr{A}(D_2)$, $\lambda \in \mathbf{C}$ we have

$$T(\lambda f) = \lambda f \circ \varphi = \lambda Tf$$
$$T(f + g) = (f + g) \circ \varphi = f \circ \varphi + g \circ \varphi = Tf + Tg$$
$$T(fg) = fg \circ \varphi = (f \circ \varphi)(g \circ \varphi) = TfTg.$$

The proof is complete.

Corollary. *If $\varphi \in A(D)$ the mapping $f \to f \circ \varphi$ is an algebraic automorphism of $\mathscr{A}(D)$.*

If the domain D_1 has a simpler structure than D_2 we may first solve problems for the algebra $\mathscr{A}(D_1)$ and then transfer the results to $\mathscr{A}(D_2)$ *via* the above algebraic isomorphism. In particular if D_2 is a simply connected domain, $D_2 \neq \mathbf{C}$, problems on $\mathscr{A}(D_2)$ can sometimes be solved by considering the algebra $\mathscr{A}(\Delta(0, 1))$.

The above proposition is essentially routine and trivial. The converse below is much more exciting and by no means trivial. We need to recall some algebraic terminology. A non-empty subset M of $\mathscr{A}(D)$ is an *ideal* if $f, g \in M$, $\lambda \in \mathbf{C}$ impy $f + g$, $\lambda f \in M$ and if $fh \in M$ whenever $f \in M$, $h \in \mathscr{A}(D)$. An ideal M is a *maximal ideal* if $M \neq \mathscr{A}(D)$ and if $\mathscr{A}(D)$ is the only ideal that properly contains M.

Theorem 9.3. *Let T be an algebraic isomorphism between $\mathscr{A}(D_2)$ and $\mathscr{A}(D_1)$. If $\mathbf{u}$ is the identity mapping on D_2, then $T\mathbf{u}: D_1 \to D_2$ is an isomorphism.*

Proof. Given $f \in \mathscr{A}(D_2)$, $w \in f(D)$ iff $f - w\mathbf{1}$ has a zero on D, i.e. iff $f - w\mathbf{1}$ has no multiplicative inverse in $\mathscr{A}(D_2)$. We show now that Tf has the same range as f. Since T is a non-zero algebraic isomorphism we have $T\mathbf{1} = \mathbf{1}$. If $f, g \in \mathscr{A}(D_2)$ with $(f - w\mathbf{1})g = \mathbf{1}$ then $T(f - w\mathbf{1})Tg = T\mathbf{1}$ and so $(Tf - w\mathbf{1})Tg = 1$, i.e. $Tf - w\mathbf{1}$ has a multiplicative inverse. A similar argument applied to T^{-1} shows that $f - w\mathbf{1}$ has a multiplicative inverse if $Tf - w\mathbf{1}$ has. It follows from above that f and Tf have the same range. In particular if $\varphi = T\mathbf{u}$ then φ has the same range as $\mathbf{u}$, namely D_2. Since $\varphi \in \mathscr{A}(D_1)$ it is now sufficient to show that φ is one-to-one.

Suppose that φ is not one-to-one so that there exist $z_1, z_2 \in D_1$, $z_1 \neq z_2$ with $\varphi(z_1) = \varphi(z_2) = w$. Let $M = \{f: f \in \mathscr{A}(D_2), f(w) = 0\}$. It is trivial to verify that M is an ideal of $\mathscr{A}(D_2)$ and clearly $M \neq \mathscr{A}(D_2)$. Suppose J is an ideal of $\mathscr{A}(D_2)$ that contains M strictly. Then there is $g \in J$ such that $g(w) \neq 0$. Since $g - g(w)\mathbf{1} \in M \subset J$ it follows that $g(w)\mathbf{1} \in J$ and so $\mathbf{1} \in J$. Since J is an ideal we have $J = \mathscr{A}(D_2)$ and so M is a maximal ideal. Moreover $M = \{(\mathbf{u} - w\mathbf{1})f: f \in \mathscr{A}(D_2)\}$, since $f(w) = 0$ iff there is $g \in \mathscr{A}(D_2)$ with $f(z) = (z - w)g(z)$ $(z \in D)$. Let $N = \{Tf: f \in M\}$. Since T is an algebraic isomorphism it follows that N is an ideal, and it is clear that N is in fact a maximal ideal since M is a maximal ideal. On the other hand

$$\begin{aligned} N &= \{(T\mathbf{u} - wT\mathbf{1})Tg: g \in \mathscr{A}(D_2)\} \\ &= \{(\varphi - w\mathbf{1})h: h \in \mathscr{A}(D_1)\} \end{aligned}$$

so that N is contained in the proper ideal $\{f: f \in \mathscr{A}(D_1), f(z_1) = 0\}$. But N is strictly contained in this ideal for if $g(z) = z - z_1$ $(z \in D_1)$ then $g \notin N$. Therefore N is not a maximal ideal. This contradiction shows that φ must be one-to-one as required.

This last theorem and the previous proposition show that two domains are isomorphic if and only if the associated algebras of analytic functions are isomorphic. This suggests that one might obtain an algebraic proof of the Riemann mapping theorem. Unfortunately no simple proof has yet been obtained along these lines.

PROBLEMS 9

1. An oriented arc Γ is *regular* at $z_0 \in \Gamma$ if Γ is smooth in some neighbourhood of z_0 and if there is some representative function γ such that $z_0 = \gamma(t_0)$, $\gamma'(t_0) \neq 0$. Let Γ_1, Γ_2 be oriented arcs that intersect at z_0 and are regular at z_0. The *angle* between Γ_1 and Γ_2 at z_0, $<(\Gamma_1, \Gamma_2; z_0)$, is defined up to equivalence modulo 2π by

$$<(\Gamma_1, \Gamma_2; z_0) \equiv \arg \gamma_1'(t_1) - \arg \gamma_2'(t_2)$$

where $z_0 = \gamma_1(t_1) = \gamma_2(t_2)$.

(i) Show that the definition is independent of the choice of those representative functions γ_1, γ_2 such that $\gamma_1'(t_1) \neq 0$, $\gamma_2'(t_2) \neq 0$.

(ii) Let $f \in \mathscr{A}(D)$ and let Γ be an oriented arc in D regular at z_0 where $f'(z_0) \neq 0$. Show that $f(\Gamma)$ contains an oriented arc regular at $f(z_0)$.

(iii) Let $f \in \mathscr{A}(D)$ and let $z_0 \in D$ with $f'(z_0) \neq 0$. Let Γ_1, Γ_2 be oriented arcs intersecting at z_0 and regular at z_0. If $\theta = <(\Gamma_1, \Gamma_2; z_0)$ show that θ is also the angle between the regular subarcs of $f(\Gamma_1)$, $f(\Gamma_2)$ that intersect at z_0, i.e. f preserves angles at z_0.

(iv) Let $f \in \mathscr{A}(D)$ and let $z_0 \in D$ be a zero of f' of order k. We then say that z_0 is a *critical point* of f. Show that angles through z_0 are multiplied (modulo 2π) by a factor $k + 1$.

2. The lines $\{z: \operatorname{Re} z = a\}$ $\left(a \in \left(-\frac{\pi}{2}, \frac{\pi}{2}\right)\right)$ and $\{z: \operatorname{Im} z = b\}$ $(b \in \mathbf{R})$ are mutually orthogonal. Determine their mutually orthogonal images under the mapping sin. Discuss the critical points of the mapping sin.

3. Give an example of an isomorphism of a starlike domain with a domain that is not starlike. (*Hint.* Consider the map $z \to z^n$ on a slice of a disc.)

4. (i) Show that $\Delta'(0, 1)$ is not isomorphic with $\Delta(0, 1)$. (*Hint.* If f is such an isomorphism, f is analytic and bounded on $\Delta'(0, 1)$ and so has an extension $\tilde{f}$ that is analytic on $\Delta(0, 1)$. Show that $\tilde{f}(0) \in \Delta(0, 1)$ and derive a contradiction.)

(ii) Show that any two open annuli are homeomorphic (including the case of a punctured disc).

(iii) Show that $\Delta'(0, 1)$ is not isomorphic with $A(0; R, 1)$ for any R with $1 < R < +\infty$. (*Hint.* Employ a similar argument to that used in part (i).)

(iv) Show that $\Delta'(0, 1)$ is isomorphic with $\nabla(0, 1)$.

5. Given $\varphi \in I(D_1, D_2)$ show that the mapping $f \to f \circ \varphi$ is a continuous mapping between the metric spaces $\mathscr{A}(D_2)$, $\mathscr{A}(D_1)$.

6. Given $\psi \in A(D)$ define T_ψ by

$$T_\psi f = f \circ \psi \qquad (f \in \mathscr{A}(D)).$$

Show that T_ψ is an algebraic automorphism of $\mathscr{A}(D)$ and that

$$T_{\psi \circ \chi} = T_\chi T_\psi \qquad (\psi, \chi \in A(D)).$$

9.2 The automorphisms of a domain

In this section we shall answer questions 2 and 3 of the introduction. In particular, given any domain D, we show that the set $A(D)$ of all automorphisms of D is a group under the operation of composition of mappings. In order to determine this group for simply connected domains we shall see that it is sufficient (modulo the Riemann mapping theorem!) to determine the cases $D = \mathbf{C}$ and $D = \Delta(0, 1)$. It turns out to be useful on the way to define automorphisms of $\mathbf{C}^\infty$ and to determine all such mappings.

Proposition 9.4. (i) *$A(D)$ is a group under the operation of composition of mappings.*

(ii) *Given $f \in I(D_1, D_2)$,*

$$I(D_1, D_2) = \{f \circ \psi : \psi \in A(D_1)\}$$

and the mapping $\psi \to f \circ \psi \circ f^{-1}$ is a group isomorphism of $A(D_1)$ with $A(D_2)$.

Proof. (i) This is routine verification.

(ii) The first statement follows by the method of Lemma 5.2. Define S on $A(D_1)$ by

$$S\psi = f \circ \psi \circ f^{-1}.$$

Clearly $S\psi \in A(D_2)$ and if $S\psi = S\chi$ then

$$\psi = f^{-1} \circ S\psi \circ f = f^{-1} \circ S\chi \circ f = \chi$$

so that S maps $A(D_1)$ one-to-one into $A(D_2)$. Given $\psi_2 \in A(D_2)$ we have $f^{-1} \circ \psi_2 \circ f \in A(D_1)$ and

$$S(f^{-1} \circ \psi_2 \circ f) = f \circ \psi_2 \circ f^{-1} \circ f^{-1} = \psi_2$$

so that S maps $A(D_1)$ onto $A(D_2)$. Given $\psi, \chi \in A(D)$ we have

$$S(\psi \circ \chi) = f \circ \psi \circ \chi \circ f^{-1} = f \circ \chi \circ f^{-1} \circ f \circ \chi \circ f^{-1} = (S\psi) \circ (S\chi).$$

We have thus shown that S is a group isomorphism of $A(D_1)$ with $A(D_2)$.

The second part of the above result answers question 2 of the introduction. It also asserts that isomorphic domains have groups of automorphisms that are (group) isomorphic. In view of the Riemann mapping theorem we can now determine $A(D)$ for any simple connected domain by first determining $A(\mathbf{C})$ and $A(\Delta(0, 1))$. Before determining these groups explicitly it is convenient to extend the definition of automorphism to include the case of the extended complex plane.

An *automorphism* of $\mathbf{C}^\infty$ is a homeomorphism f of $\mathbf{C}^\infty$ with itself such that f is differentiable except at $f^{-1}(\infty)$ and f^{-1} is differentiable except at $f(\infty)$. The set of all automorphisms of $\mathbf{C}^\infty$ is denoted by $A(\mathbf{C}^\infty)$. It is clear that $\alpha = f^{-1}(\infty)$ is an isolated singularity of f and since f is continuous at α we have $\lim_{z \to \alpha} f(z) = \infty$. It follows from Theorem 7.11 (iii) that α is a pole of f and similarly $f(\infty)$ is a pole of f^{-1}. Since f is thus meromorphic on $\mathbf{C}^\infty$ it follows from Theorem 8.14 that f is a rational function. Using Problem 8.56 we see that the pole $\alpha = f^{-1}(\infty)$ is simple. If $\alpha \in \mathbf{C}$ we then see that f is of the form

$$f(z) = \frac{\lambda}{z - \alpha} + \beta.$$

If $\alpha = \infty$ we obtain f of the form

$$f(z) = \beta z + \gamma.$$

We have now effectively determined $A(\mathbf{C}^\infty)$, but it is instructive to develop the theory of $A(\mathbf{C}^\infty)$ by a more algebraic method. To begin with we should like $A(\mathbf{C}^\infty)$ to be a group.

Proposition 9.5. *$A(\mathbf{C}^\infty)$ is a group under the operation of composition of mappings.*

Proof. It is clear that the identity mapping belongs to $A(\mathbf{C}^\infty)$. From the symmetric definition of an automorphism of $\mathbf{C}^\infty$ we have that $f \in A(\mathbf{C}^\infty)$ implies $f^{-1} \in A(\mathbf{C}^\infty)$. It is thus sufficient to show that $f, g \in A(\mathbf{C}^\infty)$ imply $f \circ g \in A(\mathbf{C}^\infty)$. Clearly $f \circ g$ is a homeomorphism of $\mathbf{C}^\infty$ with itself and the singularities of $f \circ g$ can occur only at $(f \circ g)^{-1}(\infty)$ and $g^{-1}(\infty)$. We have already observed that the point $(f \circ g)^{-1}(\infty)$ must be a pole of $f \circ g$. Let $\beta = g^{-1}(\infty)$ and suppose that $\beta \neq (f \circ g)^{-1}(\infty)$. Then $f \circ g(\beta) \in \mathbf{C}$ and so $f \circ g$ is analytic and bounded on a punctured neighbourhood of β. Since $f \circ g$ is continuous at β it follows from Proposition 7.9 that $f \circ g$ is differentiable at β. Thus $f \circ g$ is differentiable except at $(f \circ g)^{-1}(\infty)$ and similarly $(f \circ g)^{-1}$ is differentiable except at $f \circ g(\infty)$. Therefore $f \circ g \in A(\mathbf{C}^\infty)$ as required.

In the following definitions and proposition, D denotes either $\mathbf{C}^\infty$ or a domain in $\mathbf{C}$.

A subgroup G of $A(D)$ is said to be *transitive on D* if for every $z_1, z_2 \in D$ there is $\psi \in G$ such that $\psi(z_1) = z_2$. Given $z_0 \in D$ the *stabilizer of z_0 in D* is defined by

$$H(D; z_0) = \{\psi : \psi \in A(D), \psi(z_0) = z_0\}.$$

It is trivial to verify that $H(D; z_0)$ is a subgroup of $A(D)$.

Proposition 9.6. *Let G be a subgroup of $A(D)$. If G is transitive on D and if there is $z_0 \in D$ such that $H(D; z_0) \subset G$, then $G = A(D)$.*

Proof. Let $\psi \in A(D)$. Then $z_0, \psi(z_0) \in D$ and since G is transitive on D there is $\chi \in G$ such that $\chi(z_0) = \psi(z_0)$. Therefore $\chi^{-1} \circ \psi(z_0) = z_0$ so that $\chi^{-1} \circ \psi \in H(D; z_0) \subset G$. Since $\chi \in G$ it follows that $\psi = \chi \circ (\chi^{-1} \circ \psi) \in G$. Therefore $A(D) \subset G$ and so $G = A(D)$.

An *affine mapping* is a function of the form

$$f(z) = \alpha z + \beta \qquad (z \in \mathbf{C})$$

where $\alpha, \beta \in \mathbf{C}$, $\alpha \neq 0$. If we wish to regard f as a mapping of $\mathbf{C}^\infty$ into $\mathbf{C}^\infty$ then we take $f(\infty) = \infty$. It is routine to verify that an affine mapping is an automorphism of $\mathbf{C}$ (or of $\mathbf{C}^\infty$). In fact there are no other automorphisms of $\mathbf{C}$.

Theorem 9.7. *$A(\mathbf{C})$ is the set of affine mappings.*

Proof. We need only show that any function in $A(\mathbf{C})$ is an affine mapping. If $f \in A(\mathbf{C})$ then f is an entire function and so f is either a polynomial function or else has an isolated essential singularity at ∞. Suppose that f has an isolated essential singularity at ∞ and let $U = \nabla(0, 1)$. The Corollary to Theorem 7.14 gives $f(U)^- = \mathbf{C}$. Since f is a homeomorphism, $f(\Delta(0, 1))$ is open and $f(U) \subset \mathbf{C} \setminus f(\Delta(0, 1))$. It follows that $\mathbf{C} \setminus f(\Delta(0, 1))$ is closed and $f(U)^- \subset \mathbf{C} \setminus f(\Delta(0, 1))$. This contradiction shows that f must be a polynomial function. If f is of degree m then $m \geqslant 1$. Suppose that $m > 1$. By the fundamental theorem of algebra f is of the form

$$f(z) = \beta(z - \alpha_1)\ldots(z - \alpha_m) \qquad (z \in \mathbf{C}).$$

Since f is one-to-one we must have $\alpha_j = \alpha_1$ $(j = 1, \ldots, m)$ and so $f'(\alpha_1) = 0$. This contradicts Proposition 8.50 since f is one-to-one. Therefore f must be an affine mapping as required.

An *homography* is a rational function f of the form

$$f(z) = \frac{\alpha z + \beta}{\gamma z + \delta} \qquad (z \in \mathbf{C}^\infty)$$

where $\alpha, \beta, \gamma, \delta \in \mathbf{C}$ and $\alpha\delta - \beta\gamma \neq 0$. Since

$$f(z) = \frac{\alpha}{\gamma} - \frac{\alpha\delta - \beta\gamma}{\gamma(\gamma z + \delta)} \qquad (z \in \mathbf{C}^\infty)$$

it follows that an homography is not constant. In fact f is clearly one-to-one with inverse mapping given by

$$f^{-1}(z) = \frac{\delta z - \beta}{-\gamma z + \alpha} \qquad (z \in \mathbf{C}^\infty).$$

It is now readily verified that f is an automorphism of $\mathbf{C}^\infty$.

Theorem 9.8. *$A(\mathbf{C}^\infty)$ is the set of homographies.*

Proof. Let G be the set of homographies. We have already remarked that $G \subset A(\mathbf{C}^\infty)$. It is routine, though slightly tedious, to verify that G is a subgroup of $A(\mathbf{C}^\infty)$ that is transitive on $\mathbf{C}^\infty$. Given $f \in H(\mathbf{C}^\infty; \infty)$ we have $f|_\mathbf{C} \in A(\mathbf{C})$ and so by Theorem 9.7 f is an affine mapping and so is an homography. This shows that $H(\mathbf{C}^\infty; \infty) \subset G$. It follows from Proposition 9.6 that $G = A(\mathbf{C}^\infty)$ as required.

The homographies play an important role in conformal mapping and some of their properties are given in the problems. Their fundamental geometrical property is that they map circles onto circles (lines being regarded as circles). To see this we observe that if

$$f(z) = \frac{\alpha z + \beta}{\gamma z + \delta}$$

then

$$f = f_1 \circ f_2 \circ f_3 \circ f_4$$

where

$$f_1(z) = \frac{\alpha}{\gamma} + z$$

$$f_2(z) = (\beta\gamma - \alpha\delta)/\gamma z$$

$$f_3(z) = z + \delta$$

$$f_4(z) = \gamma z.$$

Since f_1, f_2, f_3, f_4 each map circles onto circles so also does f.

The main task now outstanding in this section is the determination of the group $A(\Delta(0, 1))$. It is interesting to begin by determining the group of automorphisms of the upper half plane. We denote the upper half plane by Π^+ and the lower half plane by Π^-, i.e.

$$\Pi^+ = \{z: \operatorname{Im} z > 0\}, \qquad \Pi^- = \{z: \operatorname{Im} z < 0\}.$$

Observe that an homography f maps the real line into itself if the coefficients $\alpha, \beta, \gamma, \delta$ are real. It follows (proof?) that f maps Π^+ into Π^+ or Π^+ into Π^-. This suggests that some of the homographies may give automorphisms of Π^+ and in fact we shall see that $A(\Pi^+)$ is a subgroup of the homographies. We need two preparatory lemmas.

Lemma 9.9. *If*

$$f(z) = \frac{z - i}{z + i} \qquad (z \in \Pi^+)$$

then $f \in I(\Pi^+, \Delta(0, 1))$.

Proof. It is clear that $f \in \mathscr{A}(\Pi^+)$. Recall that it is now sufficient to show that f maps Π^+ one-to-one onto $\Delta(0, 1)$. Since f is the restriction of an homography it is certainly one-to-one. Given $z \in \Pi^+$ with $z = x + iy$, $y > 0$ we have

$$|f(z)|^2 = \left|\frac{x + i(y - 1)}{x + i(y + 1)}\right|^2 = \frac{x^2 + y^2 + 1 - 2y}{x^2 + y^2 + 1 + 2y} < 1$$

so that $f(\Pi^+) \subset \Delta(0, 1)$. Given $\alpha \in \Delta(0, 1)$ let

$$\beta = i\,\frac{1 + \alpha}{1 - \alpha}.$$

Then $f(\beta) = \alpha$ and

$$\operatorname{Im} \beta = \operatorname{Re}\left(\frac{1 + \alpha}{1 - \alpha}\cdot\frac{1 - \bar{\alpha}}{1 - \bar{\alpha}}\right) = \frac{1 - |\alpha|^2}{|1 - \alpha|^2} > 0$$

so that $\beta \in \Pi^+$. Therefore f maps Π^+ onto $\Delta(0, 1)$ and the proof is complete.

Lemma 9.10. *$H(\Delta(0, 1); 0)$ is the group of rotations*

$$\Theta = \{\alpha\mathbf{u} : \alpha \in C(0, 1)\}.$$

Proof. Clearly $\Theta \subset H(\Delta(0, 1); 0)$. Given $f \in H(\Delta(0, 1); 0)$ we have $f \in \mathscr{A}(\Delta(0, 1))$, $f(0) = 0$ and $|f(z)| < 1$ $(z \in \Delta(0, 1))$. By Schwarz's lemma we have

$$|f(z)| \leqslant |z| \qquad (z \in \Delta(0, 1)).$$

The same argument applied to f^{-1} gives

$$|f^{-1}(w)| \leqslant |w| \qquad (w \in \Delta(0, 1)).$$

Given $z \in \Delta(0, 1)$ there is $w \in \Delta(0, 1)$ such that $z = f^{-1}(w)$ and so

$$|f(z)| \leqslant |z| = |f^{-1}(w)| \leqslant |w| = |f(z)| \qquad (z \in \Delta(0, 1)).$$

It follows from the maximum modulus principle that the mapping $z \to \dfrac{f(z)}{z}$ is constant and hence there is $\alpha \in C(0, 1)$ with $f(z) = \alpha z$ $(z \in \Delta(0, 1))$. The proof is complete.

Theorem 9.11. *$A(\Pi^+)$ consists of the mappings of the form*

$$f(z) = \frac{\alpha z + \beta}{\gamma z + \delta} \qquad (z \in \Pi^+)$$

where $\alpha, \beta, \gamma, \delta \in \mathbf{R}$, $\alpha\delta - \beta\gamma > 0$.

Proof. Let G be the set of mappings specified in the statement of the theorem. Given $f \in G$ it is clear that f is analytic and one-to-one on Π^+. As an homography f maps $\mathbf{R} \cup \{\infty\}$ one-to-one onto $\mathbf{R} \cup \{\infty\}$ and so the domain $f(\Pi^+)$ is contained in $\Pi^+ \cup \Pi^-$. Therefore $f(\Pi^+)$ is contained in the component Π^+ or the component Π^-. Since

$$\operatorname{Im} f(i) = \frac{\alpha\delta - \beta\gamma}{\gamma^2 + \delta^2} > 0$$

it follows that $f(\Pi^+) \subset \Pi^+$. Since f maps $\mathbf{C}^\infty$ onto $\mathbf{C}^\infty$ it follows that $f(\Pi^-) \subset \Pi^-$ and then $f(\Pi^+) = \Pi^+$. We have thus shown that $G \subset A(\Pi^+)$. It is straightforward to verify that G is a subgroup of $A(\Pi^+)$ that is transitive on Π^+. In view of Proposition 9.6 the proof will now be complete when we show that $H(\Pi^+; i) \subset G$.

Let f be the isomorphism of Π^+ with $\Delta(0, 1)$ given in Lemma 9.9. Since $f(i) = 0$ it follows from Proposition 9.4 (ii) that the group $H(\Pi^+; i)$ is isomorphic with the group $H(\Delta(0, 1); 0)$ under the mapping $\psi \to f \circ \psi \circ f^{-1}$. Lemma 9.10 now gives

$$H(\Pi^+; i) = \{f^{-1} \circ \chi \circ f : \chi \in \Theta\}.$$

By simple manipulation we verify that any function g of $H(\Pi^+; i)$ is of the form

$$g(z) = \frac{z \cos\dfrac{\theta}{2} + \sin\dfrac{\theta}{2}}{-z \sin\dfrac{\theta}{2} + \cos\dfrac{\theta}{2}} \qquad (z \in \Pi^+).$$

Therefore $H(\Pi^+; i) \subset G$ as required.

Assuming the Riemann mapping theorem we may now determine $A(D)$ for any simply connected domain D, other than $\mathbf{C}$. Indeed since Π^+ is simply connected Proposition 9.4(iii) describes $A(D)$ in terms of $A(\Pi^+)$. In practice this may not lead directly to the neatest description of $A(D)$. The student may appreciate this by trying the case $D = \Delta(0, 1)$. We give below an alternative technique for determining $A(\Delta(0, 1))$. The essence of the proof is to characterize the homographies that are homeomorphisms of $\Delta(0, 1)$ with itself.

Theorem 9.12. $A(\Delta(0, 1))$ *consists of the mappings of the form*

$$f(z) = e^{i\theta} \frac{z + \lambda}{1 + \bar{\lambda} z} \qquad (z \in \Delta(0, 1))$$

where $\theta \in \mathbf{R}$ *and* $\lambda \in \Delta(0, 1)$.

Proof. Using Theorem 9.11, Proposition 9.4(ii) and Lemma 9.9 we see that $A(\Delta(0, 1))$ consists of restrictions to $\Delta(0, 1)$ of certain homographies. We show first that $A(\Delta(0, 1))$ consists of the restrictions to $\Delta(0, 1)$ of those homographies f such that $f(\Delta(0, 1)) \subset \Delta(0, 1)$ and $f(C(0, 1)) \subset C(0, 1)$.

Let f be an homography whose restriction belongs to $A(\Delta(0, 1))$. Then $f(\Delta(0, 1)) = \Delta(0, 1)$ and since the only singularity of f is a pole it follows that f must be regular at each point of $C(0, 1)$. By continuity we have $f(C(0, 1)) \subset \bar{\Delta}(0, 1)$, and since f is one-to-one we must have $f(C(0, 1)) \subset C(0, 1)$. Conversely let f be an homography which maps $\Delta(0, 1)$ and $C(0, 1)$ into themselves. Since homographies map circles onto circles we have $f(C(0, 1)) = C(0, 1)$. Therefore $f(\nabla(0, 1)) \subset \nabla(0, 1) \cup \{\infty\}$ and $f(\Delta(0, 1)) = \Delta(0, 1)$, so that $f \in A(\Delta(0, 1))$.

If

$$f(z) = \frac{\alpha z + \beta}{\gamma z + \delta}$$

then $f(C(0, 1)) \subset C(0, 1)$ if and only if

$$(\alpha z + \beta)(\bar{\alpha}\bar{z} + \bar{\beta}) = (\gamma z + \delta)(\gamma \bar{z} + \bar{\delta}) \qquad (z \in C(0, 1))$$

i.e. $|\alpha|^2 + |\beta|^2 - |\gamma|^2 - |\delta|^2 + 2 \operatorname{Re} \{(\alpha\bar{\beta} - \gamma\bar{\delta})z\} = 0 \qquad (z \in C(0, 1)).$

This is true if and only if

$$|\alpha|^2 + |\beta|^2 - |\gamma|^2 - |\delta|^2 = 0 = \alpha\bar{\beta} - \gamma\bar{\delta}. \tag{1}$$

We may now verify that

$$1 - |f(z)|^2 = \frac{(|\delta|^2 - |\beta|^2)(1 - |z|^2)}{|\gamma z + \delta|^2}$$

so that $f(\Delta(0, 1)) \subset \Delta(0, 1)$ if and only if

$$|\delta|^2 > |\beta|^2. \tag{2}$$

Thus $f \in A(0, 1))$ if and only if conditions (1) and (2) hold. These two conditions are equivalent to (3) below.

$$\alpha \neq 0, \qquad \delta \neq 0, \qquad \frac{\alpha}{\gamma} = \frac{\bar{\delta}}{\bar{\beta}} \in \nabla(0, 1), \qquad |\alpha| = |\delta|. \tag{3}$$

We may write

$$f(z) = \frac{\alpha}{\delta}\,\frac{z + \beta/\alpha}{1 + (\bar{\beta}/\bar{\alpha})z} \qquad (z \in \Delta(0, 1))$$

so that f is of the required form. If we take $\alpha = 1$, $\beta = \lambda$, $\gamma = \bar{\lambda}\,e^{-i\theta}$, $\delta = e^{-i\theta}$ then (3) is satisfied and

$$f(z) = e^{i\theta}\,\frac{z + \lambda}{1 + \bar{\lambda}z} \qquad (z \in \Delta(0, 1)).$$

The proof is complete.

In this section we have determined the three groups $A(\mathbf{C})$, $A(\mathbf{C}^\infty)$, $A(\Pi^+)$. The group $A(\mathbf{C})$ is sometimes called the '$\alpha z + \beta$' group. The group $A(\mathbf{C}^\infty)$ is easily seen to be isomorphic with the group $SL(2, \mathbf{C})$ of all 2×2 complex matrices with determinant 1. Similarly the group $A(\Pi^+)$ is isomorphic with the group $SL(2, \mathbf{R})$ of all 2×2 real matrices with determinant 1. The groups $SL(2, \mathbf{C})$, $SL(2, \mathbf{R})$ are of fundamental importance in the theory of Lie Groups.

Remark. We state without proof the interesting fact that for every doubly connected domain D, the group $A(D)$ is isomorphic with the group $SO(2, \mathbf{R})$ of all 2×2 real orthogonal matrices with determinant 1.

PROBLEMS 9

7. Let D be a domain such that $A(D)$ is transitive on D.

(i) Given $\alpha, \beta \in D$ show that the groups $H(D; \alpha)$, $H(D; \beta)$ are isomorphic.

(ii) If D is isomorphic with D_1, show that $A(D_1)$ is transitive on D_1.

8. Show that $A(\mathbf{C})$ has an infinite number of transitive subgroups. Consider the same problem for $A(\Delta(0, 1))$ and $A(\mathbf{C}^\infty)$.

9. Show that $f \in A(\mathbf{C}^\infty)$ iff f is a homeomorphism of $\mathbf{C}^\infty$ with itself, all of whose singularities are isolated.

10. Show that an homography is uniquely determined by its values at three distinct points.

11. Determine the fixed points of an homography f, i.e. the points p such that $f(p) = p$. If f has distinct fixed points $\alpha, \beta \in \mathbf{C}$ show that f is of the form

$$\frac{f(z) - \alpha}{f(z) - \beta} = \lambda \frac{z - \alpha}{z - \beta}.$$

What happens in the other cases?

12. Every function in $A(D)$ admits a unique extension to an homography if $D = \mathbf{C}$, $\Delta(0, 1)$, Π^+. If D is isomorphic with $\Delta(0, 1)$ show that D has the above property iff every isomorphism of D with $\Delta(0, 1)$ is the restriction of an homography. Deduce that D has this property iff it is a disc or a half plane.

9.3 Mappings of the boundary

In this section we examine questions 4 and 5 of the introduction. The questions are difficult in general and we shall not attempt to prove many results. We begin by discussing question 4.

Let D_1, D_2 be domains and let f be an isomorphism of D_1 with D_2. We should like to know if f can be extended to a homeomorphism of $D^-{}_1$ with $D^-{}_2$, i.e. if f has an extension to $D^-{}_1$ that is continuous on $D^-{}_1$ and maps $b(D_1)$ one-to-one onto $b(D_2)$. If D_1, D_2 are bounded and if $b(D_1)$, $b(D_2)$ are simple closed curves then the answer is yes, although we shall not give the proof. Thus, for example, every isomorphism of $\Delta(0, 1)$ with the semi-disc $\Delta(0, 1) \cap \Pi^+$ admits a continuous extension to $\bar{\Delta}(0, 1)$. It is not possible to remove the condition that $b(D_1)$, $b(D_2)$ be simple closed curves. For example it is not possible to extend an isomorphism of $\Delta(0, 1)$ with $D = \Delta(0, 1) \setminus [0, 1]$ to be a homeomorphism of $\bar{\Delta}(0, 1)$ with $D^- = \bar{\Delta}(0, 1)$. Otherwise we would have the contradiction that $C(0, 1)$ is homeomorphic with $C(0, 1) \cup [0, 1]$ (see Problem 5.32). If we remove the condition that f be one-to-one the situation becomes much more complicated. We shall illustrate by considering the case of the domain $\Delta(0, 1)$.

Given $f \in \mathscr{C}(\Delta(0, 1))$ we say that f has *property* (*E*) if f can be extended to a continuous function $\tilde{f}$ on $\bar{\Delta}(0, 1)$. If f has property (E) the extension is clearly unique. If f has property (E) then f is continuous on the compact set $\bar{\Delta}(0, 1)$, and so $\tilde{f}$ is bounded on $\bar{\Delta}(0, 1)$ and $\tilde{f}$ is bounded on $\Delta(0, 1)$. In order for f to have property (E) the condition that f be bounded is necessary but not sufficient, as the following example shows.

Example 9.13. *If*

$$f(z) = \exp\left(\frac{z+1}{z-1}\right) \qquad (z \in \Delta(0, 1))$$

then f is analytic and bounded on $\Delta(0, 1)$ but does not have property (*E*).

Proof. Clearly $f \in \mathscr{A}(\Delta(0, 1))$. Given $z = x + iy \in \Delta(0, 1)$,

$$\operatorname{Re}\left(\frac{z+1}{z-1}\right) = \frac{x^2 + y^2 - 1}{(x-1)^2 + y^2} < 0$$

and so $|f(z)| < 1$. Suppose f admits a continuous extension to $\tilde{f}$ on $\bar{\Delta}(0,1)$. We then have

$$\tilde{f}(1) = \lim_{x \to 1-} \exp\left(\frac{x+1}{x-1}\right) = 0.$$

Given $\theta \in \left(0, \frac{\pi}{2}\right)$ we have

$$\begin{aligned}\tilde{f}(e^{i\theta}) &= \lim_{r \to 1-} f(re^{i\theta}) \\ &= \exp \frac{e^{i\theta} + 1}{e^{i\theta} - 1} \\ &= \exp\left(-i \cot \frac{\theta}{2}\right)\end{aligned}$$

so that $|\tilde{f}(e^{i\theta})| = 1 \ (\theta \in (0, \frac{\pi}{2}))$. Since $|\tilde{f}|$ is continuous at 1 we deduce that

$$1 = \lim_{\theta \to 0+} |\tilde{f}(e^{i\theta})| = |\tilde{f}(1)| = 0.$$

This contradiction shows that f does not have property (E).

There is a simple topological characterization of property (E). Given $f \in \mathscr{C}(\Delta(0, 1))$, f has property (E) if and only if f is uniformly continuous. For if f has property (E) then $\tilde{f}$ is continuous on the compact set $\bar{\Delta}(0, 1)$ and so uniformly continuous; and hence f is uniformly continuous on $\Delta(0, 1)$. Conversely if f is uniformly continuous on $\Delta(0, 1)$ then f admits a continuous extension to $\bar{\Delta}(0, 1)$ (see Problem 9.15). If $f \in \mathscr{A}(\Delta(0, 1))$, a sufficient condition for the uniform continuity of f is the boundedness of f'. For if $|f'| \leqslant M$ it follows easily from Problem 3.22 that

$$|f(\beta) - f(\alpha)| \leqslant 2M|\beta - \alpha| \qquad (\alpha, \beta \in \Delta(0, 1))$$

and this implies that f is uniformly continuous on $\Delta(0, 1)$. On the other hand the uniform continuity of f does not imply the boundedness of f' (see Problem 9.14(ii)). We shall not pursue this problem further but the interested student may find a wealth of material in Hoffman, *Banach Spaces of Analytic Functions*, Prentice-Hall, 1962.

Finally we consider question 5 of the introduction. To simplify matters we shall consider only the case of the unit disc $\Delta(0, 1)$. Thus we suppose that f is a continuous complex function on $C(0, 1)$ and we wish to know if f has a continuous extension to $\bar{\Delta}(0, 1)$ that is analytic on $\Delta(0, 1)$. We show that this is the case if and only if all the negative Fourier coefficients of f (see Problem 7.17) are zero. Various generalizations of this result may be found in the above mentioned book of Hoffman.

We begin with a technical lemma. For $0 < r < 1$ let

$$P_r(u) = \frac{1 - r^2}{1 - 2r \cos u + r^2} \qquad (u \in \mathbf{R}).$$

The mapping $(r, u) \to P_r(u)$ is called the *Poisson kernel*. We derive below some simple properties of this kernel.

Lemma 9.14. *The Poisson kernel has the following properties.*

(i) $P_r \geqslant 0 \qquad (0 < r < 1)$.

(ii) $\displaystyle\int_0^{2\pi} P_r(u)\, du = 2\pi \qquad (0 < r < 1)$.

(iii) $\displaystyle\lim_{r \to 1-} \sup\{P_r(u) : |u| \geqslant \delta\} = 0 \qquad (0 < \delta < \tfrac{1}{2}\pi)$

Proof. (i) This is clear since

$$1 - 2r\cos u + r^2 > 0 \qquad (0 < r < 1,\ u \in \mathbf{R}).$$

(ii) This may be established by real integration. Alternatively we may employ the residue theorem to obtain

$$\int_0^{2\pi} P_r(u)\,du = \int_{C(0,1)} \frac{1 - r^2}{1 - r\left(z + \dfrac{1}{z}\right) + r^2}\,\frac{dz}{iz}$$
$$= \frac{1}{i}\int_{C(0,1)} \frac{(1 - r^2)}{(1 - rz)(z - r)}\,dz$$
$$= 2\pi.$$

(iii) Given $0 < \delta < \pi/2$, $0 < r < 1$, we have

$$\sup\{P_r(u) : |u| \geqslant \delta\} = \frac{1 - r^2}{1 - 2r\cos\delta + r^2}$$

and it follows that

$$\lim_{r \to 1-} \sup\{P_r(u) : |u| \geqslant \delta\} = 0.$$

Theorem 9.15. *Given a continuous complex function f on $C(0, 1)$, the following statements are equivalent.*

(i) *f has a unique extension to a function $\tilde{f}$ that is continuous on $\bar{\Delta}(0, 1)$ and analytic on $\Delta(0, 1)$.*

(ii) $\displaystyle\int_0^{2\pi} e^{in\theta} f(e^{i\theta})\,d\theta = 0 \qquad (n \in \mathbf{P}).$

Proof. Let condition (i) hold. It follows from Problem 6.2 that

$$\int_{C(0,1)} z^n f(z)\,dz = 0 \qquad (n = 0, 1, 2, \ldots)$$

and therefore

$$\int_0^{2\pi} e^{in\theta} f(e^{i\theta})\,d\theta = 0 \qquad (n \in \mathbf{P}).$$

Let condition (ii) hold. Extend f to a function $\tilde{f}$ on $\bar{\Delta}(0, 1)$ by defining $\tilde{f}$ on $\Delta(0, 1)$ by

$$\tilde{f}(z) = \frac{1}{2\pi i}\int_{C(0,1)} \frac{f(w)}{w - z}\,dw \qquad (z \in \Delta(0, 1)).$$

It follows from Lemma 7.2 that $\tilde{f}$ is analytic on $\Delta(0, 1)$. Given $z \in \Delta'(0, 1)$ with $|z| = r$ we have

$$\left(1 - \frac{r^2 w}{z}\right)^{-1} = \sum_{n=0}^{\infty} \frac{r^{2n} w^n}{z^n} \qquad (w \in C(0, 1))$$

the convergence being uniform on $C(0, 1)$. Therefore

$$\int_{C(0,1)} \frac{f(w)}{w - r^{-2}z} \, dw = -\frac{r^2}{z} \int_{C(0,1)} \left(1 - \frac{r^2 w}{z}\right)^{-1} f(w) \, dw$$

$$= -\frac{r^2}{z} \sum_{n=0}^{\infty} \frac{r^{2n}}{z^n} \int_{C(0,1)} w^n f(w) \, dw$$

$$= 0$$

by condition (ii). Given $z \in \Delta'(0, 1)$ we thus have

$$\tilde{f}(z) = \frac{1}{2\pi i} \int_{C(0,1)} \left\{\frac{1}{w - z} - \frac{1}{w - r^{-2}z}\right\} f(w) \, dw$$

$$= \frac{1}{2\pi} \int_0^{2\pi} \frac{1 - r^2}{|e^{i\theta} - z|^2} f(e^{i\theta}) \, d\theta.$$

This last equation holds in particular for the constant function **1**. If $z = re^{it}$ we thus readily verify that

$$\tilde{f}(re^{it}) - f(e^{it}) = \frac{1}{2\pi} \int_0^{2\pi} P_r(t - \theta)\{f(e^{i\theta}) - f(e^{it})\} \, d\theta$$

where P_r is as in Lemma 9.14. Since the function f is bounded, say $|f| \leqslant M$, we may employ Lemma 9. 14 to obtain

$$|\tilde{f}(re^{it}) - f(e^{it})|$$

$$\leqslant \frac{1}{2\pi}\left[\int_{|t-\theta| \leqslant \delta} + \int_{|t-\theta| \geqslant \delta}\right]\{|f(e^{i\theta}) - f(e^{it})| P_r(t - \theta)\} \, d\theta$$

$$\leqslant \sup \{|f(e^{it}) - f(e^{i\theta})| : |t - \theta| \leqslant \delta\} + 2M \sup \{P_r(t - \theta) : |t - \theta| \leqslant \delta\}.$$

Since $C(0, 1)$ is compact we also have that f is uniformly continuous on $C(0, 1)$. Hence given $\epsilon > 0$ there is δ with $0 < \delta < \frac{1}{2}\pi$ such that

$$|f(e^{it}) - f(e^{i\theta})| < \frac{\epsilon}{2} \qquad (|t - \theta| \leqslant \delta).$$

For this δ we may by Lemma 9.14(iii) choose $r_0 < 1$ such that

$$\sup\{P_r(t-\theta): |t-\theta| \geqslant \delta\} < \frac{\epsilon}{4M} \qquad (r_0 < r < 1).$$

It follows that

$$\lim_{r\to 1-} \tilde{f}(r\,\mathrm{e}^{it}) = f(\mathrm{e}^{it}) \qquad \text{(uniformly in } t). \qquad (*)$$

It is obvious that $\tilde{f}$ is continuous on $\Delta(0, 1)$. Given $\mathrm{e}^{i\theta} \in C(0, 1)$ let $r_n\,\mathrm{e}^{it_n} \to \mathrm{e}^{i\theta}$ with $r_n\,\mathrm{e}^{it_n} \in \Delta(0, 1)$. Then $r_n \to 1$ and $\mathrm{e}^{it_n} \to \mathrm{e}^{i\theta}$. Given $\epsilon > 0$ it follows from $(*)$ that there is $n_1 \in \mathbf{P}$ such that

$$|\tilde{f}(r_n\,\mathrm{e}^{it_n}) - \tilde{f}(\mathrm{e}^{it_n})| < \frac{\epsilon}{2} \qquad (n > n_1).$$

Since f is continuous on $C(0, 1)$ we may choose $n_2 \in \mathbf{P}$ such that

$$|\tilde{f}(\mathrm{e}^{it_n}) - \tilde{f}(\mathrm{e}^{i\theta})| < \frac{\epsilon}{2} \quad (n > n_2).$$

If $n_0 = \max(n_1, n_2)$ we have

$$|\tilde{f}(r_n\,\mathrm{e}^{it_n}) - \tilde{f}(\mathrm{e}^{i\theta})| < \epsilon \qquad (n > n_0)$$

so that $\tilde{f}(r_n\,\mathrm{e}^{it_n}) \to \tilde{f}(\mathrm{e}^{i\theta})$. This shows that $\tilde{f}$ is continuous at each point of $C(0, 1)$ and so $\tilde{f}$ is continuous on $\bar{\Delta}(0, 1)$ as required. If g is any extension of f that is continuous on $\bar{\Delta}(0, 1)$ and analytic on $\Delta(0, 1)$, then Problem 7.1 gives

$$g(z) = \tilde{f}(z) \qquad (z \in \Delta(0, 1))$$

and so $g = \tilde{f}$. The proof is complete.

PROBLEMS 9

13. Let $f \in \mathscr{A}(\Delta(0, 1))$.

(i) If f has property (E) and g is a primitive of f, show that g has property (E).

(ii) If f has property (E) show that f' need not have property (E).

14. (i) If $g(z) = (z-1)\exp\left(\dfrac{z+1}{z-1}\right)$ $(z \in \Delta(0, 1))$, show that g has property (E). Observe that g has no extension that is analytic on $\Delta(0, 1+\epsilon)$ for any $\epsilon > 0$.

(ii) If $g(z) = \sum_{n=1}^{\infty} \dfrac{z^n}{n(n+1)}$ $(z \in \Delta(0, 1))$, show that g has property (E) while g' is not bounded on $\Delta(0, 1)$.

15. Let $f: \Delta(0, 1) \to \mathbf{C}$ be uniformly continuous.

(i) Show that f is bounded.

(ii) Show that f has a continuous extension to $\bar{\Delta}(0, 1)$.

(*Hint:* Let $z \in C(0, 1)$ and let $z = \lim_{n\to\infty} z_n$ where $\{z_n\} \subset \Delta(0, 1)$. Show that $\{f(z_n)\}$ is Cauchy and define

$$\tilde{f}(z) = \lim_{n\to\infty} f(z_n).)$$

16. Let T be defined on $H^\infty(\Delta(0, 1))$ (see Problem 8.17) by

$$(Tf)(z) = \int_{[0,z]} f \qquad (z \in \Delta(0, 1)).$$

Show that each Tf has property (E) and that T is a continuous mapping from $H^\infty(\Delta(0, 1))$ into the disc algebra (see Problem 8.17). Is T onto? If not, is the range of T closed?

17. Let f be an isomorphism of $\Delta(0, 1)$ with D where $b(D)$ is a simple closed curve. If f has property (E) show that the extension $\tilde{f}$ maps $C(0, 1)$ one-to-one onto $b(D)$.

18. Let u be a continuous real function on $C(0, 1)$. Show that there is a unique continuous extension to $\bar{\Delta}(0, 1)$ that is harmonic on $\Delta(0, 1)$.

9.4 Some illustrative mappings

We give here illustrative examples of isomorphisms of the disc $\Delta(0, 1)$ with three standard domains.

Example 9.16. *Let* $D = \{z: 0 < \operatorname{Im} z < \pi\}$ *and let*

$$f(z) = \frac{\exp(z) - i}{\exp(z) + i} \qquad (z \in D^-).$$

Then f is an isomorphism of D with $\Delta(0, 1)$ and f maps $b(D)$ one-to-one onto $C(0, 1) \setminus \{1, -1\}$.

Proof. It is trivial to verify that exp is an isomorphism of D with Π^+. It now follows from Lemma 9.9 that f is an isomorphism of D with $\Delta(0, 1)$. The exponential function maps $b(D)$ one-to-one onto $\mathbf{R} \setminus \{0\}$ and the homography of Lemma 9.9 maps $\mathbf{R} \setminus \{0\}$ one-to-one onto $C(0, 1) \setminus \{1, -1\}$. This completes the proof. Note that the continuous extension of f to D^- fails to map $b(D)$ onto $b(\Delta(0, 1)) = C(0, 1)$. This does not contradict the result stated in the last section since D fails to be bounded.

Example 9.17. *Given* $0 < \lambda < 2$ *let* $D = \{z: |\arg z| < \frac{\pi\lambda}{2}\}$ *and let*

$$f(z) = p^{\lambda}\left(\frac{1+z}{1-z}\right) \qquad (z \in \Delta(0, 1)).$$

Then f *is an isomorphism of* $\Delta(0, 1)$ *with* D.

Proof. Let g be the homography given by

$$g(z) = \frac{1+z}{1-z}.$$

Then g maps $C(0, 1)$ onto $i\mathbf{R} \cup \{\infty\}$ and $g(0) = 1$. It follows (proof?) that g is an isomorphism of $\Delta(0, 1)$ with $D_1 = \{z: \operatorname{Re} z > 0\}$. Since $0 < \lambda < 2$ it is easily verified that p^{λ} is an isomorphism of D_1 with D. The proof is complete. Note that f does not even have a continuous extension to $\bar{\Delta}(0, 1)$ in this case; but this is no surprise for f is unbounded.

Example 9.18. *Let* $D = \Delta(0, 1) \cap \Pi^+$ *and let*

$$f(z) = -i\frac{1 + 2iz + z^2}{1 - 2iz + z^2} \qquad (z \in D^-).$$

Then f *is an isomorphism of* D *with* $\Delta(0, 1)$ *and* f *maps* $b(D)$ *one-to-one onto* $C(0, 1)$, *the semi-circle in the upper half-plane being mapped onto itself.*

Proof. We easily check that f is the composition of the homography g used in Example 9.17, the mapping $z \to z^2$ and the homography of Lemma 9.9. The homography g is an isomorphism of D with the quadrant $\{z: \operatorname{Re} z > 0, \operatorname{Im} z > 0\}$ (proof?) and the mapping $z \to z^2$ is an isomorphism of this quadrant with Π^+. It follows from Lemma 9.9 that f is an isomorphism of D with $\Delta(0, 1)$. The image of the boundary is easily traced under each of the three mappings. Note that the behaviour at the boundary is very satisfactory in this example.

PROBLEMS 9

19. If $f(z) = \frac{1}{2}\left(z + \frac{1}{z}\right)$ $(z \in \mathbf{C} \setminus \{0\})$ show that f is an isomorphism of $\Delta(0, 1)$ with $\mathbf{C} \setminus [-1, 1]$. Determine the inverse mapping. If $-1 \in C(\alpha, r)$ and $1 \in \Delta(\alpha, r)$ show that f is an isomorphism of $\nabla(\alpha, r)$ with the outside of an aerofoil (*Joukowski's aerofoil*).

20. Let $D = \left\{z\colon 0 < \operatorname{Im} z < \frac{\pi}{2}\right\}$ and let

$$f(z) = \tan^2\left(\frac{z}{2}\right) \qquad (z \in D).$$

Show that f is an isomorphism of D with $\Delta(0, 1) \setminus [-1, 0]$ and use it to produce an isomorphism of $\Delta(0, 1)$ with $\Delta(0, 1) \setminus [-1, 0]$. Discuss the behaviour at the boundary in both cases. (Cf. Problem 5.32.)

10

ANALYTIC CONTINUATION

In the previous chapter we considered briefly the problem of extending a function f in $\mathscr{A}(D)$ to a continuous function on D^-. In this chapter we are concerned to extend f to a function that is analytic on a domain larger than D. This leads to the concept of a complete analytic function and thence to complex analytic manifolds and Riemann surfaces.

10.1 Direct analytic continuations

We consider here the first step of extending a function f in $\mathscr{A}(D)$ to a function analytic on a larger domain. We shall see in particular that this is not always possible.

A *function element* is an ordered pair (f, D) where D is a domain in $\mathbf{C}$ and $f \in \mathscr{A}(D)$. Thus if (f_1, D_1), (f_2, D_2) are two function elements we have $(f_1, D_1) = (f_2, D_2)$ if and only of $D_1 = D_2$ and $f_1 = f_2$. Given a function element (f, D) we say that a function element (f_1, D_1) is a *direct analytic continuation* of (f, D) if $D \cap D_1 \neq \varnothing$ and

$$f(z) = f_1(z) \qquad (z \in D \cap D_1).$$

It is evident that (f, D) is then a direct analytic continuation of (f_1, D_1).

Every function element (f, D) admits direct analytic continuations. We simply take any domain $D_1 \subset D$ and $f_1 = f|_{D_1}$. It is then clear that (f_1, D_1) is a direct analytic continuation of (f, D). We say that such direct analytic continuations are *trivial* and all other direct analytic continuations are *proper*. As a simple example of a proper direct analytic continuation take $D = \Delta(0, 1)$ and

$$f(z) = \sum_{n=0}^{\infty} z^n \qquad (z \in \Delta(0, 1)).$$

Evidently $(f, \Delta(0, 1))$ is a function element. Now let $D_1 = \mathbf{C} \setminus \{1\}$ and

$$f_1(z) = \frac{1}{1 - z} \qquad (z \in \mathbf{C} \setminus \{1\}).$$

Then (f_1, D_1) is certainly a proper direct analytic continuation of (f, D). In this case we have simply extended the function f. Indeed we have extended f as far as possible.

Two simple questions arise naturally at this point. Given a function element (f, D) does f admit any proper direct analytic continuation? This question has negative answer in general as we shall show in Example 10.2. If (f, D) does have a proper direct analytic continuation, in what sense is it unique?

Proposition 10.1. *If (f_1, D_1), (f_2, D_1) are direct analytic continuations of (f, D) then $f_1 = f_2$.*

Proof. By the definition of a direct analytic continuation the open set $D \cap D_1$ is non-empty and

$$f_1(z) = f(z) = f_2(z) \qquad (z \in D \cap D_1).$$

It follows by the Corollary to Theorem 8.4 that $f_1 = f_2$.

The result above says that if (f, D) admits a proper direct analytic continuation to a function on some domain D_1 then f admits exactly one direct analytic continuation to that domain. It is of course clear that f will then admit a proper direct analytic continuation to any domain D_2 such that $D \subset D_2 \subset D_1$.

The student may wish to know how he may construct direct analytic continuations of a given function element. There are many special techniques (and tricks!) for constructing continuations. Often it is only necessary to obtain a 'closed expression' for the sum of a power series or the value of an integral. We illustrate below two other standard techniques by considering special examples.

Suppose f is analytic on the half plane $D = \{z: \operatorname{Re} z > 0\}$ and

$$f(z + 1) = zf(z) \qquad (\operatorname{Re} z > 0).$$

Now define f_1 for $\operatorname{Re} z > -1$ by

$$f_1(z) = \frac{f(z+1)}{z} \qquad (\operatorname{Re} z > -1).$$

Then f_1 is certainly differentiable except possibly at 0. In fact if $f(1) \neq 0$ then f_1 has a simple pole at 0. Evidently f_1 agrees with f for $\operatorname{Re} z > 0$. Thus if $D_1 = \{z: \operatorname{Re} z > -1\} \setminus \{0\}$ then (f_1, D_1) is a direct analytic continuation of (f, D). Further

$$f(z + 2) = (z + 1)f(z + 1) = z(z + 1)f(z) \qquad (\operatorname{Re} z > 0).$$

We may thus define f_2 for $\text{Re}\, z > -2$ by

$$f_2(z) = \frac{f(z+2)}{z(z+1)} \qquad (\text{Re}\, z > -2).$$

Arguing as above we see that this gives a direct analytic continuation (f_2, D_2) of (f, D), where $D_2 = \{z: \text{Re}\, z > -2\} \setminus \{0, -1\}$. Evidently f_1, f_2 agree for $\text{Re}\, z > -1$ by Proposition 8.1 (or direct verification). Continuing in this way we may extend f to the whole of $\mathbf{C}$ and the extended function will be meromorphic on $\mathbf{C}$. Note that this method will clearly work for many kinds of functional relationships.

Suppose now that $f \in \mathscr{A}(\Delta(0, 1))$. Given $\alpha \in \Delta'(0, 1)$, f is regular at α and so in some neighbourhood of α we have by Taylor's theorem

$$f(z) = \sum_{n=0}^{\infty} \alpha_n (z - \alpha)^n.$$

Suppose the above power series has radius of convergence ρ_α. Then $\rho_\alpha \geqslant 1 - |\alpha|$ by Proposition 8.2. If $\rho_\alpha > 1 - |\alpha|$ and we define

$$f_\alpha(z) = \sum_{n=0}^{\infty} \alpha_n (z - \alpha)^n \qquad (z \in \Delta(\alpha, \rho_\alpha))$$

then $(f_\alpha, \Delta(\alpha, \rho_\alpha))$ is evidently a proper direct analytic continuation of $(f, \Delta(0, 1))$. If $\rho_\alpha = 1 - |\alpha|$ we pick another point $\beta \in \Delta'(0, 1)$ and hope to obtain a proper direct analytic continuation starting from β. It may happen that $\rho_\alpha = 1 - |\alpha|$ for each $\alpha \in \Delta(0, 1)$. This situation prompts the following definition and example.

Given a function element (f, D) we say that $b(D)$ is a *natural boundary* for (f, D) if (f, D) admits no proper direct analytic continuation. A domain D is said to be a *domain of holomorphy* if there is a function element (f, D) for which $b(D)$ is a natural boundary.

Example 10.2. *If*

$$f(z) = \sum_{n=0}^{\infty} z^{2^n} \qquad (z \in \Delta(0, 1))$$

then $C(0, 1)$ *is a natural boundary for* $(f, \Delta(0, 1))$ *and so* $\Delta(0, 1)$ *is a domain of holomorphy.*

Proof. It is clear that $f \in \mathscr{A}(\Delta(0, 1))$. Suppose that $(f, \Delta(0, 1))$ admits a proper direct analytic continuation. Then f must have a continuous extension to some subarc of $C(0, 1)$ (why?). Any such arc must contain a point of the form $e^{i\theta}$ where $\theta = 2p\pi/2^q$ and $p, q \in \mathbf{P}$. We shall show below that

$$\lim_{r \to 1-} |f(re^{i\theta})| = +\infty.$$

This contradiction will then complete the proof.

If $z = re^{i\theta}$ we have

$$f(z) = \sum_{n=0}^{q-1} z^{2^n} + \sum_{n=q}^{\infty} r^{2^n} \exp(2^{n-q+1}p\pi i)$$

$$= g(z) + \sum_{n=q}^{\infty} r^{2^n}.$$

It is clear that g is bounded on $\Delta(0, 1)$. Let

$$h(r) = \sum_{n=q}^{\infty} r^{2^n} \qquad (0 < r < 1).$$

Suppose that $\lim_{r \to 1-} h(r) = a$. Then for each $N \in \mathbf{P}$ we have

$$\sum_{n=q}^{q+N} r^{2^n} \leqslant a \qquad (0 < r < 1)$$

and therefore

$$N + 1 = \lim_{r \to 1-} \sum_{n=q}^{q+N} r^{2^n} \leqslant a.$$

This contradiction shows that $\lim_{r \to 1-} h(r) = +\infty$ and therefore

$$\lim_{r \to 1-} |f(re^{i\theta})| = +\infty,$$

as required.

Let us now return to our original question about extending a given function in $\mathscr{A}(D)$ to a larger domain. If (f, D) has a proper direct analytic continuation (f_1, D_1) we may of course extend f to a function g on the domain $D \cup D_1$ by taking

$$g(z) = \begin{cases} f(z) & \text{if} \quad z \in D, \\ f_1(z) & \text{if} \quad z \in D_1. \end{cases}$$

It is then clear that g is analytic on $D \cup D_1$. It would seem that we could now take a proper direct analytic continuation (f_2, D_2) of (f_1, D_1) and so build an extension of f on the domain $D \cup D_1 \cup D_2$. To get such an extension we must have f_2 agreeing with f on $D \cap D_2$. This may in fact fail as the next example shows.

Example 10.3. *Let* $D_1 = \Delta(1, 1)$, $D_2 = \Delta(e^{(2\pi i)/3}, 1)$, $D_3 = \Delta(e^{(4\pi i)/3}, 1)$, $f_1 = \log|_{D1}$, $f_2 = \log_{\frac{1}{2}\pi}|_{D2}$, $f_3 = \log_{\pi}|_{D3}$. *Then* (f_{j+1}, D_{j+1}) *is a direct analytic continuation of* (f_j, D_j) $(j = 1, 2)$ *but* $f_1(z) \neq f_3(z)$ $(z \in D_1 \cap D_3)$.

Proof. It is clear that each (f_j, D_j) is a function element. Let $A = \{re^{(\pi i)/3} : 0 < r < 1\}$ so that $A \subset D_1 \cap D_2$. For each $z \in A$

$$\begin{aligned} f_1(z) &= \log|z| + i \arg(z) \\ &= \log|z| + \frac{i\pi}{3} \\ &= \log|z| + i \arg_{\pi/2}(z) \\ &= f_2(z). \end{aligned}$$

By the Corollary to Theorem 8.4 we now have

$$f_1(z) = f_2(z) \qquad (z \in D_1 \cap D_2)$$

and hence (f_2, D_2) is a direct analytic continuation of (f_1, D_1). A similar argument shows that (f_3, D_3) is a direct analytic continuation of (f_2, D_2). Now let $B = \{re^{(5\pi i)/3} : 0 < r < 1\}$. Then $B \subset D_3 \cap D_1$ and if $z \in B$

$$f_1(z) = \log|z| - \frac{\pi i}{3}$$

$$\begin{aligned} f_3(z) &= \log|z| + \frac{5\pi i}{3} \\ &= f_1(z) + 2\pi i. \end{aligned}$$

By the Corollary to Theorem 8.4 this now gives

$$f_3(z) = f_1(z) + 2\pi i \qquad (z \in D_3 \cap D_1)$$

and the proof is complete.

The above example indicates that we must pursue our general extensions with some care. We cannot in general hope to end up with a single function analytic on a suitably large domain. Rather we must expect to end up with a collection of function elements that are suitably related. It is this new type of object that we study in the next section.

PROBLEMS 10

1. Let $f \in \mathscr{A}(\Pi^+)$ and let h be an entire function. If f is such that

$$f(z + i) = h(z)f(z) \qquad (z \in \Pi^+)$$

show that f can be extended to a meromorphic function on **C**. What are the poles of the extended function?

2. (i) Let g be a homeomorphism of D^- with $\bar{\Delta}(0, 1)$ such that $g|_D$ is an isomorphism of D with $\Delta(0, 1)$. If f is as in Example 10.2 show that $b(D)$ is a natural boundary for $(f \circ g, D)$, so that D is a domain of holomorphy. Show also that Π^+ is a domain of holomorphy.

2. (ii) Let

$$\lambda_n = \begin{cases} 3^n & \text{if } n \text{ is odd,} \\ 2 \cdot 3^n & \text{if } n \text{ is even.} \end{cases}$$

Show that $\sum_{n=1}^{\infty} \frac{(-1)^n}{n} z^{\lambda n}$ converges on a dense subset of $C(0, 1)$ and diverges on a dense subset of $C(0, 1)$. (Vijayaraghavan)

3. Given $f \in \mathscr{A}(D)$ let $g(z) = \overline{f(\bar{z})}$ $(z \in \bar{D})$. Show that $g \in \mathscr{A}(\bar{D})$. If $D \cap \bar{D}$ is a domain and contains a non-empty interval of $\mathbf{R}$ on which f takes real values, show that $(g, \bar{D})$ is a direct analytic continuation of (f, D).

10.2 General analytic functions

In this section we introduce the concepts of general analytic functions and complete analytic functions. As far as this section is concerned these objects are not functions in the usual sense but special collections of function elements. (For this reason some students may prefer to think of the object as a general-analytic-function and not as a function which is in some sense general and analytic.)

A *chain* of function elements is a finite collection of function elements $\{(f_j, D_j): j = 1, 2, \ldots, n\}$ $(n > 1)$ such that (f_{j+1}, D_{j+1}) is a direct continuation of (f_j, D_j) for $j = 1, \ldots, n - 1$. Thus the three function elements of Example 10.3 clearly form a chain. The members of a chain are called *links*; we say also that the first and last function elements of the chain are *linked* by the remaining function elements of the chain. Thus in Example 10.3, (f_1, D_1) and (f_3, D_3) are linked by (f_2, D_2).

Given a function element (f, D) we say that a function element (f_1, D_1) is an *analytic continuation* of (f, D) if there is a chain of function elements linking (f, D) to (f_1, D_1). It then follows (by renumbering the chain) that (f, D) is an analytic continuation of (f_1, D_1). Any direct analytic continuation of (f, D) is clearly an analytic continuation of (f, D). In Example 10.3, (f_3, D_3) is an analytic continuation of (f_1, D_1); it is not a direct analytic continuation even though $D_1 \cap D_3 \neq \varnothing$.

A *general analytic function* is a non-empty family $\mathbf{f}$ of function elements such that any two function elements of $\mathbf{f}$ are analytic continuations of each other by a chain whose links are members of $\mathbf{f}$. In a trivial sense, the set consisting of a single function element is a general analytic function. Less trivially we may verify that the three function elements of Example 10.3 form a general analytic function. A *complete analytic function* is a general analytic function **f** which contains all direct analytic continuations of each of its function elements; in other words given $(f, D) \in \mathbf{f}$ and given that (f_1, D_1) is a direct analytic continuation of (f, D) we have $(f_1, D_1) \in \mathbf{f}$. It follows by finite induction that a complete analytic function contains all analytic continuations of each of its function elements. This means roughly that the complete analytic functions are the largest general analytic functions. This statement is made more precise in the theorem below. As an example of a complete analytic function let f be any entire function and let $\mathscr{D}$ be the set of all domains in **C**. The student may easily verify that

$$\mathbf{f} = \{(f|_D, D): D \in \mathscr{D}\}$$

is a complete analytic function.

Theorem 10.4. *Given any general analytic function* **g** *there is a unique complete analytic function* **f** *such that* $\mathbf{g} \subset \mathbf{f}$. *In particular, any function element* (f, D) *determines a complete analytic function* **f** *such that* $(f, D) \in \mathbf{f}$.

Proof. Let $\{\mathbf{f}_\lambda: \lambda \in \Lambda\}$ be the set of all general analytic functions such that $\mathbf{g} \subset \mathbf{f}_\lambda$. This set is non-empty since it contains **g**. Let

$$\mathbf{f} = \bigcup\{\mathbf{f}_\lambda: \lambda \in \Lambda\}.$$

We shall show that **f** is a complete analytic function. Given (f_1, D_1), $(f_2, D_2) \in \mathbf{f}$ there are λ_1, λ_2 such that $(f_1, D_1) \in \mathbf{f}_{\lambda_1}$ $(f_2, D_2) \in \mathbf{f}_{\lambda_2}$. Choose any element (g, D) of **g**. Since $(g, D) \in \mathbf{f}_{\lambda_1}$ there is a chain in $\mathbf{f}_{\lambda_1}$ linking (f_1, D_1) to (g, D). Similarly there is a chain in $\mathbf{f}_{\lambda_2}$ linking (g, D) to (f_2, D_2). Combining these two chains in the obvious way we obtain a chain in $\mathbf{f}_{\lambda_1} \cup \mathbf{f}_{\lambda_2}$, and so in **f**, linking (f_1, D_1) to (f_2, D_2). This shows that **f** is a general analytic function. Given $(f, D) \in \mathbf{f}$ there is some $\lambda \in \Lambda$ such that $(f, D) \in \mathbf{f}_\lambda$. Let (f_1, D_1) be any direct analytical continuation of (f, D). By an argument similar to the one used above we easily verify that $\mathbf{f}_\lambda \cup \{(f_1, D_1)\}$ is a general analytic function containing **g** and thus is contained in **f**. In particular we have $(f_1, D_1) \in \mathbf{f}$ and so **f** is a complete analytic function. Suppose now that **h** is any complete analytic function containing **g**. Since **h** is certainly a general analytic function containing **g** we have $\mathbf{h} \subset \mathbf{f}$. Given $(f, D) \in \mathbf{f}$, (f, D) is certainly an analytic continuation of some element of **g** and so of some element of **h**. Since **h** is a complete analytic function it now follows that $(f, D) \in \mathbf{h}$. Thus $\mathbf{f} \subset \mathbf{h}$ and so $\mathbf{f} = \mathbf{h}$. We have now proved the first part of the theorem. For the final part it is sufficient to recall that the singleton $\{(f, D)\}$ is a general analytic function.

In some cases it is easy to write down the complete analytic function determined by a single function element. We have already seen how to do this for an entire function and the situation is clearly similar for a function meromorphic on **C**. As we should expect from Example 10.3 the situation is much more complicated for the function element (log, $\Delta(1, 1)$). Using the method of Example 10.3 we may easily show that the complete analytic function determined by (log, $\Delta(1, 1)$) contains $\{(\log_{2n\pi}, \Delta(1, 1)) : n \in \mathbf{Z}\}$. In fact (see Problem 10.6) it is not difficult to show that the complete analytic function determined by (log, $\Delta(1, 1)$) consists of all function elements (f, D) where $0 \notin D$ and f is a branch-of-log on D.

There is another interesting way of thinking of complete analytic functions. Given two function elements (f_1, D_1), (f_2, D_2) let us write

$$(f_1, D_1) \sim (f_2, D_2)$$

if (f_1, D_1) and (f_2, D_2) are analytic continuations of each other. It is a trivial exercise to verify that $\sim$ is an equivalence relation on the set of all function elements. Using the method of argument of Theorem 10.4 we easily show that the associated $\sim$-equivalence classes are simply the complete analytic functions. This suggests that it is perhaps more significant to study complete analytic functions than general analytic functions. Indeed we shall concentrate more on complete analytic functions; results about general analytic functions usually follow as simple corollaries.

It should be clear to the student that any complete analytic function **f** contains a very large number of function elements. Despite this, we can always get from one function element to another by means of a (finite) chain. It is this 'finiteness' condition (amongst others) that makes complete analytic functions reasonably tractable objects. Moreover we need not concern ourselves unduly with function elements (f, D) in which D is a complicated domain. The next result indicates that complete analytic functions are quite adequately described by their function elements of the form $(f, \Delta(\alpha, r))$.

Proposition 10.5. *Let* **f** *be a complete analytic function and let* (f_1, D_1), $(f_2, D_2) \in \mathbf{f}$. *Given* $\Delta(\alpha_1, r_1) \subset D_1$, $\Delta(\alpha_2, r_2) \subset D_2$ *there is a chain of function elements in* **f** *of the form* $(f, \Delta(\alpha, r))$ *linking* $(f_1, \Delta(\alpha_1, r_1))$ *to* $(f_2, \Delta(\alpha_2, r_2))$.

Proof. Since **f** is a complete analytic function the function elements $(f_1, \Delta(\alpha_1, r_1))$, $(f_2, \Delta(\alpha_2, r_2))$ are in **f**, and can therefore be linked by a chain of function elements in **f**. We shall suppose that they can be linked by a single function element (f_3, D_3); our argument works equally well for the general case, only the notation becomes more complicated.

Since $\Delta(\alpha_1, r_1) \cap D_3 \neq \varnothing$, $\Delta(\alpha_1, r_1) \cup D_3$ is a domain. Similarly since $\Delta(\alpha_2, r_2) \cap D_3 \neq \varnothing$, $D = \Delta(\alpha_1, r_1) \cup D_3 \cup \Delta(\alpha_2, r_2)$ is a domain. Hence there is a polygonal line Γ in D joining α_1 to α_2. The set Γ is clearly compact. (Even though it may intersect itself the set Γ is still the continuous image of a compact set.) By the usual argument we may thus cover Γ with a finite number of discs contained in D. Indeed we may suppose (why?) that these discs consist of $\Delta(\alpha_1, r_1)$, $\Delta(\alpha_2, r_2)$ and a finite number of discs in D_3. By restricting f_3 to these discs in D_3 and indexing them in suitable order we clearly obtain a chain of the required form linking $(f_1, \Delta(\alpha_1, r_1))$ to $(f_2, \Delta(\alpha_2, r_2))$.

A further way of thinking of complete analytic functions is to regard them as 'many valued functions'. We shall not follow this approach but it is useful to give some meaning to the 'values' of a complete analytic function.

Let $\mathbf{f}$ be a complete analytic function and let α belong to the domain of some function element of $\mathbf{f}$. We say that β is a '*value*' of $\mathbf{f}$ at α if there is $(f, D) \in \mathbf{f}$ such that $\alpha \in D$ and $f(\alpha) = \beta$. For example if $\mathbf{f}$ is the complete analytic function determined by $(\log, \Delta(1, 1))$ then for each $n \in \mathbf{Z}$, $2n\pi i$ is a 'value' of $\mathbf{f}$ at 1. In fact these are all the 'values' of $\mathbf{f}$ at 1. Since the above complete analytic function appears so frequently for illustrative purposes we shall denote it by **log.** It is not difficult to see that **log** has a countable infinity of 'values' at each point of $\mathbf{C} \setminus \{0\}$. This is already a complicated situation but the next result shows that it can never be worse.

Proposition 10.6. *For any complete analytic function* $\mathbf{f}$ *the set of 'values' at any point is at most countable.*

Proof. Choose any $(f, \Delta(\alpha, r)) \in \mathbf{f}$. Since $\mathbf{f}$ is determined by this function element, the set of 'values' of $\mathbf{f}$ at α is simply the set of values at α of all analytic continuations of $(f, \Delta(\alpha, r))$. Suppose that $(f_\lambda, D_\lambda) \in \mathbf{f}$ with $\alpha \in D_\lambda$. Then there is $r_\lambda > 0$ such that $\Delta(\alpha, r_\lambda) \subset D_\lambda$ and $(f_\lambda, \Delta(\alpha, r_\lambda)) \in \mathbf{f}$. We have already seen in Proposition 10.5 that $(f, \Delta(\alpha, r))$ can be linked to $(f_\lambda, \Delta(\alpha, r_\lambda))$ by a chain of function elements of the form $(g, \Delta(\beta, \rho))$. We shall show below that this chain can be chosen such that the points β have rational real and imaginary parts and each ρ is rational. The set of all finite collections of such discs forms a countable set. Thus there is at most a countable number of ways of forming an analytic continuation of $(f, \Delta(\alpha, r))$ of the form $(f_\lambda, \Delta(\alpha, r_\lambda))$. In other words $\mathbf{f}$ has at most a countable number of 'values' at α.

Suppose that $\{(f_j, \Delta(\beta_j, \rho_j)) : j = 1, 2, \ldots, n\}$ are the links of a chain joining $(f, \Delta(\alpha, r))$ to $(f_\lambda, \Delta(\alpha, r_\lambda))$. Let

$$t_1 = \min(a, \rho_1 - a, b, \rho_1 - b)$$

where $a = \text{dist}(\beta_1, \Delta(\alpha, r))$, $b = \text{dist}(\beta_1, \Delta(\beta_2, \rho_2))$. We have $t_1 > 0$ since $\Delta(\alpha, r) \cap \Delta(\beta_1, \rho_1) \neq \varnothing$ and $\Delta(\beta_1, \rho_1) \cap \Delta(\beta_2, \rho_2) \neq \varnothing$. Recall that given any complex number z we can find points, with rational real and imaginary parts, as near z as we please. Similarly we can choose rationals in any interval of real numbers. In particular we may choose rational r_1 such that

$$\rho_1 - \tfrac{1}{2}t_1 < r_1 < \rho_1 - \tfrac{1}{4}t_1.$$

Now choose w_1 in $\Delta(\beta_1, \frac{1}{4}t_1)$ with rational real and imaginary parts. It is now a straightforward exercise in inequalities to verify that

$$\Delta(w_1, r_1) \subset \Delta(\beta_1, \rho_1), \qquad \Delta(w_1, r_1) \cap \Delta(\alpha, r) \neq \varnothing,$$
$$\Delta(w_1, r_1) \cap \Delta(\beta_2, \rho_2) \neq \varnothing.$$

We thus have that $(f_1, \Delta(w_1, r_1))$ and $(f_j, \Delta(\beta_j, \rho_j))$ $(j = 2, \ldots, n)$ are the links of a chain joining $(f, \Delta(\alpha, r))$ to $(f_\lambda, \Delta(\alpha, r))$. We now modify the other discs $\Delta(\beta_j, \rho_j)$ in the same manner, proceeding one step at a time. In this way we obtain a chain of the required form.

The set of 'values' of **log** at 1 is $\{2n\pi i : n \in \mathbf{Z}\}$ and so has no cluster points. In fact this is evidently true of the set of 'values' of **log** at any point of $\mathbf{C} \setminus \{0\}$. This is a comparatively pleasant situation. On the other hand the set of 'values' of a complete analytic function at a point may well have cluster points (see Problem 10.9).

If a general analytic function **f** is to be 'analytic' in a reasonable sense then **f** ought to have a derivative. Given any general analytic function **f** we define the derivative of **f** by

$$\mathbf{f}' = \{(f', D) : (f, D) \in \mathbf{f}\}.$$

It is easy to see that $\mathbf{f}'$ is a general analytic function. In fact given (f_1', D_1), $(f_2', D_2) \in \mathbf{f}'$ we have that (f_1, D_1), (f_2, D_2) are in **f** and so are analytic continuations of each other by a chain in **f**. It then follows that (f_1', D_1), (f_2', D_2) are analytic continuations of each other by a chain in $\mathbf{f}'$. This means that $\mathbf{f}'$ is a general analytic function. More generally we define the higher derivatives of **f** by

$$\mathbf{f}^{(n)} = \{(f^{(n)}, D) : (f, D) \in \mathbf{f}\} \qquad (n \in \mathbf{P}).$$

It is then clear that each $\mathbf{f}^{(n)}$ is a general analytic function.

There is one small point that we must note in connection with derivatives. If **f** is a complete analytic function, it need not happen that $\mathbf{f}'$ is a complete analytic function. To see this take $\mathbf{f} = \mathbf{log}$ so that $(\mathbf{j}, \Delta(1, 1)) \in \mathbf{f}'$. Evidently $(\mathbf{j}, \Delta'(0, 1))$ is a direct analytic continuation of $(\mathbf{j}, \Delta(1, 1))$ but $(\mathbf{j}, \Delta'(0, 1)) \notin \mathbf{f}'$ since $\mathbf{j}$ has no primitive on $\Delta'(0, 1)$. This means that $\mathbf{f}'$ is not a complete analytic function. For convenience we shall adopt the convention that the derivative of a complete analytic function **f** shall be the complete analytic function determined by the general analytic function $\mathbf{f}'$.

In this case we shall still denote the derivative by $\mathbf{f}'$. This is not an unreasonable convention to adopt in view of the fact that any single function element determines the associated complete analytic function.

PROBLEMS 10

4. Let $\mathbf{f}$, $\mathbf{g}$ be general analytic functions and suppose there is a mapping $\varphi: \mathbf{f} \to \mathbf{g}$ of the form

$$\varphi((f, D)) = (f_\varphi, D).$$

If φ maps direct analytic continuations in $\mathbf{f}$ onto direct analytic continuations in $\mathbf{g}$ we say that $\mathbf{f}$ is *subordinate* to $\mathbf{g}$ with respect to φ, and we define

$$\mathbf{f} +_\varphi \mathbf{g} = \{(f + f_\varphi, D): (f, D) \in \mathbf{f}\}$$
$$\mathbf{f} \times_\varphi \mathbf{g} = \{(ff_\varphi, D): (f, D) \in \mathbf{f}\}.$$

(i) Show that $\mathbf{f} +_\varphi \mathbf{g}$, $\mathbf{f} \times_\varphi \mathbf{g}$ are general analytic functions.

(ii) If $\mathbf{f}$ is also subordinate to $\mathbf{g}$ with respect to ψ show that $\mathbf{f} +_\psi \mathbf{g}$ may be different from $\mathbf{f} +_\varphi \mathbf{g}$. Show that they are equal iff φ, ψ agree at some function element of $\mathbf{f}$.

(iii) If $\mathbf{g}$ is the complete analytic function determined by an entire function show that any general analytic function $\mathbf{f}$ is subordinate to $\mathbf{g}$ and that addition and multiplication of $\mathbf{f}$ and $\mathbf{g}$ may be uniquely defined.

(iv) Show that $\mathbf{f}$ is subordinate to $\mathbf{f}^{(n)}$ by the mapping

$$(f, D) \to (f^{(n)}, D).$$

(v) Let P be a polynomial function in $n + 1$ variables whose coefficients are entire functions, and let (f, D) be such that $P(f, f', \ldots, f^{(n)}) = 0$. If (g, E) is any analytic continuation of (f, D) show that $P(g, g', \ldots, g^{(n)}) = 0$. If $\mathbf{f}$ is the complete analytic function determined by (f, D) show that $P(\mathbf{f}, \mathbf{f}', \ldots, \mathbf{f}^{(n)}) = \mathbf{0}$.

5. Let $\mathbf{f}$ be a general analytic function such that $\{(f', D): (f, D) \in \mathbf{f}\}$ is complete. Is $\mathbf{f}$ complete?

6. Let $\mathbf{f}$ be the complete analytic function determined by (f, D), and let $\mathbf{g}$ be the complete analytic function determined by (f', D). Show that $\{(f', D): (f, D) \in \mathbf{f}\}$ contains all the function elements of $\mathbf{g}$ of the form $(g, \Delta(\alpha, r))$.

7. Show that any two branches-of-log are analytic continuations of each other. Deduce that

$$\mathbf{log} = \{(f, D): f \text{ is a branch-of-log on } D\}.$$

Identify all the function elements of the form $(f, \Delta(\alpha, r))$.

8. Let $f(z) = (z - 1)\log(z)$ $(z \in \Delta(1, 1))$. Show that the complete analytic function determined by $(f, \Delta(1, 1))$ has only one 'value' at 1.

9. Let

$$f(z) = 1 + \sum_{n=1}^{\infty} \frac{\sinh(z)}{n^2(z - 2n\pi i)} \qquad (z \in \mathbf{C}).$$

Show that f is an entire function with

$$f(2n\pi i) = 1 + \frac{1}{n^2} \qquad (n \in \mathbf{P}).$$

Deduce that the set of 'values' at 1 of the complete analytic function determined by $(f \circ \log, \Delta(1, 1))$ has a cluster point.

10.3 Complex analytic manifolds

In this section we shall show how a complete analytic function can be represented as a function (in the usual sense) which is analytic (in a new sense) on a special kind of metric space. In the course of this section we shall be introducing the student to the important concept of a manifold. Since manifolds have a fairly complicated structure we shall introduce the ideas gradually step by step. Our main task of representing complete analytic functions will motivate the development.

Let $\mathbf{f}$ be a complete analytic function. We wish to represent $\mathbf{f}$ as a function on some set and our first task is to consider what we should take as the 'points' of this set. Let $E_{\mathbf{f}}$ be the set of α in $\mathbf{C}$ such that there is $(f, D) \in \mathbf{f}$ with $\alpha \in D$. Given such $\alpha \in E_{\mathbf{f}}$ the associated domain D is contained in $E_{\mathbf{f}}$. Hence $E_{\mathbf{f}}$ is an open set. It is also clear by the methods employed in the last section that α may be joined to any point of $E_{\mathbf{f}}$ by a polygonal line in $E_{\mathbf{f}}$ and hence $E_{\mathbf{f}}$ is connected by Proposition 2.1. In other words $E_{\mathbf{f}}$ is simply a domain in $\mathbf{C}$. We have already seen that $\mathbf{f}$ may have an infinite number of 'values' at each point of $E_{\mathbf{f}}$. There is thus no hope of representing $\mathbf{f}$ in any natural way as a function on $E_{\mathbf{f}}$. It seems clear that we must associate with each point α of $E_{\mathbf{f}}$ at least as many different 'points' as we have 'values' at α. In other words we might try to index the points α with the associated 'values' of $\mathbf{f}$ at α. Unfortunately this simple approach turns out to be inadequate. We have to index the points α of $E_{\mathbf{f}}$ with functions.

Each point α of $E_{\mathbf{f}}$ was associated with some function element (f, D) of $\mathbf{f}$. We now consider all such triples (α, f, D). Given another such triple (β, f_1, D_1) we write

$$(\alpha, f, D) \equiv (\beta, f_1, D_1)$$

if $\alpha = \beta$ and (f_1, D_1) is a direct analytic continuation of (f, D). It is trivial to verify that $\equiv$ is an equivalence relation on the set of all such triples. We shall take the $\equiv$-equivalence classes to be the 'points' of our new space. Since it is awkward to keep dealing with equivalence classes we shall simply denote our 'points' by (α, f, D) and agree to consider (β, f_1, D_1) as the same point when $(\alpha, f, D) \equiv (\beta, f_1, D_1)$. In particular we may take D to be a disc $\Delta(\alpha, r)$ if we so wish. We write

$$M_{\mathfrak{f}} = \{(\alpha, f, D) : (f, D) \in \mathbf{f}, \alpha \in D\}.$$

It is now simple to *represent* $\mathbf{f}$ as a function on the set $M_{\mathfrak{f}}$. We simply define $F_{\mathfrak{f}}$ on $M_{\mathfrak{f}}$ by

$$F_{\mathfrak{f}}(\alpha, f, D) = f(\alpha).$$

Then $F_{\mathfrak{f}}$ is a complex valued function on $M_{\mathfrak{f}}$ and its range is simply the set of all 'values' of $\mathbf{f}$. Let us further check the reasonableness of our definition by a simple example. If $\mathbf{f}$ is the complete analytic function determined by an entire function f then we simply take $M_{\mathfrak{f}}$ as $\{(\alpha, f, \mathbf{C}) : \alpha \in \mathbf{C}\}$ and

$$F_{\mathfrak{f}}(\alpha, f, \mathbf{C}) = f(\alpha).$$

In other words the function $F_{\mathfrak{f}}$ is merely a slightly disguised form of the function f.

Our progress is satisfactory as far as it goes, but the situation requires further development before we can do any analysis on $M_{\mathfrak{f}}$. At the moment $M_{\mathfrak{f}}$ is simply a set. Our next step is to introduce a suitable metric for the set $M_{\mathfrak{f}}$. Let $p = (\alpha, f, D)$, $q = (\beta, g, E)$ be any two points of $M_{\mathfrak{f}}$. Then (f, D), (g, E) are analytic continuations of each other and so may be linked by a chain in $\mathbf{f}$, say (f_j, D_j) $(j = 1, 2, \ldots, n)$. Let

$$A = D \cup E \cup \bigcup\{D_j : j = 1, 2, \ldots, n\}.$$

It is clear that A is a domain and so there is a polygonal line Γ in A joining α to β. We shall say that any such polygonal line *joins* p to q. We now define the function d on $M_{\mathfrak{f}} \times M_{\mathfrak{f}}$ by

$$d(p, q) = \inf\{l_\Gamma : \Gamma \text{ is a polygonal line joining } p \text{ to } q\}.$$

(A polygonal line need not be an arc, but l_Γ has the obvious meaning; namely if $\Gamma = \bigcup\{[\alpha_j, \beta_j] : j = 1, \ldots, n\}$ then $l_\Gamma = \sum_{j=1}^{n} |\alpha_j - \beta_j|$.) If Γ is any polygonal line in $\mathbf{C}$ joining p to q we must certainly have $l_\Gamma \geqslant |\alpha - \beta|$. It follows immediately that the function d satisfies

$$d(p, q) \geqslant |\alpha - \beta|.$$

By way of illustration suppose that $\mathbf{f}$ is the complete analytic function determined by an entire function f. In this case we may join $p = (\alpha, f, \mathbf{C})$ to $q = (\beta, f, \mathbf{C})$ by the line segment $[\alpha, \beta]$ and so we have

$$d(p, q) = |\alpha - \beta|.$$

It is then clear in this case that d is a metric on $M_{\mathfrak{f}}$ and moreover $(M_{\mathfrak{f}}, d)$ is simply a mildly disguised version of the metric space $\mathbf{C}$. This indicates that our definition is not unreasonable. We now prove that d is always a metric on $M_{\mathfrak{f}}$.

Theorem 10.7. *Given any complete analytic function* $\mathbf{f}$, $(M_{\mathfrak{f}}\ d)$ *is a metric space.*

Proof. It is clear that $d(p, q) \geqslant 0$ $(p, q \in M_{\mathfrak{f}})$ and that $d(p, p) = 0$. Suppose that $d(p, q) = 0$ where $p = (\alpha, f, \Delta(\alpha, r))$, $q = (\beta, g, \Delta(\beta, t))$. Since

$$0 \leqslant |\alpha - \beta| \leqslant d(p, q) = 0$$

it follows that $\alpha = \beta$. Let $\rho = \min(r, t)$ so that $\rho > 0$. Since $d(p, q) = 0$ we may choose a polygonal line Γ joining p to q such that $l_\Gamma < \rho$. It follows that $\Gamma \subset \Delta(\alpha, \rho)$. We may now choose a chain of function elements $\{(f_j, D_j): j = 1, 2, \ldots, n\}$ linking $(f, \Delta(\alpha, r))$ to $(g, \Delta(\alpha, t))$ such that each $D_j \subset \Delta(\alpha, \rho)$. This means that each f_j is simply the restriction of f to D_j and so f and g agree on $\Delta(\alpha, \rho)$, i.e. $p = q$. It is clear from the definition of d that $d(p, q) = d(q, p)$ $(p, q \in M_{\mathfrak{f}})$. It remains to prove the triangle inequality.

Let p, q, r be any points of $M_{\mathfrak{f}}$. Let Γ be a polygonal line joining p to r and let Γ' be a polygonal line joining r to q. Then $\Gamma \cup \Gamma'$ is a polygonal line joining p to q and so we have

$$d(p, q) \leqslant l_{\Gamma \cup \Gamma'} \leqslant l_\Gamma + l_{\Gamma'}.$$

This is true for any such pair Γ, Γ' and so

$$\begin{aligned} d(p, q) &\leqslant \inf\{l_\Gamma : \Gamma \text{ joins } p \text{ to } r\} \\ &\quad + \inf\{l_{\Gamma'} : \Gamma' \text{ joins } r \text{ to } q\} \\ &= d(p, r) + d(r, q). \end{aligned}$$

The proof is now complete.

Our next obvious step is to investigate the properties of the metric space (M_f, d). Is it complete, or connected, or compact? In answering such questions for a metric space it is often helpful to have a clear picture of what the open balls look like. In fact it is usually enough to know what the open balls look like when the radius is small enough. The next result shows that they simply look like open discs in the complex plane.

Theorem 10.8. *Let* **f** *be a complete analytic function and let* $p \in M_{\mathbf{f}}$ *with* $p = (\alpha, f, \Delta(\alpha, r))$. *For* $0 < \epsilon < r$, $B(p, \epsilon)$ *is isometric with* $\Delta(\alpha, \epsilon)$.

Proof. Let $q \in B(p, \epsilon)$ and let $q = (\beta, g, \Delta(\beta, t))$. We have $|\alpha - \beta| \leqslant d(p, q) < \epsilon$ so that $\beta \in \Delta(\alpha, \epsilon)$. Since $d(p, q) < \epsilon$ we may choose a polygonal line Γ joining p to q such that $l_\Gamma < \epsilon$. It follows that $\Gamma \subset \Delta(\alpha, \epsilon)$ and so by the argument used in Theorem 10.7, $(g, \Delta(\beta, t))$ is a direct analytic continuation of $(f, \Delta(\alpha, r))$. Thus $(\beta, g, \Delta(\beta, t)) = (\beta, f, \Delta(\alpha, r))$ and so

$$B(p, \epsilon) \subset \{(\beta, f, \Delta(\alpha, r)) : \beta \in \Delta(\alpha, \epsilon)\}.$$

For any $q = (\beta, f, \Delta(\alpha, r))$ we may join p to q by the line segment $[\alpha, \beta]$ so that $d(p, q) = |\alpha - \beta|$. This means that

$$B(p, \epsilon) = \{(\beta, f, \Delta(\alpha, r)) : \beta \in \Delta(\alpha, \epsilon)\}$$

and that the mapping $(\beta, f, \Delta(\alpha, r)) \rightarrow \beta$ is an isometry of $B(p, \epsilon)$ with $\Delta(\alpha, \epsilon)$.

We note further that $M_{\mathbf{f}}$ is a connected metric space. This is readily proved from the definition of connectedness; alternatively we may use the method of Theorem 2.1 since we can join points of $M_{\mathbf{f}}$ by polygonal lines. The metric space $M_{\mathbf{f}}$ is never compact and need not be complete (see Problem 10.13). Before proceeding further with the general theory it is helpful to form some intuitive geometrical pictures of $M_{\mathbf{f}}$ for various complete analytic functions **f**.

If **f** is determined by a function f meromorphic on **C** then $M_{\mathbf{f}}$ simply looks like **C** (minus the poles of f). If **f** is determined by the function element $(f, \Delta(0, 1))$ of Example 10.2, then $M_{\mathbf{f}}$ simply looks like $\Delta(0, 1)$.

Suppose now that **f** is determined by the function element $(p^{\frac{1}{2}}, \Delta(1, 1))$. It is not difficult to see that the function elements of **f** of the form $(f, \Delta(\alpha, r))$ are simply $(p_\theta^{\frac{1}{2}}, \Delta(\alpha, |\alpha|))$ where $\alpha \in \mathbf{C} \setminus \{0\}$ and $\theta = \arg(\alpha) + 2n\pi$. For any disc $\Delta(\alpha, |\alpha|)$ there are essentially only two different functions, namely $p_\theta^{\frac{1}{2}}$ and $p_{\theta+2\pi}^{\frac{1}{2}}$. Thus each point α of $\mathbf{C} \setminus \{0\}$ is associated with two points of $M_{\mathbf{f}}$. This suggests that $M_{\mathbf{f}}$ should look something like two copies of $\mathbf{C} \setminus \{0\}$. Since $M_{\mathbf{f}}$ is connected these copies must somehow be joined together. To see how we should perform the joining let us note how points of $M_{\mathbf{f}}$ are joined (in the sense introduced above). To join $(1, p^{\frac{1}{2}}, \Delta(1, 1))$ to $(1, p_{2\pi}^{\frac{1}{2}}, \Delta(1, 1))$ we must take a polygonal line which goes round the origin once in the positive sense. If we go round the origin once more in the positive sense we come to the point $(1, p_{4\pi}^{\frac{1}{2}}, \Delta(1, 1))$ i.e. $(1, p^{\frac{1}{2}}, \Delta(1, 1))$; in other words we return to our starting point. A similar situation will obtain for circuits starting from any point of $M_{\mathbf{f}}$. This suggests the following construction. Take two copies of $\mathbf{C} \setminus \{0\}$ placed one above the other and cut them along the positive real axis. Take the lower flap of the bottom plane and paste it to the upper flap of the top plane; now take the lower flap of the top plane and paste it to the upper flap of the bottom plane. This last operation is of course impossible to realize, but we have at least formed some idea as to what $M_{\mathfrak{f}}$ looks like in this case.

Remark. The above space $M_{\mathfrak{f}}$ is of course not compact. If however we add on two other points corresponding roughly to 0 and ∞ then it is possible to turn this extended space into a compact connected metric space (by a method akin to that used for $\mathbf{C}^{\infty}$). It is this extended space which is usually referred to as the *Riemann surface* associated with the function $p^{\frac{1}{2}}$. It is a remarkable fact that the extended space is actually homeomorphic with $\mathbf{C}^{\infty}$.

Suppose now that $\mathbf{f} = \mathbf{log}$. We may form a picture of $M_{\mathfrak{f}}$ by a method similar to the one used above. Only, in this case we must take a copy of $\mathbf{C} \setminus \{0\}$ for each $n \in \mathbf{Z}$ and paste them all together suitably to form an infinite continuous spiral staircase.

The student may amuse himself by producing models for $M_{\mathfrak{f}}$ in many other cases (see Problem 10.12). To return to the general case, Theorem 10.8 tells us that we may regard $M_{\mathfrak{f}}$ as being made up of lots of open balls each looking like a disc in $\mathbf{C}$. Let us examine this situation a little more closely. Given $p \in M_f$ let φ_p be the homeomorphism of $B(p, \epsilon)$ with $\Delta(\alpha, \epsilon)$ given in Theorem 10.8. Thus

$$\varphi_p(\beta, f, \Delta(\alpha, \epsilon)) = \beta \qquad (\beta \in \Delta(\alpha, \epsilon)).$$

We say that the mappings φ_p $(p \in M_{\mathfrak{f}})$ form a family of *coordinate mappings* for $M_{\mathfrak{f}}$. There are clearly many other families of homeomorphisms of the balls $B(p, \epsilon)$ with open discs in $\mathbf{C}$, but we shall use only the above mappings.

It is significant to know how the coordinate mappings compare on overlapping balls. We make the comparison as follows. Let $p, q \in M_{\mathfrak{f}}$ with associated coordinate mappings $\varphi_p\colon B(p, \epsilon) \to \mathbf{C}$, $\varphi_p\colon B(q, \delta) \to \mathbf{C}$ and $B(p, \epsilon) \cap B(q, \delta) \neq \varnothing$. If

$$U_{pq} = \varphi_p(B(p, \epsilon) \cap B(q, \delta))$$

then U_{pq} is a connected open subset of $\Delta(\alpha, \epsilon)$ and we may define the mapping $\psi_{pq}\colon U_{pq} \to \mathbf{C}$ by

$$\psi_{pq} = \varphi_q \circ \varphi_p^{-1}.$$

The mappings ψ_{pq} are called *coordinate transformations.* In this case it is trivial to verify that each ψ_{pq} is simply a restriction of the identity mapping. In particular, the coordinate transformations are conformal mappings.

We have now developed sufficient motivation to introduce the concept of a manifold. The student should understand that we shall only be considering very special cases of manifolds; our definitions are by no means the most general possible. For our purposes we shall define a *manifold* (of complex dimension 1) to be a connected metric space M such that for each $p \in M$ there is a homeomorphism φ_p of some $B(p, \epsilon)$ with an open disc in **C**. The mappings $\{\varphi_p\}$ are called *coordinate mappings* for M and the mappings

$$\psi_{pq} = \varphi_q \circ \varphi_p^{-1}$$

are called the associated *coordinate transformations*. Manifolds are classfied into various types according to the behaviour of the associated family of coordinate transformations. We are only interested in one case; we say that M is a *complex analytic manifold* if each coordinate transformation is a conformal mapping.

Remark. It is customary to introduce equivalence relations on the families of coordinate mappings for a manifold M, but we shall not trouble to do so here. We simply associate one family of coordinate mappings with M.

Given a complex analytic manifold M it is clear what we mean by a continuous complex function on M (since M is a metric space). What meaning can we give to a function being analytic on M? The obvious method is to exploit the coordinate mappings. Given $p \in M$ we say that $F: M \to \mathbf{C}$ is *differentiable* at p if $F \circ \varphi_p^{-1}$ is differentiable at α (in the usual sense), where $\alpha = \varphi_p(p)$. We say that F is *analytic* on an open subset of M if F is differentiable at each point of the open subset. Suppose now that N is another complex analytic manifold with coordinate mappings $\{\varphi_q^1\}$. Given $G: M \to N$ we say that G is *analytic* on M if for each $p \in M$ the complex function $\varphi_q^1 \circ G \circ \varphi_p^{-1}$ is differentiable at p, where $q = G(p)$. The situation is illustrated in Figure 10.1 below.

It is clear that any domain in **C** is a complex analytic manifold (with the identity mapping for each coordinate mapping). It is also clear that for any complete analytic function **f**, M_f is a complex analytic manifold (with coordinate mappings as introduced earlier). We mention one other example of a complex analytic manifold. The extended complex plane $(\mathbf{C}^\infty, d^*)$ is a complex analytic manifold (take φ_∞ to be **j**). We leave the student to verify the delightful fact that a meromorphic function on **C** is analytic as a mapping between the complex analytic manifolds **C** and $\mathbf{C}^\infty$!

We may now readily complete our main task of representing complete analytic functions.

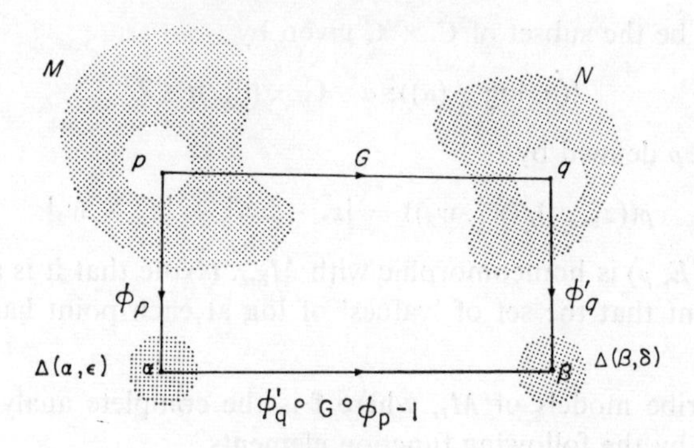

Figure 10.1

Theorem 10.9. *For each complete analytic function* **f**, *the representing function* $F_{\mathfrak{f}}$ *is analytic on the complex analytic manifold* $M_{\mathfrak{f}}$.

Proof. We have already noted that $M_{\mathfrak{f}}$ is a complex analytic manifold. Let $p \in M_{\mathfrak{f}}$ and suppose $p = (\alpha, f, \Delta(\alpha, r))$. We easily verify that

$$F_{\mathfrak{f}} \circ \varphi_p^{-1}(z) = f(z) \qquad (z \in \Delta(\alpha, r)).$$

Since f is analytic on $\Delta(\alpha, r)$ it follows that $F_{\mathfrak{f}}$ is differentiable at p. Since p was arbitrary this proves that $F_{\mathfrak{f}}$ is analytic on $M_{\mathfrak{f}}$.

If **g** is any general analytic function such that $\mathbf{g} \subset \mathbf{f}$ then the representing function for **g** is simply the restriction of $F_{\mathfrak{f}}$ to some open connected subset of $M_{\mathfrak{f}}$. This then completes the limited task we have set ourselves. It should be clear to the student that we have only touched on the beginning of a vast subject. We have not even got as far as associating a *Riemann surface* with each complete analytic function **f**. To do this we have to add to $M_{\mathfrak{f}}$ various other points. For example the origin clearly plays an important role for the complete analytic function **log** and ought to have some associated point attached to $M_{\mathbf{log}}$. The same is also true for isolated singularities of function elements of **f** and the point at infinity. When such points are added to $M_{\mathfrak{f}}$ and the metric suitably extended, we then obtain the Riemann surface associated with **f**.

PROBLEMS 10

10. For each non-zero integer n show that

$$d((1, \log, \Delta(1, 1)), (1, \log_{2n\pi}, \Delta(1, 1)) = 1.$$

11. Let E be the subset of $\mathbf{C} \times \mathbf{C}$ given by

$$\{(\alpha, \log_{2n\pi}(\alpha)) \colon \alpha \in \mathbf{C} \setminus \{0\}, n \in \mathbf{Z}\}$$

with metric ρ defined by

$$\rho((z_1, z_2), (w_1, w_2)) = |z_1 - w_1| + |z_2 - w_2|.$$

Show that (E, ρ) is homeomorphic with $M_{\log}$. (Note that it is essential for the argument that the set of 'values' of log at each point has no cluster points.)

12. Describe models of $M_{\mathbf{f}}$, where $\mathbf{f}$ is the complete analytic function determined by the following function elements.

(i) $f(z) = p^{\frac{1}{n}}(z) \qquad (z \in \Delta(1, 1)) \qquad (n \in \mathbf{P})$.

(ii) $f(z) = p^{\frac{1}{2}}(z(z - 1)) \qquad (z \in \Delta(\tfrac{1}{2}, \tfrac{1}{2}))$.

(iii) $f(z) = \log(1 - z^2) \qquad (z \in \Delta(0, 1))$.

13. Given any complete analytic function $\mathbf{f}$ show that M_f is never compact and need not be complete.

14. Given any complete analytic function $\mathbf{f}$ discuss the relation between the manifolds $M_{\mathbf{f}}$ and $M_{\mathbf{f}'}$.

15. Show that $(\mathbf{C}^\infty, d^*)$ is a complex analytic manifold and that the analytic mappings from $\mathbf{C}$ to $\mathbf{C}^\infty$ are precisely the meromorphic functions on $\mathbf{C}$. What are the analytic mappings from $\mathbf{C}^\infty$ to itself?

10.4 The gamma and zeta functions

In this section we discuss two very important special functions of analysis, namely Euler's gamma function and Riemann's zeta function. These functions appear in this chapter since they involve ideas of analytic continuation.

The gamma function arose as a generalization of the factorial function on the positive integers. There are many ways of defining the gamma function. We shall define the function in a half plane by an infinite integral and then extend to the whole of the plane by analytic continuation. We shall then obtain some of the properties of the gamma function.

Recall that t^{z-1} is a short-hand notation for $\exp\{(z - 1)\log t\}$. In particular we have $|t^{z-1}| = t^{x-1}$ $(t > 0)$, where $x = \operatorname{Re} z$. Now let $D = \{z \colon \operatorname{Re} z > 1\}$. Define H on $[0, \infty) \times D$ by

$$H(t, z) = \mathrm{e}^{-t} t^{z-1} \qquad (t \geqslant 0, z \in D).$$

For each $n \in \mathbf{P}$, H is continuous on $[0, n] \times \mathbf{D}$ (regarded as a subset of $\mathbf{R}^3$). For each $t \in [0, n]$ the mapping $z \to H(t, z)$ is analytic on D. Define f_n on D by

$$f_n(z) = \int_0^n t^{z-1}\,\mathrm{e}^{-t}\,\mathrm{d}t \qquad (z \in D).$$

It follows from Problem 7.9 that $f_n \in \mathscr{A}(D)$. It is easy to see that for each $z \in D$ there is $M > 0$ such that

$$|t^{z-1}\,\mathrm{e}^{-t}| \leqslant \frac{M}{1 + t^2} \qquad (t > 0).$$

We may thus define $\mathbf{\Gamma}$ on D by

$$\mathbf{\Gamma}(z) = \int_0^\infty t^{z-1}\,\mathrm{e}^{-t}\,\mathrm{d}t \qquad (z \in D).$$

If K is any compact subset of D there is some $m \in \mathbf{P}$ such that

$$|t^{z-1}\,\mathrm{e}^{-t}| \leqslant t^m\,\mathrm{e}^{-t} \qquad (z \in K).$$

It follows that

$$|f_n(z) - \mathbf{\Gamma}(z)| \leqslant \int_n^\infty t^m\,\mathrm{e}^{-t}\,\mathrm{d}t \qquad (z \in K)$$

and so $\{f_n\}$ converges to $\mathbf{\Gamma}$ uniformly on K. Since K was any compact subset of D it follows from Theorem 8.19 that $\mathbf{\Gamma}$ is analytic on D.

A simple integration by parts shows that

$$\mathbf{\Gamma}(z + 1) = z\mathbf{\Gamma}(z) \qquad (z \in D).$$

We may now use the method of page 265 to continue $\mathbf{\Gamma}$ into the whole of the plane. This extended function is called Euler's *gamma function* and is denoted by $\mathbf{\Gamma}$. We then have

$$\mathbf{\Gamma}(z + 1) = z\mathbf{\Gamma}(z) \qquad (z \in \mathbf{C}).$$

More generally for each $n \in \mathbf{P}$ we have

$$\mathbf{\Gamma}(z + n) = z(z + 1)\ldots(z + n - 1)\mathbf{\Gamma}(z) \qquad (z \in \mathbf{C}).$$

It is easy to verify by integration that $\mathbf{\Gamma}(1) = 1$, and so it follows from above that

$$\mathbf{\Gamma}(n) = (n - 1)! \qquad (n \in \mathbf{P}).$$

Using the equation

$$\mathbf{\Gamma}(z) = \frac{\mathbf{\Gamma}(z + n)}{z(z + 1)\ldots(z + n - 1)}$$

we deduce that the only singularities of $\mathbf{\Gamma}$ are simple poles at $0, -1, -2, -3, \ldots$. The residue at $-n$ is given by

$$\begin{aligned}
\operatorname{Res}(\mathbf{\Gamma}; -n) &= \lim_{z \to -n} (z+n)\mathbf{\Gamma}(z) \\
&= \lim_{z \to -n} \frac{\mathbf{\Gamma}(z+n+1)}{z(z+1)\ldots(z+n-1)} \\
&= \frac{\mathbf{\Gamma}(1)}{(-n)(-n+1)\ldots(-n+n-1)} \\
&= \frac{(-1)^n}{n!}.
\end{aligned}$$

We may now summarize our discussion thus far in the following theorem.

Theorem 10.10. *The function* $\mathbf{\Gamma}$ *is meromorphic on* $\mathbf{C}$ *and satisfies the equation*

$$\mathbf{\Gamma}(z+1) = z\mathbf{\Gamma}(z) \qquad (z \in \mathbf{C}).$$

The poles of $\mathbf{\Gamma}$ *occur at* $-n$ $(n = 0, 1, 2, \ldots)$, *are all simple and satisfy*

$$\operatorname{Res}(\mathbf{\Gamma}; -n) = \frac{(-1)^n}{n!}.$$

The gamma function is clearly real valued on $\mathbf{R}$ (except for the poles). It is thus a smooth extension of the factorial function on $\mathbf{P}$. It has besides this many interesting properties. Some of these properties are given in the exercises. We shall establish here only one other property of the gamma function. As a simple corollary we shall see that $\mathbf{\Gamma}$ has no zeros.

Theorem 10.11.

$$\mathbf{\Gamma}(z)\mathbf{\Gamma}(1-z) = \frac{\pi}{\sin(\pi z)} \qquad (z \in \mathbf{C}).$$

Proof. Let

$$f(z) = \mathbf{\Gamma}(z)\mathbf{\Gamma}(1-z) - 1/z \qquad (z \in \mathbf{C}).$$

It is readily seen that f is meromorphic on $\mathbf{C}$ with simple poles at $\pm n$ $(n \in \mathbf{P})$ and that $f(0) = 0$. We shall use the Mittag-Leffler theorem to show that

$$f(z) = 2z \sum_{n=1}^{\infty} \frac{(-1)^n}{z^2 - n^2} \qquad (z \in \mathbf{C}).$$

The proof will then be complete by reference to page 219.

We first compute the residues of f. Given $n \in \mathbf{P}$ we have

$$\begin{aligned}\operatorname{Res}(f; -n) &= \lim_{z\to -n} (z+n)\{\Gamma(z)\Gamma(1-z) - 1/z\}\\ &= \operatorname{Res}(\Gamma; -n)\Gamma(n+1)\\ &= (-1)^n\end{aligned}$$

$$\begin{aligned}\operatorname{Res}(f; n) &= \lim_{z\to n} (z-n)\{\Gamma(z)\Gamma(1-z) - 1/z\}\\ &= \lim_{w\to -(n-1)} (1-w-n)\Gamma(w)\Gamma(1-w)\\ &= -\operatorname{Res}(\Gamma; -(n-1))\Gamma(n)\\ &= (-1)^n\end{aligned}$$

Let Γ_n be the perimeter of the rectangle with vertices $\pm(n+\frac{1}{2}) \pm in$ ($n \in \mathbf{P}$). If we show that f is uniformly bounded on the contours Γ_n it will follow from Theorem 8.36 that

$$\begin{aligned}f(z) &= \lim_{n\to\infty} \sum_{j=1}^{n} \left\{(-1)^j\left[\frac{1}{z-j} + \frac{1}{j}\right] + (-1)^j\left[\frac{1}{z+j} - \frac{1}{j}\right]\right\}\\ &= 2z \sum_{n=1}^{\infty} \frac{(-1)^n}{z^2 - n^2}\end{aligned}$$

as required.

It is clear that we may neglect the term $-1/z$ in $f(z)$. It is clear from the initial definition of Γ that

$$|\Gamma(x+iy)| \leqslant \Gamma(x) \qquad (x \geqslant 1, y \in \mathbf{R}).$$

It is also readily checked that for each $m \in \mathbf{Z}$

$$\Gamma(z)\Gamma(1-z) = \frac{(-1)^m\Gamma(z-m)\Gamma(m+3-z)}{(m+1-z)(m+2-z)} \qquad (z \in \mathbf{C}). \quad (*)$$

If $z = n + \frac{1}{2} + iy$ ($n \in \mathbf{P}$), take $m = n - 1$ in (*) and we obtain

$$\begin{aligned}|\Gamma(z)\Gamma(1-z)| &= \left|\frac{\Gamma(\frac{3}{2}+iy)\Gamma(\frac{3}{2}-iy)}{(\frac{1}{2}+iy)(\frac{1}{2}-iy)}\right|\\ &\leqslant 4(\Gamma(\tfrac{3}{2}))^2 \qquad (y \in \mathbf{R}).\end{aligned}$$

If $z = -(n+\frac{1}{2}) + iy$ and we take $m = -(n+2)$ in (*) we again obtain the same inequality. Now let $z = x \pm in$ ($n \in \mathbf{P}$). Let $m \in \mathbf{Z}$ and suppose $x \in [m+1, m+2]$. We then have $\operatorname{Re}(z-m) \in [1, 2]$, $\operatorname{Re}(m+3-z) \in [1, 2]$. Since Γ is continuous on $[1, 2]$ it is bounded on $[1, 2]$, say by M. It now follows from (*) that

$$|\Gamma(z)\Gamma(1-z)| \leqslant M^2 \qquad (m \in \mathbf{Z}, n \in \mathbf{P}).$$

We have now shown that f is uniformly bounded on the contours Γ_n and so the proof is complete.

We can now see that Γ has no zeros. By the above theorem the zeros of Γ could appear only at the integers and we have already seen that this is impossible. It follows in particular that $1/\Gamma$ is an entire function with simple zeros at $0, -1, -2, \ldots$. It also follows from the theorem that $\Gamma(\frac{1}{2}) = \pi^{\frac{1}{2}}$. It is not difficult to show that the integral formula for Γ holds for $\operatorname{Re} z > 0$. In particular we have

$$\Gamma(\tfrac{1}{2}) = \int_0^\infty e^{-t}t^{-\frac{1}{2}}\,dt.$$

Put $t = x^2$ and we deduce that

$$\pi^{\frac{1}{2}} = 2\int_0^\infty e^{-x^2}\,dx.$$

This gives another proof of the well-known formula

$$\int_{-\infty}^\infty e^{-x^2}\,dx = \pi^{\frac{1}{2}}.$$

We turn now to the Riemann zeta function. Recall from Example 8.20 that we defined ζ for $\operatorname{Re} z > 1$ by

$$\zeta(z) = \sum_{n=1}^\infty n^{-z}.$$

Moreover we established that ζ is analytic on the half plane given by $\operatorname{Re} z > 1$. We shall now show that ζ can be extended to a meromorphic function on **C**. We begin with a formula relating the gamma and zeta functions.

Theorem 10.12.

$$\zeta(z) = \frac{1}{\Gamma(z)}\int_0^\infty \frac{t^{z-1}}{e^t - 1}\,dt \qquad (\operatorname{Re} z > 1).$$

Proof. Make the substitution $t = nu$ ($n \in \mathbf{P}$) in the integral formula for Γ and we obtain for $\operatorname{Re} z > 1$

$$\frac{1}{n^z} = \frac{1}{\Gamma(z)}\int_0^\infty e^{-nu}\,u^{z-1}\,du$$

and hence

$$\sum_{n=1}^N \frac{1}{n^z} = \frac{1}{\Gamma(z)}\int_0^\infty \sum_{n=1}^N e^{-nu}\,u^{z-1}\,du.$$

Since $\exp - \mathbf{1}$ has a simple zero at 0 we have

$$\lim_{u\to 0+} \frac{u^{z-1}}{e^u - 1} = 0 \qquad (\operatorname{Re} z > 2).$$

It therefore follows for Re $z > 1$ that

$$\sum_{n=1}^{N} \frac{1}{n^z} = \frac{1}{\Gamma(z)} \int_0^\infty \frac{u^{z-1}}{e^u - 1}\, du - \frac{1}{\Gamma(z)} \int_0^\infty \frac{u^{z-1} e^{-Nu}}{e^u - 1}\, du.$$

To complete the proof it will be sufficient to show that the second term on the right converges to 0 as $N \to \infty$. Let $a = \text{Re}\,(z - 1)$ so that $a > 0$. Given $\delta > 0$ we have

$$\int_0^\infty \frac{u^{z-1} e^{-Nu}\, du}{e^u - 1} \leqslant \int_0^\delta \frac{u^a}{e^u - 1}\, du + e^{-N\delta} \int_\delta^\infty \frac{u^a}{e^u - 1}\, du.$$

Given $\epsilon > 0$ we may choose $\delta > 0$ such that the first integral on the right is less than $\epsilon/2$. With this fixed δ we may now choose N_0 such that the second term on the right is less than $\epsilon/2$ whenever $N > N_0$. This completes the proof.

We shall now show how the formula in Theorem 10.12 enables us to extend ζ to the whole plane. Let us define P and Q for Re $z > 1$ by

$$P(z) = \int_0^1 \frac{t^{z-1}}{e^t - 1} dt$$

$$Q(z) = \int_1^\infty \frac{t^{z-1}}{e^t - 1}\, dt.$$

Using the methods employed for Γ we may easily show that the formula for $Q(z)$ actually defines an entire function. Recall from Example 7.13 that

$$\frac{t}{e^t - 1} = \sum_{n=0}^{\infty} \frac{B_n t^n}{n!} \qquad (t \in [0, 1])$$

the convergence being uniform on [0, 1]. We deduce for Re $z > 2$ that

$$\begin{aligned} P(z) &= \sum_{n=0}^{\infty} \int_0^1 \frac{B_n t^{n+z-2}}{n!} dt \\ &= \sum_{n=0}^{\infty} \frac{B_n}{n!(n + z - 1)} \\ &= \frac{1}{z - 1} - \frac{1}{2z} + \sum_{n=1}^{\infty} \frac{B_{2n}}{(2n)!(2n + z - 1)}. \end{aligned}$$

Using the fact that

$$\overline{\lim_{n \to \infty}} \left| \frac{B_{2n}}{(2n)!} \right|^{1/2n} = \frac{1}{2\pi} < 1$$

we may show that the last equation above gives a direct analytic continuation of P to a function meromorphic on $\mathbf{C}$ with simple poles at $1, 0, -1, -3, -5, \ldots$. Since $1/\Gamma$ is an entire function we may now extend ζ to the whole of $\mathbf{C}$ by

$$\zeta(z) = P(z)/\Gamma(z) + Q(z)/\Gamma(z) \qquad (z \in \mathbf{C}).$$

This defines the Riemann *zeta function* on the whole of $\mathbf{C}$ and it is evidently meromorphic on $\mathbf{C}$. Slightly more information is given in the theorem below.

Theorem 10.13. *The Riemann zeta function is meromorphic on* $\mathbf{C}$ *with a single pole. This pole, at the point* 1, *is simple and has residue* 1.

Proof. Since Q and $1/\Gamma$ are entire functions, the singularities of ζ can only be those of P/Γ. We have seen that P has simple poles at $1, 0, -1, -3, -5, \ldots$. Since $1/\Gamma$ has simple zeros at $0, -1, -2, \ldots$ it follows that ζ is regular at $0, -1, -3, -5, \ldots$. Since $\Gamma(1) = 1$ we have that 1 is a simple pole for ζ and this is the only singularity of ζ in $\mathbf{C}$. Since P has residue 1 at 1 it is clear that

$$\operatorname{Res}(\zeta; 1) = 1/\Gamma(1) = 1.$$

This completes the proof.

We shall not develop here any deeper theory of the Riemann zeta function; but we give some indication as to why it is important in number theory, and we discuss the nature of the zeros of ζ. The student may easily show that for $\operatorname{Re} z > 1$

$$(1 - 2^{-z})\zeta(z) = 1 + \frac{1}{3^z} + \frac{1}{5^z} + \frac{1}{7^z} + \cdots.$$

By a similar argument he may show that

$$(1 - 2^{-z})(1 - 3^{-z})\zeta(z) = 1 + \frac{1}{5^z} + \frac{1}{7^z} + \frac{1}{11^z} + \frac{1}{13^z} + \cdots.$$

Now let $\{p_n : n \in \mathbf{P}\}$ denote the set of prime numbers. By continuing the above process we may establish that for $\operatorname{Re} z > 1$

$$\frac{1}{\zeta(z)} = \prod_{n=1}^{\infty} (1 - p_n^{-z}).$$

This equation (due to Euler) gives the basic connection between number theory and the Riemann zeta function. Some other relations are given in the exercises at the end of the chapter.

We turn finally to the question of the zeros of ζ. The distribution of these zeros turns out to be of considerable significance in the theory of numbers. It is clear from our earlier considerations that ζ has zeros at $-2, -4, -6, \ldots$. It was shown by Riemann that ζ satisfies the equation

$$\zeta(1-z) = 2^{1-z}\pi^{-z}\cos(\pi z/2)\zeta(z) \qquad (z \in \mathbf{C}).$$

This equation enables us to deduce the behaviour of ζ for $\operatorname{Re} z < \frac{1}{2}$ from its behaviour for $\operatorname{Re} z > \frac{1}{2}$. In particular it may be used to prove that the only zeros of ζ outside the strip $0 \leqslant \operatorname{Re} z \leqslant 1$ occur at $-2, -4, -6, \ldots$. It is known that ζ has an infinite number of zeros in the strip $0 \leqslant \operatorname{Re} z \leqslant 1$, that none lie on $\operatorname{Re} z = 0$ or $\operatorname{Re} z = 1$, and that an infinite number lie on the line $\operatorname{Re} z = \frac{1}{2}$. There is a conjecture of long standing that all the zeros in this strip lie on the line $\operatorname{Re} z = \frac{1}{2}$. The conjecture has not yet been proved or disproved and is known as the *Riemann hypothesis*.

PROBLEMS 10

16. Show that $\mathbf{\Gamma}$ has exactly one absolute minimum point in $\{x : x > 0\}$ and that it lies in $(1, 2)$.

17. Let

$$F(z) = \mathbf{\Gamma}(z) - \sum_{n=0}^{\infty} \frac{(-1)^n}{n!(z+n)} \qquad (z \in \mathbf{C}).$$

Show that F is an entire function and that

$$F(z) = \int_1^{\infty} t^{z-1} \mathrm{e}^{-t}\, \mathrm{d}t \qquad (\operatorname{Re} z > 1).$$

18. Show that

$$\mathbf{\Gamma}(\bar{z}) = \overline{\mathbf{\Gamma}(z)} \qquad (z \in \mathbf{C})$$

and deduce that

$$|\mathbf{\Gamma}(iy)| = \pi/(y \sinh(\pi y)) \qquad (y \in \mathbf{R} \setminus \{0\}).$$

19. Let $\mu: \mathbf{P} \to \mathbf{R}$ be defined by $\mu(1) = 1$ and

$$\mu(n) = \begin{cases} (-1)^r & \text{if } n \text{ is the product of } r \text{ distinct prime factors,} \\ 0 & \text{otherwise.} \end{cases}$$

Show that for $\operatorname{Re} z > 1$

$$\frac{1}{\zeta(z)} = \sum_{n=1}^{\infty} \frac{\mu(n)}{n^z}$$

$$\frac{\zeta(z)}{\zeta(2z)} = \sum_{n=1}^{\infty} \frac{|\mu(n)|}{n^z}.$$

20. Let $\mathrm{d}(n)$ denote the number of divisors of n and let $\sigma_k(n)$ denote the sum of the kth powers of all the divisors of n. Show that

$$(\zeta(z))^2 = \sum_{n=1}^{\infty} \frac{\mathrm{d}(n)}{n^z} \qquad (\mathrm{Re}\, z > 1)$$

$$\zeta(z)\zeta(z-k) = \sum_{n=1}^{\infty} \frac{\sigma_k(n)}{n^z} \qquad (\mathrm{Re}\, z > \max(1, k+1)).$$

Appendix

RIEMANN-STIELTJES INTEGRATION

In this appendix we define the Riemann-Stieltjes integral which we used in Chapter 5. We also prove those results on integration which were only stated in Chapter 5, namely Theorem 5.9 and the formula for the length of a piecewise smooth arc or contour. It will be necessary for us to have some basic facts about complex functions of bounded variations. Some of these facts were given in Problem 3.7, but for the sake of completeness we begin by proving all the results that we need.

A.1 Functions of bounded variation

Let $a, b \in \mathbf{R}$ with $a < b$. A *subdivision* of $[a, b]$ is a set of the form $P = \{t_j : j = 0, 1, \ldots, n\}$ where

$$a = t_0 < t_1 < \ldots < t_n = b.$$

The set of all subdivisions of $[a, b]$ is denoted by $\mathscr{P}$. If we wish to indicate which interval we have in mind we shall write $\mathscr{P}[a, b]$. Given $f: [a, b] \to \mathbf{C}$ and $P \in \mathscr{P}$ we define

$$T(f, P) = \sum_{j=1}^{n} |f(t_j) - f(t_{j-1})|$$

$$V(f; a, b) = \sup \{T(f, P): P \in \mathscr{P}\}.$$

We say that f is of *bounded variation* on $[a, b]$ if $V(f; a, b) < +\infty$. We then call $V(f; a, b)$ the *total variation* of f on $[a, b]$. (If f represents an arc or contour Γ the total variation of f is simply the length of Γ.) We denote by $\mathscr{BV}([a, b])$ the set of all functions of bounded variation on $[a, b]$. By way of example note that if f is real valued and monotonic then $f \in \mathscr{BV}([a, b])$ and $V(f; a, b) = |f(b) - f(a)|$. It follows from Proposition 1 below that linear combinations of monotonic functions belong to $\mathscr{BV}([a, b])$. We then show in Proposition 2 that we obtain all the functions of $\mathscr{BV}([a, b])$ in this way.

Proposition 1. *$\mathscr{BV}([a, b])$ is a linear space.*

Proof. Let $f, g \in \mathscr{BV}([a, b])$ and let $P \in \mathscr{P}$. Then

$$\begin{aligned} T(f+g, P) &\leqslant T(f, P) + T(g, P) \\ &\leqslant V(f; a, b) + V(g; a, b). \end{aligned}$$

It follows that

$$V(f+g; a, b) \leqslant V(f; a, b) + V(g; a, b)$$

and so $f + g \in \mathscr{BV}([a, b])$. Given $\alpha \in \mathbf{C}$ we easily see that $V(\alpha f; a, b) = |\alpha| V(f; a, b)$ so that $\alpha f \in \mathscr{BV}([a, b])$. This completes the proof.

We now make some simple observations. Since $\{a, b\}$ is a subdivision of $[a, b]$, if $f \in \mathscr{BV}([a, b])$ we have

$$|f(b) - f(a)| \leqslant V(f; a, b).$$

Suppose now that $[c, d] \subset [a, b]$. Given $P \in \mathscr{P}[c, d]$ we may clearly extend P to be a subdivision P' of $[a, b]$ by adding the points a and b. It follows that for each $K \in \mathscr{P}[c, d]$

$$|f(c) - f(a)| + T(f, P) + |f(d) - f(b)| \leqslant T(f, P') \leqslant V(f; a, b)$$

Therefore $f|_{[c, d]} \in \mathscr{BV}([c, d])$ and

$$V(f; c, d) \leqslant V(f; a, b).$$

If $a < x < b$ we have in particular that f is of bounded variation on $[a, x]$. Further if $x \leqslant y \leqslant b$ we have that

$$V(f; a, x) \leqslant V(f; a, y).$$

We need one other observation before the next result. Given $f \in \mathscr{BV}([a, b])$ and $c \in (a, b)$ we have

$$V(f; a, b) = V(f; a, c) + V(f; c, b).$$

To see this let $P_1 \in \mathscr{P}[a, c]$, $P_2 \in \mathscr{P}[c, b]$ so that $P_1 \cup P_2 \in \mathscr{P}[a, b]$. We then have

$$T(f, P_1) + T(f, P_2) = T(f, P_1 \cup P_2) \leqslant V(f; a, b)$$

and this gives

$$V(f; a, c) + V(f; c, b) \leqslant V(f; a, b).$$

Conversely, given $P \in \mathscr{P}[a, b]$ we may adjoin c to P (if necessary) to form P^*. Evidently there exist $P_1 \in \mathscr{P}[a, c]$ and $P_2 \in \mathscr{P}[c, b]$ such that $P^* = P_1 \cup P_2$. Then

$$\begin{aligned} T(f, P) \leqslant T(f, P^*) &= T(f, P_1) + T(f, P_2) \\ &\leqslant V(f; a, c) + V(f; c, b) \end{aligned}$$

and so it follows that

$$V(f; a, b) \leqslant V(f; a, c) + V(f; c, b).$$

Proposition 2. *Given* $f \in \mathscr{BV}([a, b])$ *there exist four non-negative non-decreasing functions* f_k *such that*

$$f = f_1 - f_2 + i(f_3 - f_4).$$

Proof. It follows easily from the inequalities $|\text{Re}\, z| \leqslant |z|$, $|\text{Im}\, z| \leqslant |z|$ that $\text{Re} f, \text{Im} f \in \mathscr{BV}([a, b])$. It is thus sufficient to show that if $f \in \mathscr{BV}([a, b])$ is real valued then there exist non-negative non-decreasing functions g, h such that $f = g - h$. We may choose $M > 0$ such that $f(a) + M \geqslant 0$. Define g and h on $[a, b])$ by

$$\begin{aligned} g(x) &= f(a) + M + V(f; a, x) \\ h(x) &= f(a) + M + V(f; a, x) - f(x). \end{aligned}$$

It is then clear that g is non-negative and non-decreasing and $f = g - h$. We have $h(a) = M > 0$ and if $a \leqslant x < y \leqslant b$ then

$$\begin{aligned} h(y) - h(x) &= V(f; a, y) - V(f; a, x) - (f(y) - f(x)) \\ &= V(f; x, y) - (f(y) - f(x)) \\ &\geqslant 0. \end{aligned}$$

This shows that h is both non-decreasing and non-negative.

The student shoud observe that the decomposition of a real function of bounded variation into the difference of two non-negative non-decreasing functions is not unique. For example if $f = g - h$ is one such decomposition and if $M > 0$ then $f = (g + M) - (h + M)$ is another such decomposition. We shall return to this remark in the next section.

A.2 The Riemann-Stieltjes integral

We begin by defining the integral for a special case and we then go on to define the integral for the case we wish to consider. Let $a, b \in \mathbf{R}$ with $a < b$ and let f be a bounded real function on $[a, b]$ and g a non-decreasing function on $[a, b]$. Let $P = \{t_j : j = 0, 1, \ldots, n\}$ be an element of $\mathscr{P}[a, b]$. We define the *norm* of P by

$$\nu(P) = \max\{t_j - t_{j-1} : j = 1, \ldots, n\}.$$

For $j = 1, \ldots, n$ we define

$$M_j = \sup\{f(t) : t_{j-1} < t < t_j\}$$
$$m_j = \inf\{f(t) : t_{j-1} < t < t_j\}.$$

For each $P \in \mathscr{P}$ we define

$$S(f, g; P) = \sum_{j=1}^{n} M_j\{g(t_j) - g(t_{j-1})\}$$

$$s(f, g; P) = \sum_{j=1}^{n} m_j\{g(t_j) - g(t_{j-1})\}.$$

We define the *upper* and *lower Riemann-Stieltjes integrals of f with respect to g* by

$$\int^{-} f\,\mathrm{d}g = \inf\{S(f, g; P) : P \in \mathscr{P}\}$$

$$\int_{-} f\,\mathrm{d}g = \sup\{s(f, g; P) : P \in \mathscr{P}\}$$

respectively. If these two integrals are equal we say that f is *Riemann-Stieltjes integrable with respect to g*. We then denote their common value by

$$\int f\,\mathrm{d}g \quad \text{or} \quad \int_a^b f\,\mathrm{d}g \quad \text{or} \quad \int_a^b f(t)\,\mathrm{d}g(t)$$

and call this the *Riemann-Stieltjes integral of f with respect to g.*

Theorem 3. *If f is a continuous real function on [a, b] and g is non-decreasing on [a, b] then f is Riemann-Stieltjes integrable with respect to g. If h is another non-decreasing function on [a, b] then*

$$\int_a^b f\,\mathrm{d}(g + h) = \int_a^b f\,\mathrm{d}g + \int_a^b f\,\mathrm{d}h.$$

Proof. Since f is continuous on the compact set $[a, b]$, f is bounded, say $|f| \leqslant M$. Given $P \in \mathscr{P}$ we easily verify that

$$-M\{g(b) - g(a)\} \leqslant S(f, g; P)$$
$$s(f, g; P) \leqslant M\{g(b) - g(a)\}.$$

It follows that

$$-M\{g(b)-g(a)\} \leqslant \int^{-} f\,\mathrm{d}g$$

$$\int_{-} f\,\mathrm{d}g \leqslant M\{g(b)-g(a)\}.$$

It is clear that for any $P \in \mathscr{P}$

$$s(f, g; P) \leqslant S(f, g; P).$$

Given $P_1, P_2 \in \mathscr{P}$, let P_3 be the subdivision of $[a, b]$ formed from $P_1 \cup P_2$. We now verify that

$$s(f, g; P_1) \leqslant s(f, g; P_3) \leqslant S(f, g; P_3) \leqslant S(f, g; P_2)$$

and it follows from this that

$$\int_{-} f\,\mathrm{d}g \leqslant \int^{-} f\,\mathrm{d}g.$$

Since f is continuous on the compact set $[a, b]$, it is uniformly continuous. Given $\epsilon > 0$ there is thus $\delta > 0$ such that

$$t_1, t_2 \in [a, b],\ |t_1 - t_2| < \delta \quad \Rightarrow \quad |f(t_1) - f(t_2)| < \epsilon.$$

Given $P \in \mathscr{P}$ with $\nu(P) < \delta$ we then have

$$M_j - m_j \leqslant \epsilon \qquad (j = 1, \ldots, n)$$

and so

$$S(f, g; P) - s(f, g; P) \leqslant \epsilon\{g(b) - g(a)\}.$$

Since

$$s(f, g; P) \leqslant \int_{-} f\,\mathrm{d}g \leqslant \int^{-} f\,\mathrm{d}g \leqslant S(f, g; P)$$

we thus have

$$\left|\int^{-} f\,\mathrm{d}g - \int_{-} f\,\mathrm{d}g\right| \leqslant \epsilon\{g(b) - g(a)\}$$

for each $\epsilon > 0$ and so

$$\int^{-} f\,\mathrm{d}g = \int_{-} f\,\mathrm{d}g.$$

This proves that f is Riemann-Stieltjes integrable with respect to g.

If h is a non-decreasing function on $[a, b]$ then so is $g + h$ and thus f is Riemann-Stieltjes integrable with respect to h and $g + h$. It is clear that

$$S(f, g + h; P) = S(f, g; P) + S(f, h; P) \qquad (P \in \mathscr{P})$$

and therefore

$$\int f \,\mathrm{d}(g + h) \geqslant \int f \,\mathrm{d}g + \int f \,\mathrm{d}h.$$

Similarly

$$s(f, g + h; P) = s(f, g; P) + s(f, h; P) \qquad (P \in \mathscr{P})$$

and therefore

$$\int f \,\mathrm{d}(g + h) \leqslant \int f \,\mathrm{d}g + \int f \,\mathrm{d}h.$$

This shows that

$$\int f \,\mathrm{d}(g + h) = \int f \,\mathrm{d}g + \int f \,\mathrm{d}h.$$

Suppose now that g is a real function of bounded variation on $[a, b]$. Then there exist non-decreasing functions g_1, g_2 such that $g = g_1 - g_2$. Suppose $g = h_1 - h_2$ is another decomposition of g with h_1, h_2 non-decreasing. We then have $g_1 + h_2 = g_2 + h_1$ and so

$$\begin{aligned}\int f \,\mathrm{d}g_1 + \int f \,\mathrm{d}h_2 &= \int f \,\mathrm{d}(g_1 + h_2)\\ &= \int f \,\mathrm{d}(g_2 + h_1)\\ &= \int f \,\mathrm{d}g_2 + \int f \,\mathrm{d}h_1.\end{aligned}$$

It follows that

$$\int f \,\mathrm{d}g_1 - \int f \,\mathrm{d}g_2 = \int f \,\mathrm{d}h_1 - \int f \,\mathrm{d}h_2.$$

We now define the *Riemann-Stieltjes integral of f with respect to g* by

$$\int f \,\mathrm{d}g = \int f \,\mathrm{d}g_1 - \int f \,\mathrm{d}g_2$$

where $g = g_1 - g_2$ is any decomposition of g into the difference of two non-decreasing functions.

Suppose now that f is a continuous complex function on $[a, b]$ and g is a complex function of bounded variation on $[a, b]$. If $f = f_1 + if_2$, $g = g_1 + ig_2$ we define the *Riemann-Stieltjes integral of f with respect to g* by

$$\int f \, dg = \int f_1 \, dg_1 - \int f_2 \, dg_2 + i \int f_1 \, dg_2 + i \int f_2 \, dg_1.$$

Observe that if g is given by

$$g(t) = t \qquad (t \in [a, b])$$

then $\int f \, dg$ is simply the usual complex Riemann integral.

A.3 Proof of Theorem 5.9

We are now ready to prove Theorem 5.9. We state the results again for convenience.

Theorem 4. *Let $f \in \mathscr{C}([a, b])$ and let $g \in \mathscr{BV}([a, b])$.*

(i) *If $\{t_j : j = 0, \ldots, n\}$ is a subdivision of $[a, b]$ then*

$$\int_a^b f \, dg = \sum_{j=1}^{n} \int_{t_{j-1}}^{t_j} f \, dg.$$

(ii) *If φ is a continuous strictly increasing function from $[c, d]$ onto $[a, b]$ then*

$$\int_a^b f \, dg = \int_c^d f \circ \varphi \, d(g \circ \varphi).$$

(iii) *If $g \in \mathscr{C}([a, b])$ and is differentiable on (a, b) except at a finite number of points with g' bounded and continuous then*

$$\int_a^b f \, dg = \int_a^b f g'$$

where the second integral is the complex Riemann integral.

Proof. In view of the definition of the Riemann-Stieltjes integral for complex valued functions it will clearly be sufficient to establish the proof in the case in which f and g are real valued with g non-decreasing.

(i) We give the proof for the subdivision $\{a, c, b\}$; only the details are more complicated in the general case. Given $P_1 \in \mathscr{P}[a, c]$, $P_2 \in \mathscr{P}[c, b]$ we have that $P_1 \cup P_2 \in \mathscr{P}[a, b]$ and

$$\int_a^b f \, dg \leqslant S(f, g; P_1 \cup P_2) = S(f, g; P_1) + S(f, g; P_2).$$

Therefore

$$\int_a^b f\,\mathrm{d}g \leqslant \int_a^c f\,\mathrm{d}g + \int_c^b f\,\mathrm{d}g.$$

Similarly we have

$$\int_a^b f\,\mathrm{d}g \geqslant s(f, g; P_1 \cup P_2) = s(f, g; P_1) + s(f, g; P_2)$$

and so

$$\int_a^b f\,\mathrm{d}g \geqslant \int_a^c f\,\mathrm{d}g + \int_c^b f\,\mathrm{d}g.$$

This completes the proof of (i).

(ii) Since φ is continuous and strictly increasing we have that $f \circ \varphi$ is continuous and $g \circ \varphi$ is non-decreasing. It follows that $f \circ \varphi$ is Riemann-Stieltjes integrable with respect to $g \circ \varphi$. Given $P \in \mathscr{P}[c, d]$ with $P = \{t_j : j = 0, \ldots, n\}$ let $\varphi(P) = \{\varphi(t_j) : j = 0, \ldots, n\}$. It is clear that $\varphi(P) \in \mathscr{P}[a, b]$. Conversely given $\{x_j :\ j = 0, \ldots, n\} \in \mathscr{P}[a, b]$ we have $\{\varphi^{-1}(x_j) : j = 0, \ldots, n\} \in \mathscr{P}[c, d]$. This shows that $P \to \varphi(P)$ is a one-to-one mapping of $\mathscr{P}[c, d]$ onto $\mathscr{P}[a, b]$. Since

$$S(f, g; \varphi(P)) = S(f \circ \varphi, g \circ \varphi; P) \qquad (P \in \mathscr{P}[c, d])$$

we now conclude that

$$\begin{aligned} \int_a^b f\,\mathrm{d}g &= \inf\{S(f, g; \varphi(P)) : P \in \mathscr{P}[c, d]\} \\ &= \inf\{S(f \circ \varphi, g \circ \varphi; P) : P \in \mathscr{P}[c, d]\} \\ &= \int_c^d f \circ \varphi\,\mathrm{d}(g \circ \varphi). \end{aligned}$$

(iii) In view of part (i) we may clearly suppose that g is actually differentiable on (a, b) with g' bounded and continuous. This means that fg' is bounded on $[a, b]$ and continuous except possibly at a and b. In particular fg' is Riemann integrable on $[a, b]$. Given $[t_{j-1}, t_j] \subset [a, b]$ it follows from the first mean value theorem that there is $u_j \in (t_{j-1}, t_j)$ with

$$g(t_j) - g(t_{j-1}) = g'(u_j)(t_j - t_{j-1}).$$

It follows that

$$S(f, g; P) \leqslant S(fg', \mathbf{u}; P) \qquad (P \in \mathscr{P})$$

and therefore

$$\int_a^b f\,\mathrm{d}g \leqslant \int_a^b fg'.$$

Similarly we have

$$s(f, g; P) \geqslant s(fg', \mathbf{u}, P) \qquad (P \in \mathscr{P})$$

and therefore

$$\int_a^b f \, \mathrm{d}g \geqslant \int_a^b fg'.$$

This completes the proof.

A.4 The formula for arc length

It now remains to prove the formula for the arc length which was given on page 127. The formula will follow as a special case of our final result.

Theorem A.5. *Let $g \in \mathscr{C}([a, b])$ be differentiable on $[a, b]$ except at a finite number of points with g' bounded and continuous. Then*

$$V(g; a, b) = \int_a^b |g'|.$$

Proof. Given $c \in (a, b)$, recall from page 293 that

$$V(g; a, b) = V(g; a, c) + V(g; c, b).$$

It is thus clearly sufficient to give the proof for the case in which g is actually differentiable on (a, b). Given $P = \{t_j : j = 0, \ldots, n\}$ we have

$$|g(t_j) - g(t_{j-1})| = \left| \int_{t_{j-1}}^{t_j} g' \right| \leqslant \int_{t_{j-1}}^{t_j} |g'|$$

and thus

$$T(g, P) \leqslant \int_a^b |g'|.$$

It follows that

$$V(g; a, b) \leqslant \int_a^b |g'|.$$

We also have $u_j \in (t_{j-1}, t_j)$ such that

$$g(t_j) - g(t_{j-1}) = g'(u_j)\,(t_j - t_{j-1}).$$

It follows that

$$T(g, P) \geqslant s(|g'|, \mathbf{u}; P)$$

and so

$$V(f; a, b) \geqslant \int_a^b |g'|.$$

This completes the proof.

PROBLEMS

1. Given $f \in \mathscr{BV}([a, b])$ show that

$$\sup |f| \leqslant |f(a)| + V(f; a, b).$$

Given $f, g \in \mathscr{BV}([a, b])$ show that $fg \in \mathscr{BV}([a, b])$ with

$$V(fg; a, b) \leqslant \sup |f| \, V(g; a, b) + \sup |g| \, V(f; a, b).$$

2. Given $f \in \mathscr{BV}([a, b])$ with $\inf(f) > 0$ show that $1/f \in \mathscr{BV}([a, b])$ with

$$V(1/f; a, b) \leqslant \frac{V(f; a, b)}{(\inf(f))^2}$$

3. Given $f \in \mathscr{BV}([a, b])$ let

$$||f|| = |f(a)| + 2V(f; a, b).$$

Show that $\mathscr{BV}([a, b])$ is a Banach algebra with respect to $|| \cdot ||$ (see Problem 2.6).

4. Give examples of the following situations.

(i) $f \in \mathscr{C}([a, b]), f \notin \mathscr{BV}([a, b])$.

(ii) $f \in \mathscr{BV}([a, b]), f \notin \mathscr{C}([a, b])$.

5. Given $f \in \mathscr{C}([a, b]) \cap \mathscr{BV}([a, b])$ show that the mapping $x \to V(f; a, x)$ is continuous on $[a, b]$.

6. Given $f \in \mathscr{BV}([a, b])$ show that f has one-sided limits at each point of $[a, b]$. Let E be the set of points in $[a, b]$ for which one of the following inequalities holds:

$$\lim_{h \to 0+} f(t + h) \neq f(t), \qquad \lim_{h \to 0+} f(t - h) \neq f(t),$$

$$\lim_{h \to 0+} f(t + h) \neq \lim_{h \to 0+} f(t - h).$$

Show that E is at most countable. It follows that f is continuous except on a countable set.

7. Let g be a *step function* on $[a, b]$, i.e. there is $\{t_j : j = 0, \ldots, n\} \in \mathscr{P}[a, b]$ such that g is constant on each (t_{j-1}, t_j), say β_j, and $g(t_j) = \gamma_j$. Show that $g \in \mathscr{BV}([a, b])$ and determine $V(g; a, x)$ for $x \in [a, b]$. Show that any bounded function f is Riemann-Stieltjes integral with respect to g. Evaluate $\int_a^b f \, \mathrm{d}g$.

8. Let $f_1, f_2 \in \mathscr{C}([a, b])$, $\lambda_1, \lambda_2 \in \mathbf{C}$, $g_1, g_2 \in \mathscr{BV}([a, b])$. Show that

$$\int_a^b (\lambda_1 f_1 + \lambda_2 f_2) \, \mathrm{d}g_1 = \lambda_1 \int_a^b f_1 \, \mathrm{d}g_1 + \lambda_2 \int_a^b f_2 \, \mathrm{d}g_1$$

$$\int_a^b f_1 \, \mathrm{d}(\lambda_1 g_1 + \lambda_2 g_2) = \lambda_1 \int_a^b f_1 \, \mathrm{d}g_1 + \lambda_2 \int_a^b f_1 \, \mathrm{d}g_2.$$

What change arises in $\int_a^b f_1 \, \mathrm{d}g_1$ if (i) the value of f_1 is altered at one point; (ii) the value of g_1 is altered at one point?

9. Given $f \in \mathscr{C}([a, b])$, $g \in \mathscr{BV}([a, b])$ show that

$$\left| \int_a^b f \, \mathrm{d}g \right| \leqslant \int_a^b |f(t)| \, \mathrm{d}V(g; a, t) \leqslant \sup |f| \; V(g; a, b).$$

10. Given $f, g \in \mathscr{BV}([a, b]) \cap \mathscr{C}([a, b])$ show that

$$\int_a^b f \, \mathrm{d}g = f(b)g(b) - f(a)g(a) - \int_a^b g \, \mathrm{d}f.$$

Given $n \in \mathbf{P}$ evaluate

$$\int_a^b g^n \, \mathrm{d}g.$$

SUGGESTIONS FOR FURTHER STUDY

In the course of the book we have attempted to introduce the student to the basic ideas of complex analysis. We hope that the student's appetite has been whetted to read deeper and wider in this subject. Almost every theorem that we have proved in this book was known in some form to analysts at the end of the 19th century; but complex analysis has not stood still since 1900 and this century has seen many developments in the theory. We give here some descriptive remarks about a few of these developments.

It is in order to begin by mentioning some of the more elementary aspects of the classical theory that we have omitted. Apart from the gamma and zeta functions we have not discussed many of the special functions of complex analysis. In particular we have neglected the rich theory of doubly periodic functions (or elliptic functions). Other standard functions such as those of Legendre and Bessel appear only in veiled references in the problems. We have also omitted the elementary theory of asymptotic expansions.

The student may recall that several theorems were discussed in the text without proof, e.g. Runge's theorem, Montel's theorem and the Riemann mapping theorem. He may now wish to take up these topics in some of the references suggested in the bibliography e.g. Ahlfors, Cartan, Hille, Saks and Zygmund. He may also wish to study some advanced topics such as the Nevanlinna theory of meromorphic functions, analytic number theory, cluster set theory, and Riemann surfaces.

At several points in the book we attempted to show the value of interrelating the algebraic and analytic aspects of a problem. This is one of the themes that has received much attention in several branches of a large subject known as *functional analysis*. To give a simple illustration let A be the disc algebra (see Problem 8.17), i.e. the complex Banach algebra of all complex functions that are continuous on $\bar{\Delta}(0, 1)$ and analytic on $\Delta(0, 1)$. The disc algebra is clearly a closed subalgebra of $\mathscr{C}(\bar{\Delta}(0, 1))$. Wermer has proved the interesting fact that A is actually a maximal closed subalgebra of $\mathscr{C}(\bar{\Delta}(0, 1))$, i.e. that $\mathscr{C}(\bar{\Delta}(0, 1))$ is the only closed subalgebra that properly contains A. Conversely certain maximal closed subalgebras of $\mathscr{C}(\bar{\Delta}(0, 1))$ admit properties of an analytic nature. Now let B be the algebra $H^{\infty}(\Delta(0, 1))$ (see Problem 8.17), i.e. the complex Banach algebra of all complex functions that are analytic and bounded on $\Delta(0, 1)$. We may regard A as a closed subalgebra of B by identifying A with those functions in B that have property (E) of

Chapter 9. The algebra B is much larger than the algebra A and its properties are much more complicated. It follows from a theorem of Gelfand (for general commutative Banach algebras) that B can be represented as a subalgebra of the continuous functions on a certain compact topological space, called the carrier space, which contains a homeomorphic image of the open disc $\Delta(0, 1)$. It follows from a theorem of Carleson that the image of the disc $\Delta(0, 1)$ is even dense in the carrier space (this remarkable result is called the *corona theorem*). The carrier space consists of the set of maximal ideals of B endowed with a suitable topology. Certain analytic problems are intimately related to properties of this set of maximal ideals. For example, a sequence $\{z_n\}$ in $\Delta(0, 1)$ is called an *interpolation sequence* if for each bounded sequence $\{w_n\}$ of complex numbers there exists $f \in B$ such that $f(z_n) = w_n$ $(n \in \mathbf{P})$. The characterization of such sequences is intimately related to the corona theorem. Many topics of similar nature are being investigated at the present time (Carleson's corona theorem dates from 1963) in a subject known as *function algebras*.

In elementary complex analysis we have been studying functions of the form $f: D \to \mathbf{C}$. We can generalize in two directions. In the first place we may allow our functions to take their values in a more general space than the complex numbers. This programme has been carried through in functional analysis where one considers functions with values in a complex Banach algebra, for example. It is quite simple to define analytic in this context and it is then possible to obtain extensions of the classical theory. For example, Liouville's theorem extends in the expected manner, and as such it formed a significant tool in Gelfand's original development of the theory of commutative Banach algebras. This generalized complex analysis also forms a basic tool for the standard treatment of the topic known as *spectral theory*.

We may also generalize in the opposite direction and consider complex valued functions of *several complex variables*. The standard definition of *analytic* in this case is given in terms of power series expansions in several complex variables. Although some theorems generalize (for example the important maximum modulus principle) the student should be forewarned that the theory of functions of several complex variables is quite different from the theory for a single variable. For example, in two or more variables an isolated singularity is automatically removable. Moreover there are quite innocent-looking open connected subjects of $\mathbf{C}^2$ such that every function that is analytic on the set admits an extension to a function that is analytic on a strictly larger open connected set. We have indicated in this book that the technical difficulties in obtaining the most general theorems for functions of a single complex variable are usually of a topological nature. These difficulties are multiplied for several complex variables, not least by the fact that it is not possible (for the common mortal) to visualize more than three real dimensions. Indeed it is not possible to understand even the statements of some of the deeper theorems for several complex variables without a knowledge of some of the language of *algebraic topology*. Advances in mathematical research and teaching have brought elementary complex function theory within the scope of an ever-widening audience, and it is to be hoped that the same may be done in the future for functions of several complex variables.

BIBLIOGRAPHY

This selected bibliography is in three parts. The first consists of collateral reading. Almost all the books in this section contain much more material than the present book. The second section consists of advanced topics in complex analysis. The final section gives a few references to generalizations. I have made no attempt to give a complete bibliography, and the selection reflects my own personal tastes.

A. Collateral

Ahlfors, L. V., *Complex Analysis*, (2nd ed.), McGraw-Hill, New York, (1966).
Boas, R. P., Jr., *Entire Functions*, Academic Press, London, (1954).
Cartan, H., *Elementary Theory of Analytic Functions of One or Several Variables*, (translated from the French), Hermann, Paris, (1963).
Gleason, A. M., *Fundamentals of Abstract Analysis*, Addison-Wesley, Reading, Mass., (1966).
Hille, E., *Analytic Function Theory*, 2 volumes, Blaisdell, New York, (1963).
Markushevich, A. I., *Theory of Functions of a Complex Variable*, 3 volumes, (revised English edition), Prentice-Hall, London, (1965).
Nehari, Z., *Introduction to Complex Analysis*, Allyn and Bacon, Boston, (1961).
Rudin, W., *Real and Complex Analysis*, McGraw-Hill, New York, (1966).
Saks, S. and Zygmund, A., *Analytic Functions* (2nd English ed.), Polish Scientific Publishers, Warsaw, (1965).
Titchmarsh, E. C., *The Theory of Functions*, (2nd ed.) Oxford University Press, Oxford, (1939).

B. Advanced

Collingwood, E. F. and Lohwater, A. J., *The Theory of Cluster Sets*, Cambridge University Press, Cambridge, (1966).
Hayman, W. F., *Meromorphic Functions*, Clarendon Press, Oxford, (1964).
Heins, M., *Selected Topics in The Classical Theory of Functions of a Complex Variable*, Holt, Rinehart and Winston, London, (1962).
Springer, G., *Introduction to Riemann Surfaces*, Addison-Wesley, Reading, Mass., (1957).

Titchmarsh, E. C., *The Theory of the Riemann Zeta-Function*, Oxford University Press, Oxford, (1951).

Whyburn, G. T., *Topological Analysis*, (revised ed.), Princeton, New Jersey, (1964).

C. Generalizations

Gunning, R. C. and Rossi, H., *Analytic Functions of Several Complex Variables*, Prentice-Hall, London, (1965).

Hoffman, K., *Banach Spaces of Analytic Functions*, Prentice-Hall, London (1962).

Simmons, G. F., *Introduction to Topology and Modern Analysis*, McGraw-Hill, New York (1963).

Seminars on Analytic Functions, Institute for Advanced Study, Princeton University, New Jersey, Vol. II, Seminar V, (*Analytic Functions as Related to Banach Algebras*), (1958).

INDEX OF SPECIAL SYMBOLS

SUBJECT INDEX